Residential Wiring

to the 1999 NEC®

Craftsman Book Company
6058 Corte del Cedro / P.O. Box 6500
Carlsbad, CA 92018

Library of Congress Cataloging-in-Publication Data

Markell, Jeff.
Residential wiring to the 1999 NEC / by Jeff Markell.
p. cm.
New ed. of: Residential wiring to the 1996 NEC.
Includes index.
ISBN 1-57218-073-0
1. Electric wiring, Interior--Handbooks, manuals, etc.
2. Electric wiring--Insurance requirements. 3. National Fire Protection Association. National electrical code (1999)
I. Title.
TK3285.M173 1999
621.319'24--dc21 99-19966
CIP

First edition ©1984 by Reston Publishing Company, Inc.
ISBN 0-8359-6661-5

Second edition ©1987 by Craftsman Book Company
ISBN 0-934041-19-9

Third edition ©1993 by Craftsman Book Company

Fourth edition ©1996 by Craftsman Book Company

Fifth edition ©1999 by Craftsman Book Company

CONTENTS

PREFACE

This book was written for anyone who intends to make a living wiring residential buildings. If you can read, understand and follow the instructions in this manual, you should have no trouble installing safe, modern, efficient electrical systems in homes and apartments.

As an electrician, you need to know how to use a wide variety of tools and materials. This manual describes the tools that should be in every electrician's tool box and suggests how they can be used to best advantage. I'll also explain what you should know about electrical materials: wire, cable, conduit, fixtures, boxes, switches, breakers and panels. There's a correct tool and a right material for every purpose. Sometimes selecting the right tools and materials isn't easy. After reading this book, you should have little trouble choosing both tools and materials appropriate for the work you do.

This manual isn't a book of electrical theory. But every professional electrician needs some background on how electricity is generated and distributed. And, of course, you should know how Ohm's Law and Watt's Law are used to design electrical systems. The first two chapters cover these important subjects.

If you've worked as an electrician for some time, you know that nearly everything an electrician does is governed by the National Electrical Code. For our purposes, the only right way is the Code way. Until you're comfortable with the Code, doing everything the Code way can be a nuisance. Once you understand the Code and the reasons for Code requirements, you may have a different perspective. Most experienced electricians would agree that the NEC protects everyone (including electricians and electrical contractors) and is a good guide to professional practice — even if the building inspector didn't enforce it.

This book will help you learn and understand the NEC. Throughout the chapters you'll find references to specific Code sections. These references are shown in parentheses. For example, (NEC 310-2b) refers to an article numbered 310 and to section 310-2. In that section 310-2 is a part (b). That's the portion of the code you may want to study.

This book will help you follow the Code. But this isn't a substitute for the NEC. Every professional electrician needs a copy of the current Code. Many bookstores sell the NEC. If your local bookstore doesn't have a copy, ask an inspector or another electrician where he got his. But just having the current NEC isn't enough. Many cities and counties don't adopt the model Code exactly as published by National Fire Protection Association. Instead, they supplement the Code with amendments or changes that will be enforced on jobs in that city or county. Once you have the current NEC, ask at your local building department about amendments or changes that apply in that jurisdiction. Keep those changes with your copy of the Code.

I'll explain floor plans, cable plans and wiring diagrams in detail. This is important information for every electrician. The Code has a lot to say about types of outlets, spacing of outlets, what must be switch controlled and what need not be switch controlled. The work you do will have to follow the plans and comply with the Code. The information in this book should help you understand and follow plans prepared for your jobs.

Finally, I'll explain how to diagram the circuits you're likely to find in a home or apartment. As a teacher of electrical wiring for many years, I've found that a student who can diagram a circuit correctly has a reasonably good chance of wiring it correctly as well. And a student who can't diagram a circuit probably can't install it either!

Before we begin, I want to acknowledge the firms and individuals that have supplied materials used in preparation of this manual. Among those are:

Harvey Hubbel Inc.
Slater Electric, Inc.
Appleton Electric Company
Advanced Transformer Company
Midland-Ross Corporation
General Electric Corporation
Westinghouse Electric Corporation
Pass & Seymour, Inc.
The Wiremold Company
Intermatic Incorporated
Bridgeport Fittings, Inc.
Gould Inc.
Amprobe Instrument Company
Bell Electric Company
McGraw Edison Company

Thanks to the National Fire Protection Association we have been able to reprint various reference materials directly from the National Electrical Code® as shown in the latest 1999 edition.

My particular personal thanks to Mr. L.K. Raeburn, Manager of Amfac Electric Supply Co. in San Diego. He kindly loaned much electrical material for photographic illustrations.

And finally — my profound thanks to Ms. Betty Marelli. She patiently converted my quite messy drafts into an eminently readable manuscript.

Now let's get down to business — what you need to know to wire homes and apartments.

Jeff Markell

CHAPTER 1

ELECTRICAL ENERGY

In order to understand his work satisfactorily, the electrician needs a modest background as to what electricity is and what can be done with it—at least to the extent that these matters are known at this time. Electricity has always been, and to a considerable extent still is, mysterious since it is an invisible form of energy, but as the reader may have discovered with dismay, it is by no means intangible. This is a practical book on how electrical wiring in a small building should properly be done in accordance with accepted standards of good workmanship, and in accordance with the provisions of the National Electrical Code®*. It is not a treatise on theory, so the discussion of theoretical matters will be kept to a minimum.

Historical Introduction

It might surprise some people to learn that as far back as 600 BC the Greeks were amusing and amazing themselves with elementary uses of static electricity. They had discovered, for example, that a piece of amber rubbed with cloth attracted bits of straw, hair, and the like. Their word for amber was "elektron," which is obviously the root of "electron," "electricity," "electronics," and other words containing "electro."

The ancients also discovered that certain heavy, black stones that were found every so often mysteriously attracted iron. Since they happened to be found fairly frequently in a part of Asia Minor called Magnesia they were called magnets. Naturally all manner of hocus pocus was

*National Electrical Code® and NEC® are Registered Trademarks of the National Fire Protection Association, Inc., Quincy, MA.

imagined to explain these curious phenomena, and all the early explanations of which we have records were complete nonsense.

Over many centuries it was observed that various other materials had characteristics similar to amber; they too could be rubbed, and would attract light objects. Scientific thinking developed the theory that the rubbed materials leaked a sort of "fluid" that was the cause of the attraction. This fluid was called electricity. Discontent with one "fluid," thinkers by the 18th Century hypothesized there were two fluids. One was called "vitreous," the other was "resinous." The difference was based on the nature of the substances being rubbed. By the mid-18th Century Ben Franklin went back to the one "fluid" theory. He decided that the two fluids were simply different aspects of the same thing. When an object had too much of this electric fluid it was "positive"; if it had too little, it was "negative," and if it was normal, whatever that meant, it was "neutral." While those in scientific fields were dissatisfied with this theory, it was the only one available until the early 20th Century investigations of the structure of matter began to offer a more satisfactory alternative.

The Composition of Matter

Matter is anything that has mass and occupies space. It exists in three states: solid, such as a rock; liquid, such as water; and gaseous, such as the air around us. By means of variations in temperature and pressure, matter can be changed from one state to another. Remove enough heat from a quantity of water, thus reducing the temperature, and at 32°F it will change in state from liquid to a solid, ice. Take the same quantity of water, and add heat, increasing the temperature. At 212°F it will start to vaporize, changing from a liquid to a gas.

Although a particular type of matter may change state from solid, to liquid, to gaseous, the component building blocks of which it is made remain the same, which leaves the question—of what is matter made? To find out one must divide, and subdivide, and subdivide again in order to arrive at the smallest particle that retains the characteristics of that particular type of matter be it water, steel, or foam plastic. That smallest particle is termed a "molecule." There is a different molecule that corresponds to each different kind of matter. But the molecule is by no means the end of the line. Molecules are composed of yet smaller parts termed atoms. Water, for example, is composed of molecules made of two hydrogen atoms plus one oxygen atom—H_2O. All matter, then, consists of the atoms of 100 elements combined in different compounds to form the molecules that distinguish different substances from each other. How atoms accomplish this business of combining into molecules will be discussed after we look more closely at the atom itself.

The atoms that compose molecules are quite complicated structures. Each one seems to be a miniature solar system, consisting of a nucleus surrounded by varying numbers of revolving electrons. The nucleus contains various particles such as protons, neutrons, positrons, neutrinos, mesons, and according to recent theories even a couple odd bits called "quarks" and "charms." We are primarily concerned with the bulk of the nucleus which consists of the protons and neutrons. The number of protons in the nucleus differentiate the atoms of the 100 elements from each other. The number of protons in the nucleus of an atom is its "atomic number." Hydrogen is #1, helium is #2, and so on.

Protons are positively charged, neutrons have no electrical charge, and the orbiting electrons are negatively charged. Since under normal conditions atoms are electrically neutral, an atom of any particular element will contain equal numbers of electrons and protons. The number of neutrons, as well as the various other nuclear components (neutrinos, mesons, etc.), has nothing to do with the electron-proton balance. Hydrogen has no neutrons, while the 92 protons of uranium are out numbered by 146 neutrons.

Curiously, while the magnitude of the opposite electrical charges in electrons and protons is equal to each other the difference in mass between the two is staggering. The mass of a proton is 1840 times that of an electron. A similarity appears between what is observed on a gigantic scale in the solar system, and in miniscule scale in the atom. All but a tiny part of the mass of the solar system is contained in the sun. Similarly all but a tiny part of the mass of an atom is contained in the nucleus.

The planets of the solar system are held in their orbits around the sun by a complex of factors involving the mutual attraction of their gravitational fields, and that of the sun, as well as their masses, and the velocities at which they move. Electrons are held in their orbit around the nucleus by the electrostatic attraction between their negative charges, and the positive charges of the protons in the nucleus, and again a relationship between mass and velocity is involved.

At this point the parallel between the atom and the solar system breaks down. The planets of the solar system each differ greatly from each other as to mass, composition, orbital velocity, and other characteristics. The electrons orbiting the nucleus of an atom do not differ.

In order to maintain the centrifugal force to keep from falling into the nucleus, on the one hand, or spinning away from its nucleus, on the other, the electron must move at a constant speed. Because it has mass, it must also have a level of energy determined by the combination of its mass and its velocity. Only a very limited number of specific energy levels are possible for electrons. There are seven altogether. As an electron can only occupy an orbital path appropriate to its energy level, there are seven possible orbits.

The more complex atoms might have as many as 100 electrons, but since only seven possible energy levels exist, the electrons must group at various appropriate orbit distances from the nucleus (Figure 1-1) forming

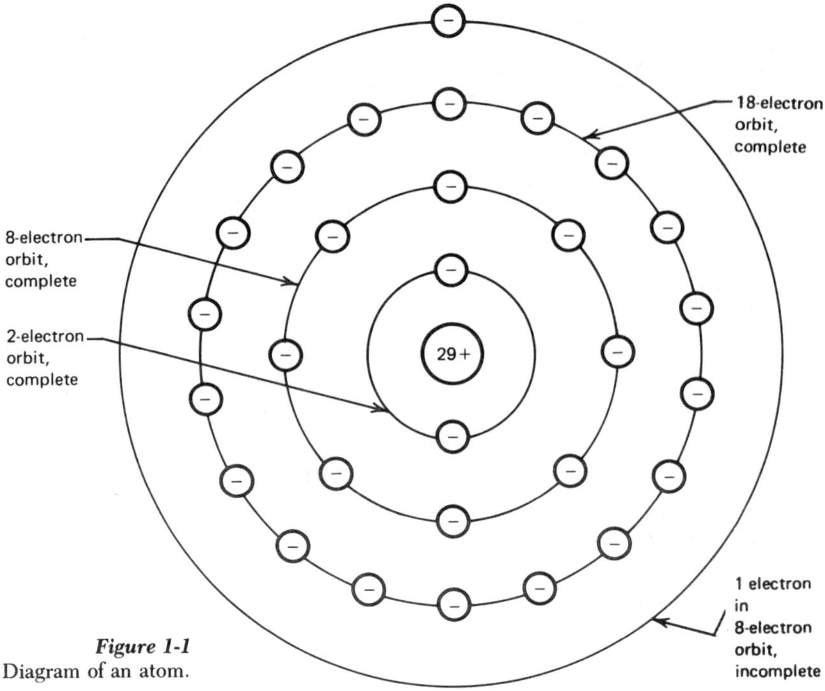

Figure 1-1
Diagram of an atom.

"shells" in layers around it. A consistently repeated pattern is found in the formation of these shells. The innermost shell (#1) can hold no more than two electrons. Any number above two starts the second shell. It can hold up to eight. When it is filled the third shell is started.

At this point the picture becomes a bit more complicated. The third shell (#3) can hold up to 18 electrons, however, the *outermost* shell of any atom regardless of which one, may hold no *more* than eight. Thus when Shell #3 is the outside one, and has eight electrons the next electron must take up an orbit in shell #4. Only after shell #4 has one or two occupants can the rest of the 18 possible spaces in #3 be filled. By the time shell #4 has eight electrons, shell #3 will already have its allotted 18. When #4 is the outermost shell and is holding eight electrons shell #5 starts. Shell #4 can hold 32 electrons. When it has its 32, and shell #5 is up to eight, shell #6 is started. Shell #6 is completed and shell #7 started in a similar manner. With all elements it is the spare electrons of the outer shell, whatever shell number that happens to be, that take part in any of the various chemical and electrical phenomena. These are termed "valence electrons."

Formation of Molecules

Regardless of its shell number the outermost shell of any atom can contain no more than eight valence electrons. Any atom that has all eight is stable and does not normally combine with other atoms. The atoms with valence electrons anywhere between one and seven, trying to attain stability, are candidates for combination with other atoms to form molecules. The process of molecule formation is termed atomic bonding and occurs in any one of three ways: ionic bonding, covalent bonding, or metallic bonding.

Ionic Bonding

An atom by itself will contain matching numbers of electrons and protons which, since they have opposite electrical charges, results in a neutral charge for the atom as a whole. However, this matter of valence electrons gets in the way. An atom with more than four but less than eight valence electrons is unstable. It is looking to obtain whatever number of valence electrons are missing to fill its outer shell to eight. Conversely, an atom with less than four valence electrons is also unstable and willing to unload its excess.

Thus where an atom with one valence electron meets another one with seven, there is a tendency for the one to join the seven, thus stabilizing both atoms. However, in the process something else happens. The atom that lost an electron now has a net positive electrical charge of one. The atom that picked up an electron has also picked up a net negative electrical charge of one. An atom that is no longer electrically neutral but has a net positive or negative charge has become an "ion." Ions with opposite electrical charges are attracted to each other tending to combine via "ionic bonding" to form molecules.

Covalent Bonding

Hydrogen with atomic number 1 has only a single electron in the #1 shell. It is unstable because that shell is incomplete without two electrons. One way it stabilizes is to join with another hydrogen atom (Figure 1-2) to form a hydrogen molecule in which the two component atoms share their two electrons. This is an example of "covalent bonding."

Metallic Bonding

Since it is the most commonly used material for electrical wires, copper is a good example of "metallic bonding." The atom in this case has 29 electrons. Two complete shell #1, another 8 complete shell #2, and the next 18 fill out shell #3. That is a total of 28. The 29th is a lone valence electron in shell #4. This lone electron is loosely held and has a tendency to wander off becoming a "free electron." The copper atom has become a positive ion, and so have a lot of others that have also lost their single

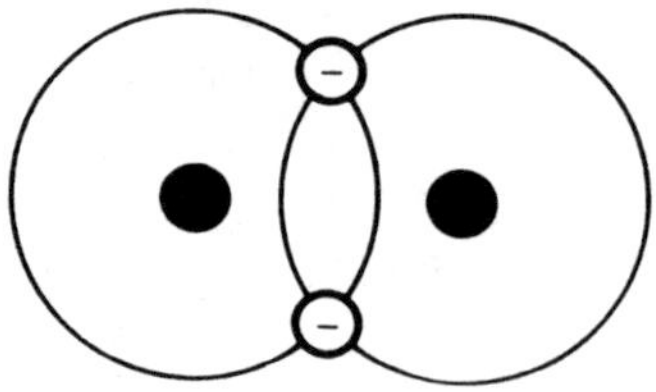

Figure 1-2
Covalent bond of two hydrogen atoms.

valence electrons. Although like charges repel, the copper ions do not simply fly apart as one might expect. They are immersed in a sort of soup of free electrons. The mutual attraction between the positive copper ions and the negatively charged electron mass around them holds the whole business together by "metallic bonding."

That same soup of free electrons, unattached to specific atoms, flows as an electrical current through a metal when it is connected to a source of electrical pressure. We will measure that pressure in volts and measure the current it creates in amperes.

The more free electrons available in a given material, the more readily they will move in response to a given electrical pressure, and the more free electrons there are in a material, the less "resistance" it will have to the flow of those electrons as electrical current.

Materials containing large numbers of free electrons, and therefore offering little resistance to the flow of electron current, are termed "conductors." Those with very few free electrons will necessarily have a high resistance to the flow of electron current since there are but few available electrons to participate in the process. These materials, due to their high resistance, are good insulators.

Metals in general, because of metallic bonding, have many free electrons and are good conductors. Glass, rubber, wood, cloth, and plastics having few free electrons are good insulators. A few materials exist that fall into the cracks between conductors and insulators. They have some of the characteristics of both, hence are called semiconductors. Silicon and germanium are two. These types of materials are used in various electronic devices, but not directly in building wiring; thus they will not concern us.

Static Electricity

Under normal conditions the atoms of a substance are neutrally charged since the negative charges of the orbiting electrons are exactly balanced by the positive charge of the protons in the nucleus. When two electrically unbalanced atoms bond ionically to form a molecule, the molecule then also becomes neutral since the net positive charge of one atom has been off set by the net negative charge of the other.

Figure 1-3
Amber rubbed with cloth results in a positive static charge.

However, when some outside influence forces many atoms of a material either to gain or lose an electron that material becomes either negatively or positively charged. This charge collects on the object's surface and tends to stay there until conduced away. The pieces of amber rubbed with cloth by the Greeks in 600 BC were charged in this way (Figure 1-3). When you walk across a thick carpet, and then see a small spark when you touch a door knob, you were charged in the same way. This type of surface charge is called a static charge.

It is interesting to note that one important use of static electricity is to precipitate the particulate matter from exhaust gases of industrial plants (Figure 1-4). As the gases enter the precipitation chamber the dust, soot, and other particles are attracted to a positively charged plate. The moment they touch it, they become positively charged and are strongly repelled, and drop to the bottom of the chamber where the charge is grounded.

While this and a few other constructive uses of static electricity exist, the static form is generally useless because it is essentially an instantaneous rather than a steady, dependable force.

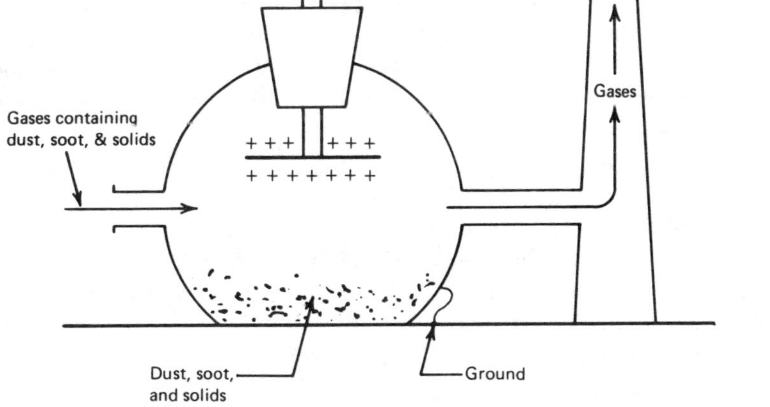

Figure 1-4
Smoke precipitator.

Current Electricity

When a neutral atom loses an electron it becomes a positive ion. A neutral atom that gains an electron becomes a negative ion. Between any two charged particles a force field exists in which like charges are repelled, and unlike charges are attracted. This force field is called an "electrostatic field." In response to the force being exerted by the field, charged particles move. This movement constitutes an electrical current. In a solid conductor the only mobile particles are free electrons that have escaped from the outer shell of an atom leaving it as a positive ion. In liquids and gases the positive ions are also free to move, an effect we shall encounter in connection with certain types of lighting equipment.

When an excess of electrons causing a negative charge is built up at one end of a conductor, and a deficiency of electrons causing a positive charge is built up at the other end, the pressure caused by the field existing between the two ends will cause the loose electrons in the conductor to flow from the area of excess to the area of deficiency, if permitted to do so. As the electron differential between the area of excess and deficiency increases or decreases, the pressure differential between them varies as well.

The difference in electrical pressure between two points is measured in units called volts. The volt is named after an 18th-century Italian experimenter named Alessandro Volta, the inventor of the battery. One volt is defined as the pressure necessary to force one ampere of electrical current through a resistance of one ohm. This definition is not too helpful until we understand what is meant by ampere and ohm.

The ampere is named after Andre Marie Ampere, also a late 18th-century electrical experimenter. His experiments dealt in part with the flow of current in a conductor. In consequence the ampere, the unit used to measure current flow, is named in his honor. Since an electrical current consists of a flow of electrons through a conductor, then the measurement of that flow is a count of the electrons passing a designated metering point in a specific length of time. As a comparison, amperage measures the flow of electricity per second the same way gallons per minute measures the flow of water. An electrical flow of 6,250,000,000,000,000,000 electrons per second equals one ampere for that is what a pressure of one volt will push through a resistance of one ohm.

For an electrical pressure (voltage) to push a current (amperage) through any substance, that voltage must be sufficient to overcome the resistance of the substance. All substances, to greater or lesser degrees, are resistant to the flow of electrical current. Conductors such as metals have low resistance. Various insulators such as plastics, paper, glass, or rubber have high resistance, but no material exists that has no resistance. Since the resistances of different materials vary so widely, it is necessary to have a means for measuring these differences. International agreement was reached long ago to standardize on a unit termed the ohm as the measure of resistance. The ohm is named after another late 18th- and early 19-century investigator of electrical phenomena, Georg Simon Ohm. It was Ohm who recognized resistance as an inherent property of all materials. He also worked out the law bearing his name that explains the relationship between voltage, amperage, and resistance.

Ohm's Law

Ohm's Law states an absolutely fixed relationship exists between current, voltage, and resistance such that the current flowing in a circuit is directly proportional to the applied voltage, and inversely proportional to the resistance. Expressed in words, this sounds rather complicated, but it can be reduced to a very simple and easily understood mathematical formula. This formula can be stated three ways, but first for mathematical purposes the following symbols are used:

Electrical current in amperes = I
Electrical pressure in volts = E
Electrical resistance in ohms (Ω) = R

1. The current in amperes is equal to the pressure in volts divided by the resistance in ohms.

$$I = \frac{E}{R}$$

2. The resistance in ohms is equal to the pressure in volts divided by the current in amperes.

$$R = \frac{E}{I}$$

3. The pressure in volts is equal to the current in amperes multiplied by the resistance in ohms.

$$E = I \times R$$

With any two factors known, the third can easily be calculated by either division of multiplication. All that is necessary is to keep track of when to multiply and when to divide. The following diagram is often helpful:

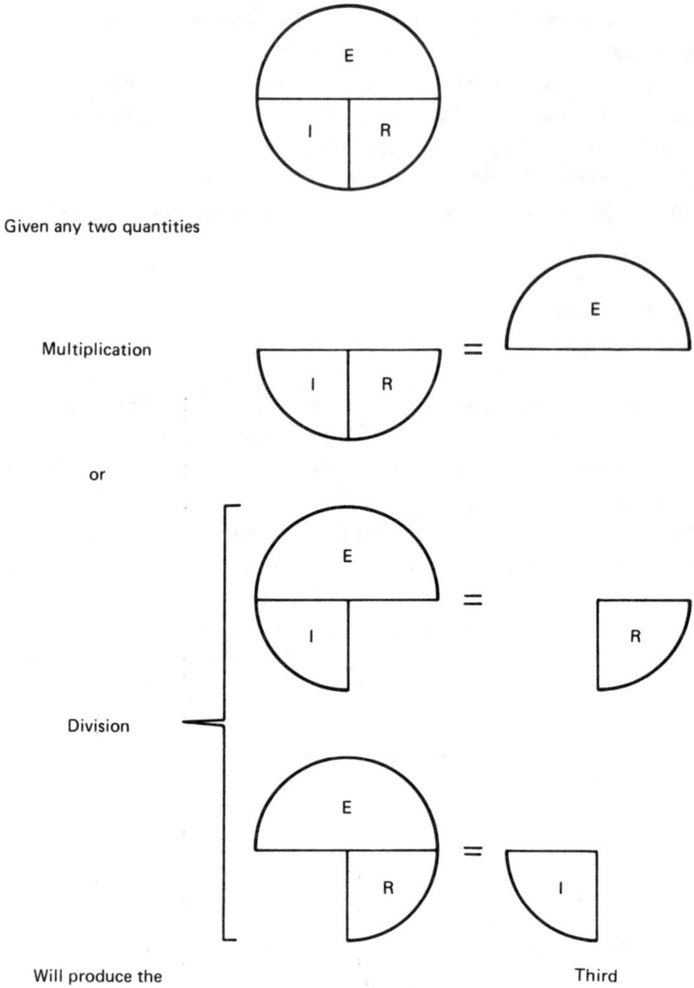

Figure 1-5
Ohm's law.

EXAMPLE Using Ohm's Law a device with a resistance of 18 ohms will be connected to a 120V circuit. What amperage will it draw?

$$I = \frac{E}{R} = \frac{120}{18} = 6.6667 \text{ A}$$

EXAMPLE A countertop appliance draws 4A at 120V. What is its internal resistance?

$$R = \frac{E}{I} = \frac{120}{4} = 30$$

EXAMPLE An electric dryer has a resistance of 10.67 and draws 22.5 A. What voltage should be supplied?

$$E = I \times R = 10.67 \times 22.5 = 240V$$

An ohmmeter can be used to read resistances directly only when the circuit is off. However, many electrical devices that show very little resistance when off and cold, increase in resistance dramatically when turned on and hot. A broiler, a toaster, or a tungsten light bulb are examples of this. A 60 watt tungsten light bulb has a cold resistance of only 5 ohms. A resistance of 5 Ω with a pressure of 120V would mean that

$$I = \frac{120}{5} = 24 \text{ amperes}$$

a current of 24 amperes would be drawn by that bulb. What happens in fact is completely different. The filament heats instantaneously which increases its resistance instantaneously from 5 Ω to 240 Ω. This resistance, however, can be found only by computation rather than direct measurement. This particular computation requires the use of another formula in addition to Ohm's Law. This one was formulated by James Watt and is known as Watt's Law.

Watt's Law

This law is named after the same James Watt who invented the reciprocating steam engine. After doing so he found it difficult to sell to a skeptical public until he could work out a way to compare its power to perform work with that of a horse. The horsepower ratings for not only steam engines, but also gasoline, diesel, and electric motors are derived from his basic formulations.

In the same way that a fixed relationship exists in Ohm's law between voltage, amperage, and resistance, so in Watt's Law a fixed relationship exists between power, expressed in watts, and amperage, and voltage. Watt's Law states that the power available in watts is equal to the amperage multipled by the voltage.

$$P = I \times E$$

There are two other common versions of this formula. One is the current in amperes is equal to the power in watts divided by the voltage.

$$I = \frac{P}{E}$$

The other version is the voltage is equal to the power in watts divided by the amperage.

$$E = \frac{P}{I}$$

As with Ohm's Law when any two quantities are known the third is obtained by simple multiplication or division. A similar diagram may be drawn for Watt's Law

Watt's Law provides a simple method for converting watts to equivalent amperage, and vice versa (Figure 1-6). This type of computation is needed, as will be seen in Chapter 8, to determine connected loads on circuits. Loads must be known accurately in order to insure that proper wire and breaker sizes are specified. Load computations are also used in

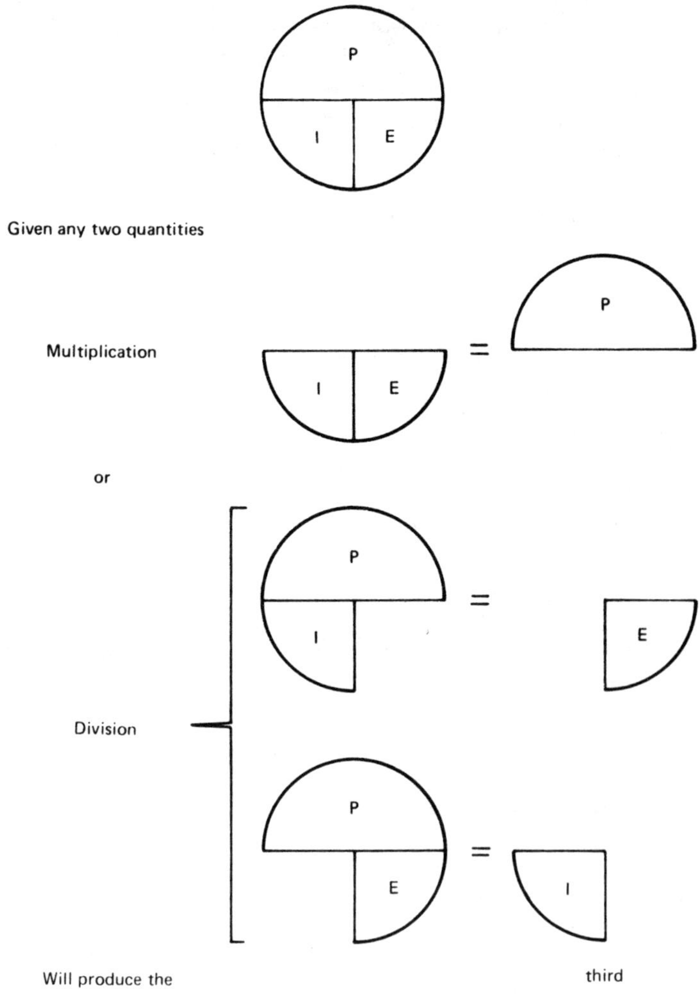

Figure 1-6
Watt's law.

troubleshooting to determine when a circuit is overloaded, and help determine proper action when an overload is found.

EXAMPLE A 1 horsepower electric motor draws 746 watts of power at 120V. Will it operate satisfactorily on a 15 A circuit?

$$I = \frac{P}{E} = \frac{746}{120} = 6.22 \text{ A}$$

No problem!

EXAMPLE A coffee maker drawing 1000 W, a toaster at 1200 W, and an egg cooker at 600 W are plugged into a 20A 120V counter-top appliance circuit. Any two will operate satisfactorily, but as soon as the third one (no matter which one it is) is turned on the breaker trips. What is the matter?

$P = I \times E = 20 \times 120 = 2400$ watts is the maximum the circuit can handle. Above that wattage the breaker *should* trip.

Coffee maker	1000	
Toaster	1200	
Total	2200	No problem
Coffee maker	1000	
Egg maker	600	
Total	1600	No problem
Toaster	1200	
Egg maker	600	
Total	1800	No problem
Coffee maker	1000	
Toaster	1200	
Egg maker	600	
Total	2800	Overloaded by 400 watts

*National Electrical Code® and NEC® are Registered Trademarks of the National Fire Protection Association, Inc., Quincy, MA.

Electrical Measurements

The electrician wiring residences or other small buildings actually takes very few electrical measurements. However, when he does need a measurement, he must know what instrument to use and how to use it. His workhorse and most commonly used instrument is the multimeter (Figure 1-7). This instrument will give readings in AC volts, DC volts, ohms (resistance), or milliamperes. In price they vary from about $20.00 to approximately $1,000.00.

The immense precision of the expensive instruments is not necessary for building wiring. A small, inexpensive instrument is entirely satisfactory. In fact it is preferable since it is compact and can be carried in a pocket while its owner squirms in and out of unlikely nooks and corners in the normal course of completing his work. If it is accidentally smashed in the process, he hasn't lost much.

The multimeter will have a central function selector switch to shift between the various AC (alternating current), DC (direct current), ohms, and milliampere scales. In normal building wiring only AC voltage and ohms scales are used. AC voltages in a building will be either 120 nominal, or 240 nominal. Nominal means that the actual voltage at any given time might vary anywhere between 110V and 120V, or between 220V and 240V. When reading building voltages be sure the selector switch is in fact set on to AC. If, by accident, it is set to DC, there could be 240V AC in a receptacle, but the multimeter will give a reading of 0 volts because there is no *DC voltage* at that point. Your instrument will gladly give you

Figure 1-7
Multimeter.

accurate information, but you must ask it the right questions. If it is asked whether there is DC voltage in AC outlet, it will correctly answer "0." If asked whether there is AC voltage in a car battery it will equally correctly answer "0."

The ohms scales will primarily be used to check circuit continuity, and to test for shorts. When using the ohms scales *be sure the power is off*. Resistances cannot be read on a live circuit, only on a dead one. If power is on, it will burn out the meter. After setting the selector to ohms, and before taking any readings, short the test probes to each other, and use the "ohms adjust" control to set the meter accurately on 0. If this is not done, misleading readings might result.

The measurement of amperage is not commonly necessary when wiring small buildings. However, when troubleshooting it is a very helpful measurement to have when tracking an overload that keeps tripping a breaker. The instrument to use for this purpose is a clamp-on field sensing ammeter (Figure 1-8). This meter when clamped around the power lead from a breaker will detect the electrical field around it and translate the intensity of that field into a measurement of the amperage flowing in that wire. In order to read amperage, this meter must be clipped around the hot wire only. For reasons mentioned in the next chapter, if it is clipped around the complete cable feeding an appliance, it will read 0 instead of the amperage being drawn by the appliance. To read the exact amperage drawn by a plug-in appliance, an adapter such as is shown in

Figure 1-8
Measuring circuit amperage in power ire, leaving breaker with clamp-on ammeter.

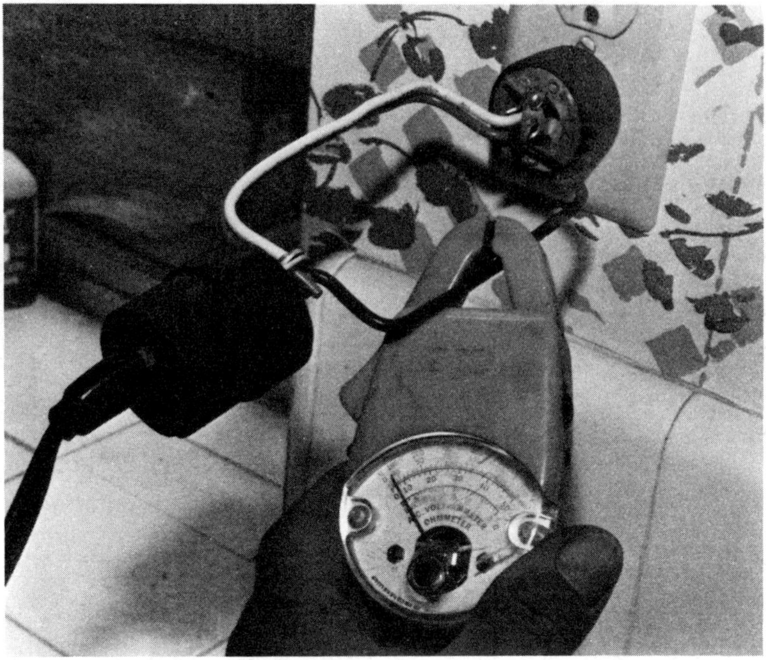

Figure 1-9
Reading amperage of appliance using author's adapter and clamp-on ammeter.

Figure 1-9 is necessary to separate the hot wire from the common in the appliance feed.

In addition to voltage, resistance, and amperage, wattage, the fourth factor in electrical computations, can also be directly measured with a wattmeter. The wiring of buildings discussed here does not require this measurement since wire sizing, breaker sizing, and circuit loading are specified and limited in the code by amperage rather than wattage. In all likelihood the only contact the average electrician will have with wattage measurement will be the installation of the meter box in the service entrance (Chapter 9). It is here that the utility company mounts their watthour meter (Figure 1-10) to record electrical power usage for billing purposes.

Basic Electrical Circuits

Since an electrical current will flow readily through a conductor, it is a simple matter to direct electrical energy from a remote source to a desired point by connecting conductors to form a low resistance path from one point to the other. Conductors so connected become an electrical circuit. The simplest electrical circuit consists of a minimum of four parts (Figure 1-11). They are a source of electrical pressure or voltage, conduc-

Figure 1-10
Watthour meter.

tors to connect the source to the use point; an electrical load or using device; and a switch or other mechanism to control that load. Since a current will only flow when the path is complete from the high pressure, or hot side, of the source back to the low pressure, or grounded side, a return conductor is necessary to complete the circuit.

Electrical loads can be connected to a power source in either of two ways (Figure 1-12). They can be connected in ***series*** or in ***parallel.*** In the case of a series circuit, there is only one path through which current can flow, and consequently the same amperage flows through all parts of the circuit. In the case of the parallel connection, there is a separate electrical path through each load with part of the amperage coming from the source passing through each path. The part of the total amperage drawn that passes through each load is proportional to its wattage and inversely proportional to its resistance.

In Figure 1-13 three loads connected in parallel on one circuit are shown. The total wattage being drawn is the sum of the three loads:

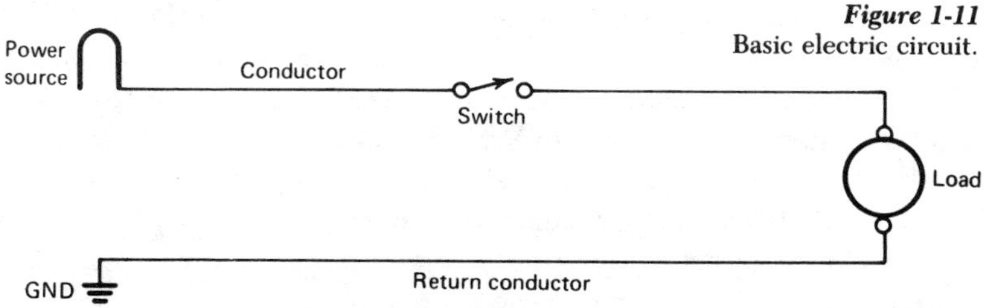

Figure 1-11
Basic electric circuit.

$$\begin{array}{r} 300\ \text{W} \\ 100\ \text{W} \\ \underline{75\ \text{W}} \\ 475\ \text{W} \end{array}$$

Using Watt's Law to determine the total amperage being drawn:

$$I = \frac{P}{E} = \frac{475}{120} = 3.958 \text{ amperes}$$

The amperage being drawn by the television alone:

$$I = \frac{P}{E} = \frac{300}{120} = 2.5 \text{ amperes}$$

Its resistance by Ohm's Law:

$$R = \frac{E}{I} = \frac{120}{2.5} = 48 \text{ ohms}$$

The fan by contrast draws an amperage:

$$I = \frac{P}{E} = \frac{75}{120} = .625 \text{ amperes}$$

But has a resistance vastly greater:

$$R = \frac{E}{I} = \frac{120}{.625} = 192 \text{ ohms}$$

Let us note something else that is happening in this parallel circuit that will help in understanding overloads. The total resistance on this circuit as shown, again using Ohm's Law, is:

$$R = \frac{E}{I} = \frac{120}{3.958} = 30 \text{ ohms}$$

Now, we already have a resistance of 48 ohms, and another of 192 ohms. We have not calculated the third one, but we know it will be somewhere between the two figures. How can the total resistance of the circuit be only 30 ohms? This brings us to the fact that in parallel circuitry there are multiple paths through which the current may pass. Regardless of how high the resistance of a particular path is, as soon as that path exists ***some*** current can pass through it—some current that could not pass, and was

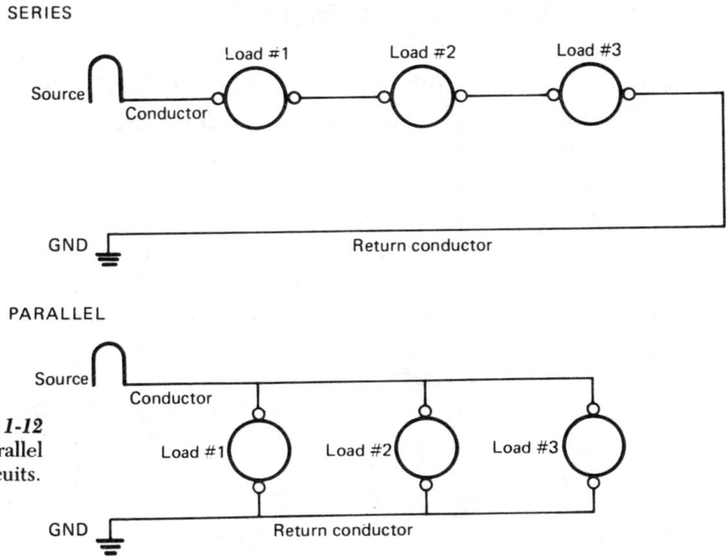

Figure 1-12
Series and parallel circuits.

not passing through other existant paths. Therefore, the more paths there are, regardless how high their resistances may be, the more current the circuit will allow to pass, because with the opening of each additional path the total resistance of the circuit has been reduced.

In building wiring all power-using devices are wired in parallel because wired this way each is independent of the others. Referring to the series circuit in Figure 1-12, if Load #2 were to break down, #1 and #3 will stop as well because only one electrical path exists and it has been broken. If #2 in the parallel circuit failed, this would have no effect on either #1 or #3, because each has independent access to the power source.

In building wiring the basic rule is: "All loads are wired in parallel, all switches are wired in series." A switch completes or breaks an electrical path, but uses no power. It presents either no resistance, or infinite resistance to the passage of an electrical current. Its purpose is merely to open or close the path to some electrical equipment.

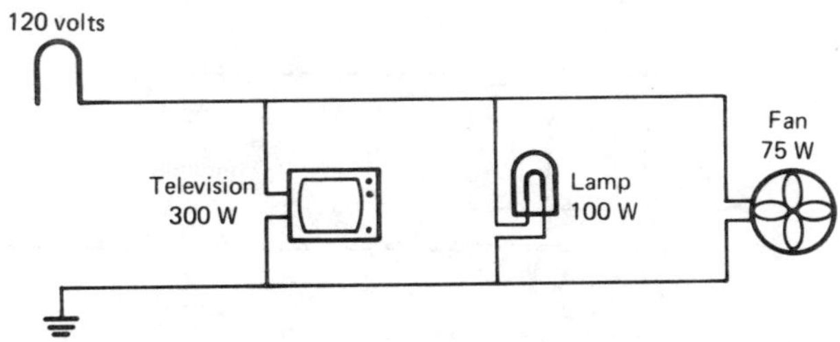

Figure 1-13
Three loads in parallel.

Effects of Electrical Energy

Electrical energy can easily be channeled so as to produce heat, magnetism, chemical reactions, and, as some know all too well, physiological effects as well. All of these effects involve the conversion of energy from one form to another. Such conversions always involve some loss.

Mechanical energy applied to an apparatus encounters resistance in the form of friction within the mechanism. In the process of transmitting power from the engine to the wheels of an automobile, some of that power, despite the best lubrication, is lost in friction. Well, actually it is not lost. It is still present in the form of heat that develops at friction points. A transformation of mechanical energy into heat energy has taken place.

Similarly electrical energy is transformed into heat in the process of overcoming the resistance in a conductor. Conductors specifically designed to maximize this transformation are used in the heating elements of certain electrical appliances, such as toasters, broilers, electric ranges, water heaters, clothes dryers, and the other useful electrical heating equipment. These are all common and well known uses of the heating effect that can be produced with electrical energy.

While not commonly thought of in this way the incandescent light bulb is another example of the heating effect of electricity. Inside the bulb an electrical current passes through a filament of tungsten wire. The resistance of the wire causes it to heat white hot, producing light. In the process considerable waste heat is produced, as direct contact with a burning bulb will quickly prove.

The fluorescent light is another example of the heating effect of electricity (Figure 1-14). In this instance filaments do not produce light, but act as heaters and ionizing electrodes. Air is pumped from a fluorescent tube after which a bit of argon gas and a few drops of mercury are introduced. The heater current passed through the filaments vaporizes the mercury. The higher ionizing voltage then ionizes first the argon, then the mercury vapor, producing ultraviolet light. The ultraviolet hits the phosphor coating of the tube, producing visible light. The fluorescent tube is a vastly more efficient light than the incandescent; a far higher

Figure 1-14
Fluorescent tube.

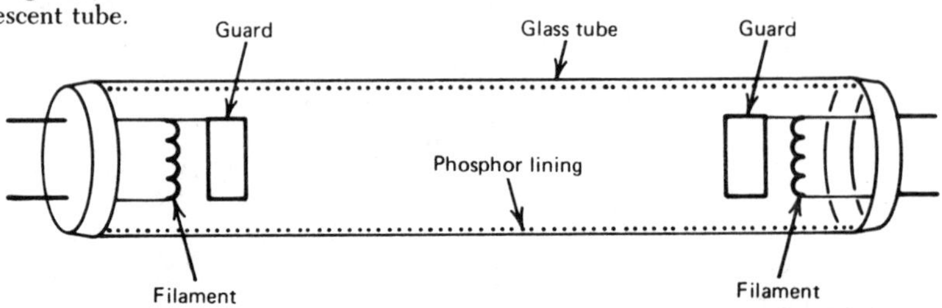

percentage of the electrical energy used appears as visible light and far less is wasted in heat. A touch of an operating fluorescent tube demonstrates the difference.

Other electrical lighting systems such as neon, metal halide, sodium vapor, and mercury vapor lamps are all examples of electrical heating effects.

In addition to producing heat, electricity can be used to produce many other useful results through magnetic effects. As will be discussed in Chapter 2, a magnetic field can produce an electrical current. The reverse is true as well. An electrical current can produce magnetic effects. The electric motor in its many forms is probably the most important use of the electromagnetic effect, but there are many others.

Examples of many other everyday items whose operation is based on magnetism are doorbells, buzzers, telephone transmitters and receivers, solenoid controls, electromagnets, dynamic stereo loudspeakers, and all material recorded on magnetic tape.

Although chemical effects of electricity are perhaps not as commonly encountered as heating or magnetic effects, it is highly likely that this book was printed using electroplated type—a very important chemical effect. You may very well have eaten your last meal with electroplated silverware.

An electro-chemical effect produces the power to turn the car engine over every time it is started, and the drycell batteries used in flashlights are all chemical effect items.

The physiological effects of electricity generally are not the ones we are most eager to encounter. However, while we tend to think of these effects as generally unpleasant, there are some extremely useful ones as well. Remember, the pacemaker on which many heart patients depend is an apparatus that regulates the heart beat electrically. The defibrillator found in all coronary care units is another very important lifesaving means of regulating heart action. In addition the medical arsenal contains a number of other pieces of electrical equipment with important uses in the saving and maintaining of life, and the improvement of comfort for ill patients.

CHAPTER 2

DISTRIBUTION OF ALTERNATING CURRENT

Electrical current as discussed in the previous chapter exists in two forms: direct current (DC) and alternating current (AC). Direct current is produced in several ways. Alternating current is produced only by magnetic mechanical generators. Direct current flows steadily in one direction, and often, but not always, at constant voltage. Alternating current rhythmically reverses its direction, and constantly increases and decreases in voltage.

DC Sources

One form of DC is thermoelectricity. When two dissimilar metals are connected (Figure 2-1) and heated, at one end a small electrical current is generated. This is the principle behind the safety thermocouple used in both gas water heaters and ovens. One end of the thermocouple is mounted in the flame of the pilot light. The small current thus generated tells the control unit that the pilot light is indeed on, making it safe to turn on the main burner.

Another form of DC is pizeoelectricity. Pressure applied to certain crystaline substances such as Rochelle salts or barium titanate produces a weak DC current. The pressure of sound waves on the diaphragm of a crystal microphone causes the crystal to produce an electrical signal. The same principle applies in the case of the crystal phonograph pick-up.

Another source of DC is the photovoltaic cell using silicon to generate an electrical current in the presence of light. Photovoltaic cells, although very expensive until recently, have been a primary source of power for the electrical equipment used in many satellites. Extensive

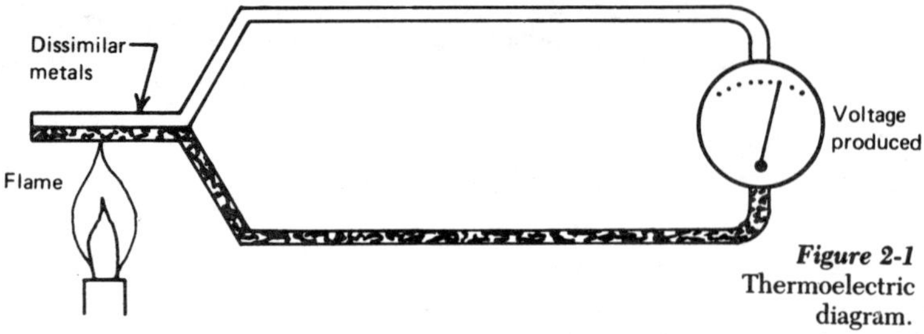

Figure 2-1
Thermoelectric diagram.

development is under way with an eye to reducing the cost and augmenting the output of photovoltaics. Several manufacturers are expressing confidence that these devices will become economically feasible in the near future. Presently, a photovoltaic powered pocket calculator is available at about the same cost as the battery powered types.

The DC source with which many are most familiar is electrochemical: drycells, and storage batteries (wet cell). The drycell, familiar in many forms, powers flashlights, portable radios, pocket calculators, and many other items. The storage battery is probably most commonly encountered as the battery in our cars, boats, or recreational vehicles.

Both the drycell and the storage battery produce electricity through chemical interactions. Although some rechargeable drycells have become available, most are not rechargeable. When the chemicals in the cell are used, the cell dies and must be replaced. By contrast the storage battery is rechargeable. Its chemical reaction is reversible so the battery can be recharged and used again. It is recharged with the use of electrical power produced by magnetic-mechanical means.

Magnetic-Mechanical Generation and Alternating Current

The electrical power produced and delivered by the utility companies is alternating current, or simply AC. Consequently the building power and lighting circuits installed by the electrician will use AC. The term alternating current refers to the voltage in this type of power (Figure 2-10) that constantly alternates over time from 0 volts to a controlled maximum voltage, back to 0, then down to the same maximum in the opposite direction, and finally back to 0 to start the cycle over again. The current used in the United States cycles at the rate of 60 times a second, or one complete cycle takes 1/60th of a second. The correct term to use when referring to cycles per second in electrical and radio frequencies is Hertz. AC current thus alternates at the rate of 60 Hertz, which is abbreviated as 60 Hz.

In order to understand why AC voltage alternates, it is necessary to see how it is generated. An understanding of AC generation is dependent on first having an acquaintance with magnetism and electromagnetic induction.

Magnetism and Electrical Induction

Ancient peoples, as mentioned earlier, discovered a peculiar black stone that could attract iron. They also found that if a bar of iron were stroked with such a stone, it too would attract other pieces of iron. Meanwhile, the Chinese had long ago discovered that if a magnet could be suspended in a manner that allowed it to move freely, one end would always point fairly close to what we call north. This was the origin of the magnetic compass, which is still standard equipment on all ships and aircraft regardless of how elaborate their electronic guidance systems are.

Near the turn of the 16th Century, William Gilbert started studying magnets seriously and found that the attracting property of magnets is concentrated at the two ends, termed "poles," with little force detectable between. This observation coupled with the far earlier Chinese discovery pointed to the idea that the entire earth is a colossal magnet with opposing magnetic poles, fairly close to the geographic poles.

It had also been noted that if the north seeking pole of one magnet were brought close to the north seeking pole of another, they repelled each other. Similarly two south seeking poles repelled each other. However, any two opposite poles, a north seeking and a south seeking pole, when brought close to each other are definitely attracted. These observations indicate very clearly that a force field exists in the vicinity of a magnet. This field can readily be demonstrated by placing a thin cardboard on top of a bar magnet, and then sprinkling the cardboard with iron filings. The filings will form a pattern resembling Figure 2-2.

The mutual repulsion of like poles, north and north or south and south, can be visually demonstrated in a similar way by placing similar poles of two magnets under a cardboard and again sprinkling iron filings. This time the pattern will look like Figure 2-3. The mutual attraction of

Figure 2-2
Bar magnet force field.

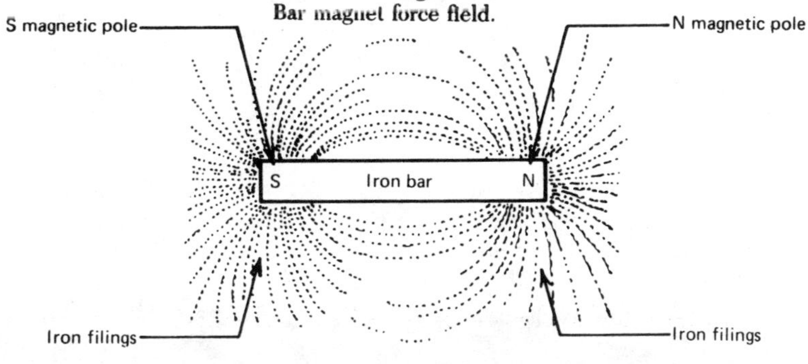

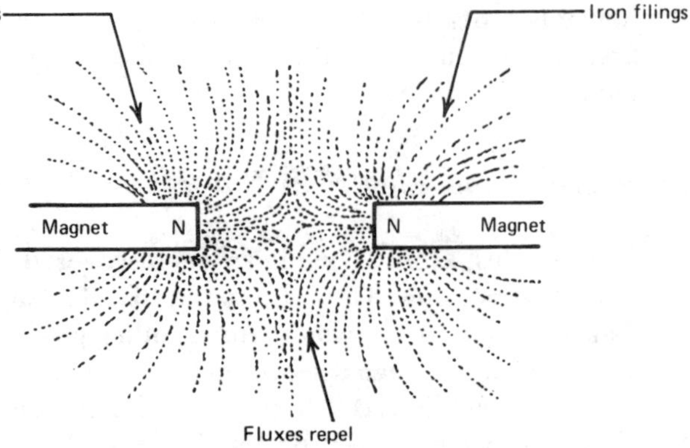

Figure 2-3
N-N or S-S repulsion pattern.

north and south poles when demonstrated in the same manner will produce the pattern of Figure 2-4.

As was mentioned in the previous chapter, positive and negative electrical charges exhibit a behavior curiously similar to that of polar magnetism. Positive repels positive, negative repels negative, but positive and negative attract each other. As we shall see by looking more closely at magnetism, these similarities are most certainly related.

Further investigations in the 19th Century by a German physicist Wilhelm Weber resulted in the molecular theory of magnetism. This theory proposes that the individual molecules of a material that can be magnetized such as iron or steel are each miniature magnets with their own north and south poles, and their own magnetic fields. When a piece of such material is unmagnetized (Figure 2-5), the molecules are arranged in random fashion with the result that the magnetic fields of the molecules cancel each other out. This neutralizes the material so no external magnetic field results. However, when that material is magnetized the molecules (Figure 2-6) align themselves in an orderly manner like a proper

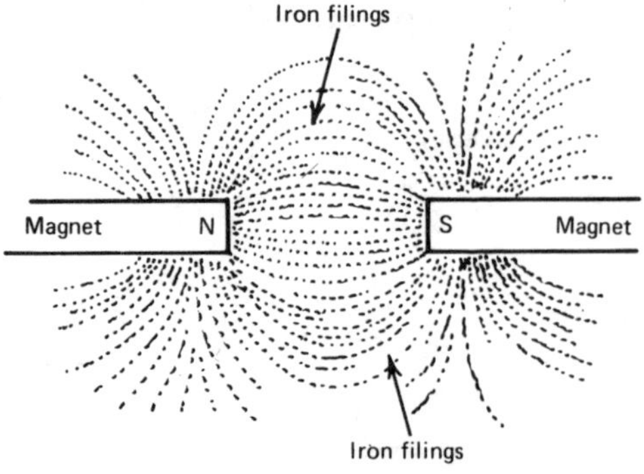

Figure 2-4
N-S attraction pattern.

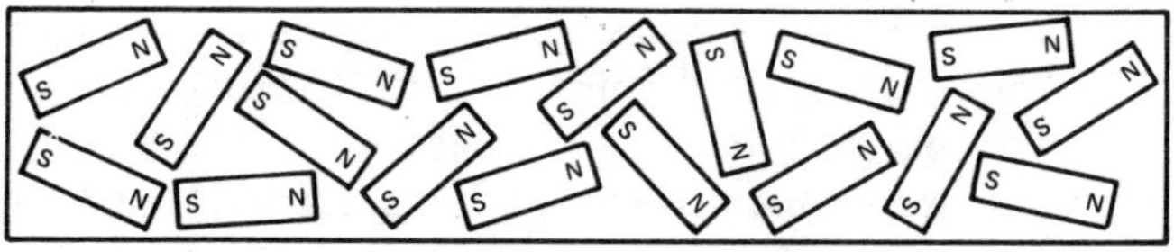

Figure 2-5 Molecular theory random pattern.

Prussian infantry company. The north pole of one molecule is facing the south pole of the next one with the result that the various individual molecular magnetic fields reenforce each other. This, in turn, produces a magnet with an external field.

Weber's model explained much of what had been observed regarding magnets and has only recently been modified in light of current atomic research. It had long been recognized that magnetism and electricity are integrally interrelated, but it had not been proposed that this interrelationship is consistant on the level of subatomic particles.

Looking back again at the model of the atom, we have noted that the electrons orbit in concetric ***shells*** around the nucleus. It has been found that in addition to rotating around the nucleus the electrons each spin on their own axis just as the planets of the solar system spin on theirs. The discovery of electron spin has led to the further discovery that magnetism appears to be related to the spin of the electrons particularly in the third shell of atoms of magnetic materials.

As mentioned earlier the electron has an electrically negative charge. When it spins, its electrical field spins with it creating a magnetic field the polarity of which depends on the direction of spin of the electron. With most materials half the electrons spin clockwise, the other half counter-clockwise. The opposing polarities thus cancel each other out, leaving no external magnetic field.

In the case of a magnetic material there is an imbalance. More electrons rotate in one direction than in the other so that the magnetic fields do not cancel out completely. The magnetic material iron is a good case in point. Its atomic number is 26 meaning it has 26 protons and 26 electrons. Those 26 electrons are arranged in shells as with any other atom. The first shell has the normal two electrons. Since they spin in opposite directions they neutralize each other magnetically.

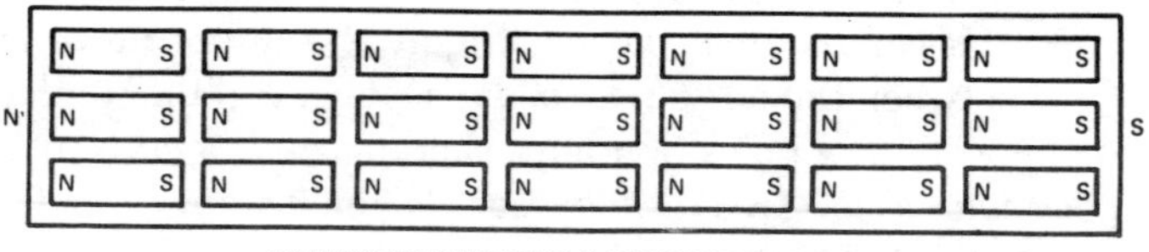

Figure 2-6 Molecular theory alignment pattern.

The second shell has its full complement of eight electrons which spin four one way, and four the opposite way so the second shell internally cancels itself out in the same way as the first. The third shell is, however, incomplete. It can hold 18 electrons, but in the case of an iron atom it only has 14. Of those 14 half of the number that would constitute a complete third shell spin one way, and the rest spin the opposite. Half of 18 means that nine spin one way with only five to oppose them. This leaves an external magnetic field resulting from the four uncanceled fields. The fourth shell has only two valence electrons spinning in opposite directions thus they again cancel each other out. The final result is that the atom as a unit has an external magnetic field due to the imbalance of four in the third shell.

The effect of the force fields of individual atoms on each other causes groups of nearby atoms to align themselves so that their various fields reenforce each other. Such a group is called a ***domain.*** In a piece of unmagnetized iron, as an example, these domains are placed in random in relation to each other similarly to Weber's molecules shown in Figure 2-5. When the material is magnetized, the domains are forced into alignment like the molecules in Figure 2-6. Actually, while Weber talked of ***molecules*** and modern physicists speak of ***domains*** the two explanations of magnetic behavior are essentially the same in terms of accounting for observed phenomena. The advantage of the modern explanation is that it shows more clearly ***why*** things happen as they do.

Electrical Induction It had long been noted that the reactions of electrical charges and magnetic poles followed the same rules. Like charges and like poles repel each other. Unlike charges and unlike poles attract each other. This similarity between electricity and magnetism quite naturally raised the question whether there is any connection between the two. The first connection was found when it was discovered that an electrical current flowing in a conductor creates a magnetic field around that conductor (Figure 2-7). The field so created is of a different shape from that of a bar magnet. It is circular and has a clockwise direction relative to the direction of flow of the current. Regardless of its shape, however, it very clearly is a magnetic field. This finding naturally raised the question whether the reverse would work. Could a magnetic field create an electrical current in a conductor? Michael Faraday found that indeed this could be done. There are two classic demonstrations of the ways this can be accomplished.

If a conductor is moved (Figure 2-8) up or down in the attraction field between the north pole of one magnet and the south pole of another, or between the poles of a horseshoe magnet, a current will flow in that conductor as long as it is in motion. When it stops, even though it remains within the field, the current will stop as well. So, current can be generated by moving a conductor in a magnetic field.

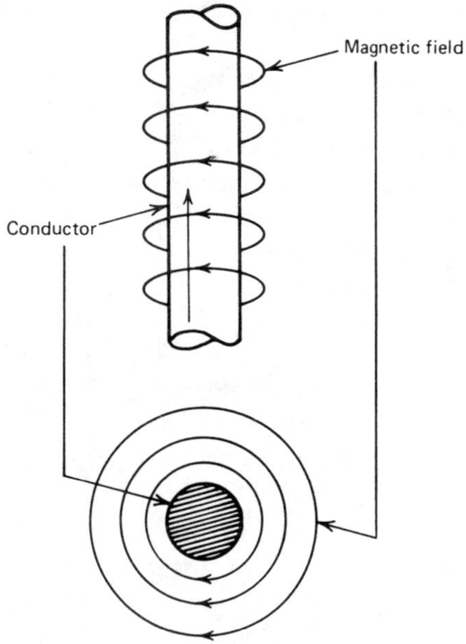

Figure 2-7
Magnetic field around a conductor.

The other way of accomplishing the generation of electricity by the use of magnetism is to coil the conductor, leave it stationary, and move the magnet relative to the coiled conductor. Again, when the magnet stops, whether inside or outside the coil, the current stops as well. Thus given the two basic components—a conductor and a magnetic field—an electrical current will be produced as long as the two are moving relative to each other.

After the discovery that moving a conductor and a magnetic field relative to each other will produce voltage it was soon noted that any increase in the speed of that movement will increase the magnitude of the

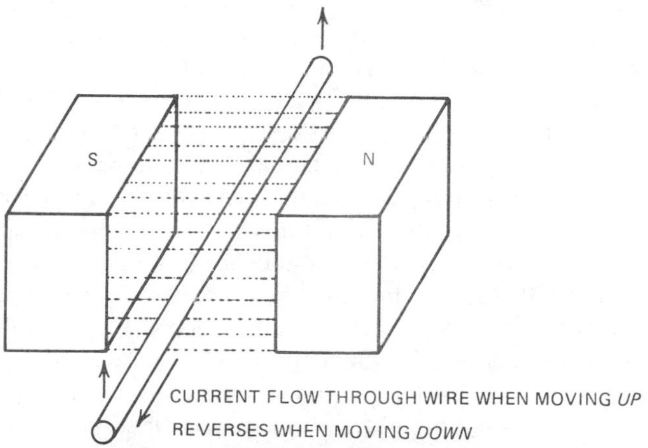

Figure 2-8
›nductor moving in a magnetic field.

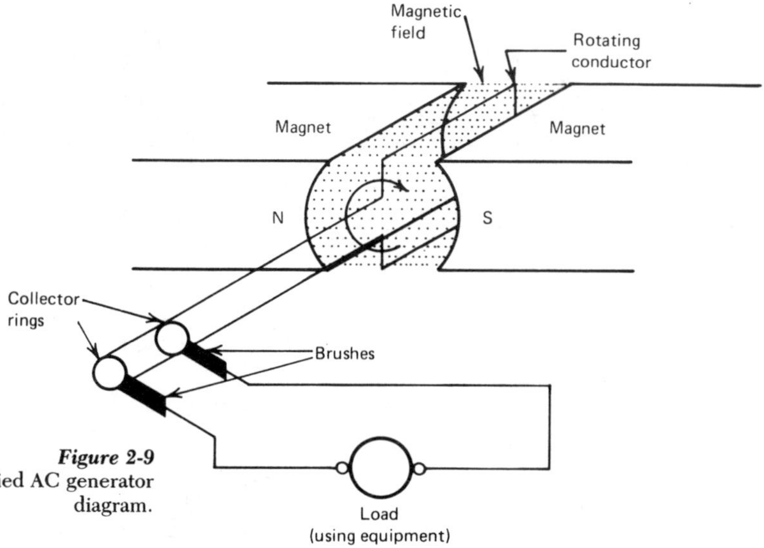

Figure 2-9
Simplified AC generator diagram.

voltage produced. It was also found that the stronger the magnetic field becomes the greater will be the voltage induced. Additionally, the greater the number of turns in a coiled conductor the greater the induced voltage will become. Thus by various means the physical energy of movement in conjunction with a magnetic field and a conductor can be used to produce electrical energy.

Induction and AC Generation

Since the movement of a conductor in a magnetic field will induce an electrical voltage, and the voltage varies with the movement, then if a way could be found to maintain constant and steady movement, it would then be possible to make a steady and constant flow of electrical energy. Let us (Figure 2-9) see how this is done.

A simplified AC generator consists of a magnet to provide the field, a loop of conductor arranged so as to rotate within that field, and a pair of collector rings to take off the electrical energy produced in the conductor as it rotates. As the loop rotates it cuts across magnetic lines of force inducing voltage. The voltage induced at any given time is a function of the number of lines of magnetic force being cut at that time. As will be seen from Figure 2-10, that number of lines of force changes continually as the conductor rotates in the field of the magnet.

At 0° the conductor loop is moving in line with the flux of the magnetic field and cutting no lines of force, thus no voltage is generated. As it turns through 45° it cuts more and more lines of force the farther it turns. As it turns through 90° it cuts the maximum possible number of force lines for this magnetic field, and thus has reached the maximum

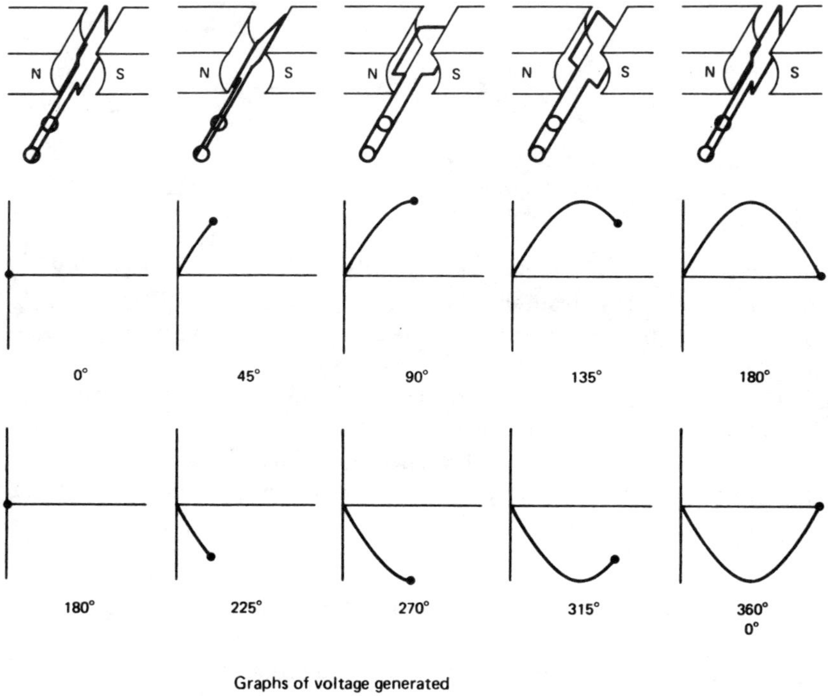

Figure 2-10 AC generator cycle.

possible voltage. Beyond 90° it cuts fewer and fewer force lines the farther it goes until it reaches 180° where it again is moving parallel to the force lines of the magnetic field, cutting no lines, and producing no voltage. Everything is as it was at 0° ***except*** for the fact that the half of the rotating loop that just moved past the south pole of the magnet is now starting past the north pole, and vice versa. This reversal of the magnetic field relative to the moving conductor will also reverse the flow of the voltage induced in that conductor. During the second half of the rotation from 180° back to 0° the curve representing the magnitude of the voltage being generated will be exactly the same as the curve for the first half of the rotation from 0° to 180° except that the ***direction*** is reversed. It is this reversal of direction in the generating process that gives its name to ***alternating current***.

The graphic curve representing the voltage changes that occur during one full rotation of the conductor in the magnetic field of a generator (Figure 2-11) is called a ***sine curve.*** Another way of stating it is to say that AC voltage follows a ***sinusoidal*** curve. One rotation of the armature (the

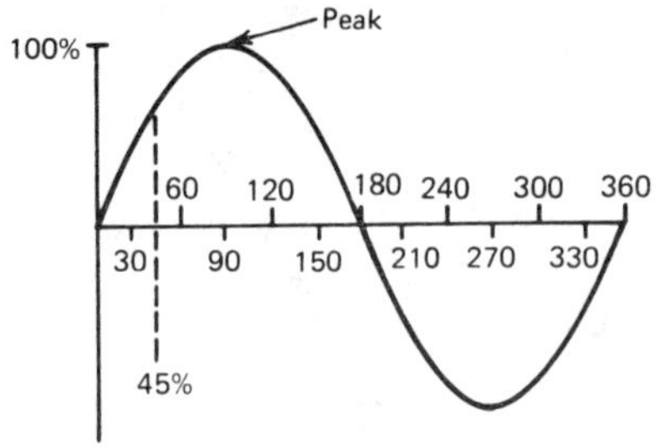

Figure 2-11
One full AC cycle (sine wave curve).

rotating part of an AC generator) takes that voltage curve (Figure 2-11) through one full cycle from 0 volts up to maximum on one side, back to 0, down through the maximum in reverse and back to 0 at a rate or ***frequency*** equal to the speed of rotation of that armature. The standard frequency of AC voltage in the United States as mentioned earlier is 60 cycles per second, or 60 Hz. This frequency is held steady in power generating plants by carefully controlling the speed at which the armatures of the large generators are turned by either steam turbines or other mechanical drive systems.

Single Phase and Three Phase Power Generation

Figure 2-11 shows in graphic form the voltage produced by a generator with a single armature coil. A generator of this type is called a ***single phase*** generator. To produce the voltage illustrated the armature coil had to be mounted on a shaft and rotated in a magnetic field. Since mechanical energy must be used to rotate that shaft why not add more coils each of which will produce a similar voltage? This would result in vastly greater electrical energy output in return for relatively little increase to the mechanical energy put in.

This is precisely what is normally done. It happens that three coils generating three phases (Figure 2-12) is presently the optimum system. Virtually all utilities generate power this way. Coil #1 produces phase A, coil #2 produces phase B, and coil #3 produces phase 3. Each phase alternates independently of the others at a frequency of 60 Hz, but they follow each other in time by 180th of a second since they necessarily pass through the force line variations of the magnetic field in sequence.

Since each of the three armature coils has two ends one might expect that a total of six wires would be required to transmit three phase current. However, using either of two different methods it is possible to join the three coils at the generator so that the three phases can be transmitted by one wire for each phase (Figure 2-13). Using the ***delta*** connection each armature coil is connected to one of the others at each end. With the ***Y*** connection, all three coils meet at a single point. In either configuration, only three wires are needed leading out.

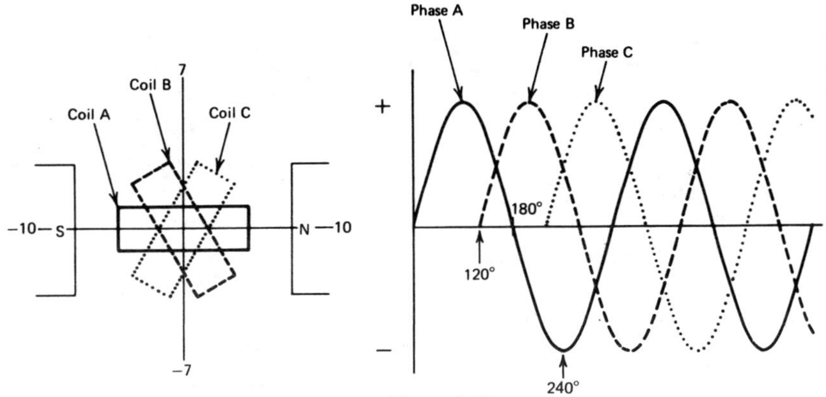

Figure 2-12
Three-phase generator and voltage wave form diagram.

While most electrical power is generated as three phase, the majority of it is finally used as a single phase. The portion used as three phase is primarily powering heavy industrial machinery. In this application three phase is far superior to single phase. An electric motor gets 120 pushes per second from single phase (60 cycles per second each cycle having two peaks). On the face of it it might seem as though that should be fast enough for any normal purpose. An average electric motor turns at about 1800 rpm, which means the armature turns at 30 revolutions per second. With 120 electrical pushes turning the armature at 30 revolutions per second the armature is getting four pushes per turn. If it is a little electric drill motor with an armature diameter of about 2″, the circumference is something over 6″ in which case it is getting a push every 1½″. However,

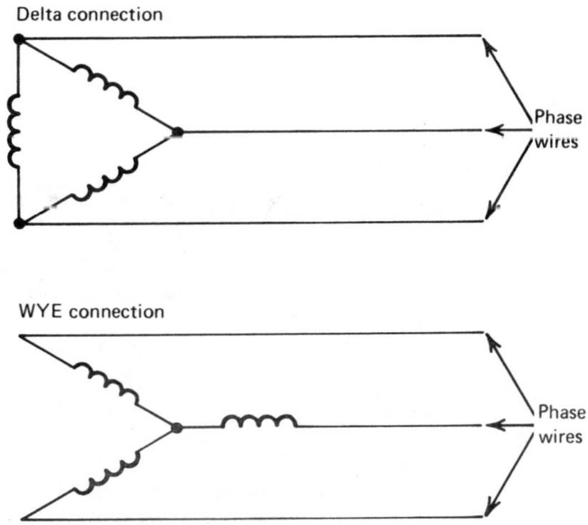

Figure 2-13
Wye and delta three-phase connection diagrams.

when the armature diameter of a somewhat larger motor gets up to the vicinity of 12″, the circumference is now better than 36″. At the same four pushes per revolution, it is now better than 9″ between them. By changing over to three phase current the larger motor is now going to get 12 pushes per revolution, or one every 3″ of its circumference. This will obviously facilitate much smoother operation.

While the larger electric motor benefits both in smoothness of operation and in efficiency through the use of three phase power, smaller equipment is much simpler to construct and operate using single phase power. For this reason although the majority of electrical power is generated and transmitted over long distances as three phase, most of it is ultimately broken down into single phase before it is used. When three phase power is split into single phases, voltage is reduced. The apparatus used to either increase or decrease voltage is a transformer.

Transformers The operation of a transformer brings us back to the subject of magnetic fields. As shown in Figure 2-7, an electrical current creates a magnetic field around its conductor. When that conductor is coiled (Figure 2-14) the resulting magnetic field closely resembles that of a simple bar magnet as shown in Figure 2-2. However, when that field is created by an alternating current in the coil, a significant difference appears. The field created by the bar magnet is constant in both polarity and intensity. The field produced by the alternating current is constant in neither respect. Since the voltage of alternating current is constantly changing, and its direction regularly reverses, the magnetic field it produces when passed through a coil is constantly changing in intensity as well as periodically reversing in polarity.

Figure 2-14
Magnetic field around a coiled conductor.

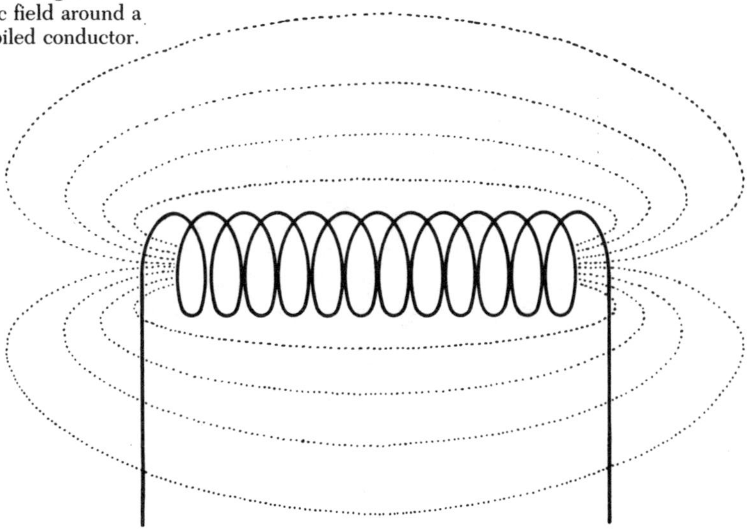

This constantly changing and reversing field around a coil has a very interesting and useful capability. As we have already seen, a conductor and a magnetic field moving relative to each other will induce a voltage in the conductor. It is the conductor cutting magnetic lines of force, due to the movement, that produces this result.

In the case of a coil through which alternating current is passing, the magnetic field around that coil builds up in intensity as the voltage goes up, then collapses as the voltage goes back down. Since the voltage is constantly changing, the field is in constant change as well. As the voltage goes up, the magnetic lines of force around the coil move outward. As the voltage decreases they contract. Thus a conductor placed within the area of influence of such a field is actually being crossed and recrossed by moving lines of magnetic force even though no physical movement is involved. Since magnetic lines of force are being cut, voltage is induced in the second conductor. When the second conductor is also coiled we have a device whereby the magnetic field of the current passing through the first or ***primary*** coil produces a voltage in the second or ***secondary*** coil.

By varying the number of turns in the two coils relative to each other, the voltage induced in the secondary coil can be made either greater or smaller than that passing through the ***primary*** coil. As an example, if the primary coil has 100 turns and the secondary has only 10 turns the voltage in the secondary will be 1/10 that of the primary. If such a primary is plugged into a 120V line, the secondary will produce 12V. The initial 120V has been transformed to 12V, and the pair of coils that together do this are called a ***transformer*** (Figure 2-15).

A transformer that changes 120V current to 12V is called a ***step-down*** transformer because the voltage is being reduced. A transformer that does the reverse is called a ***step-up*** transformer. A step-up trans-

Figure 2-15
120V to 12V step-down transformer.

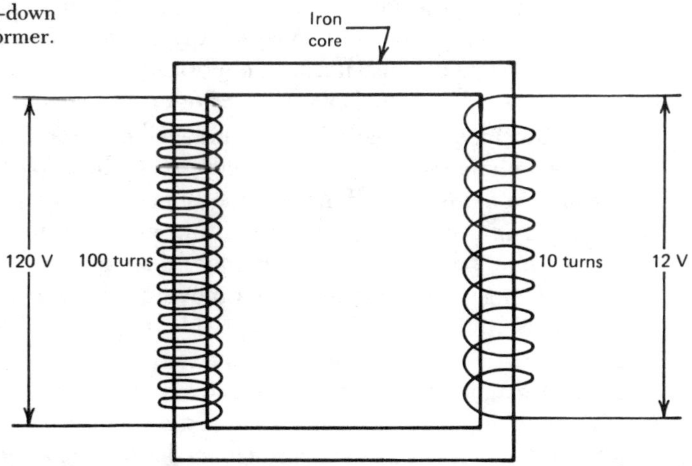

former has more turns in the secondary coil than in the primary, and again transforms the voltage up in proportion to the ratio of turns in the two coils. Take a primary coil with the same 100 turns, but this time put 200 turns in the secondary coil. In this case when the primary is connected to the same 120 line voltage the secondary will yield 240V. The efficiency of a transformer is considerably improved when both primary and secondary coils are wound around an iron core as indicated in Figure 2-15.

The fact that AC can be easily transformed up or down in voltage is one of its major advantages. It can be generated, transmitted over long distances, and finally put to various uses the voltage being changed at each step of the way to one that is optimum for the immediate purpose. We shall examine more closely what some of these voltages changes are, and why, in connection with commercial power generation and distribution. However, first it will be useful to investigate a few other aspects of AC behavior.

Inductive Reactance

As we have seen, when a coil is cut by a magnetic field, however that field is produced, voltage is induced in the coil. It could be produced by a permanent magnet as in Figure 2-8. It could also be produced by an electrical current originating at some external source, and flowing through the coil. When the current flowing through a coil starts to build up a magnetic field around that coil, the field, as it builds up, cuts across the turns of its own coil and induces a second voltage.

Such an induced voltage and the current that flows because of it are always due to a polarity opposing any change in the existing magnetic field. As the source current increases causing the magnetic field around the coil to expand, the induced voltage builds an increasing field in opposition to that of the source current. When the source current starts to decrease, causing its field to collapse, the induced current reverses to support that field. Because it acts in opposition to the source, the induced voltage is termed ***counter voltage*** or ***counter electromotive force*** (cemf).

Obviously, with steady DC this induced counter voltage is a purely momentary phenomenon occurring only when the current is turned on, and again when it is shut off. As was mentioned earlier an induced voltage is produced in a conductor that is cutting magnetic lines of force. The conductor may be moving in a stationary magnetic field (Figure 2-8) or the conductor may be stationary in a moving magnetic field or as in the case of a transformer the magnetic field itself may be expanding and collapsing relative to a stationary conductor. The critical factor for induction is that magnetic lines of force must move relative to the conductor in which a voltage is induced. In the case of the counter voltage produced by an alternating current passing through a coil it is the alternately expanding and collapsing field of the coil itself that induces its own opposing voltage.

The property of opposition to change in current flow, and consequent change in magnetic field intensity, is called ***inductance.*** A component that produces inductance, such as a coil, is called an ***inductor.*** The amount of inductance produced depends on the strength of the magnetic field of the coil, the number of turns in the coil, and the frequency with which the field changes and cuts across these turns.

In a DC circuit the only opposition to the flow of current is resistance. That resistance is made up of the individual resistances of the various components of the circuit. In the case of an AC circuit the presence of an inductor in the circuit causes the appearance of a counter voltage that further opposes the flow of current. When a force in addition to resistance opposes the flow of a current, the total combined opposition is called ***impedance.*** In electrical formulas the symbol for impedance is **Z.** Since Z, like R, opposes the flow of electrical current, it too is measured in ohms and may be substituted for R in the Ohm's Law formulas:

$$E = I \times Z \qquad I = \frac{E}{Z} \qquad Z = \frac{E}{I}$$

Impedance exceeds resistance because of a factor termed ***reactance.*** Reactance is the counter electromotive force, or counter voltage, that only appears in AC circuits as a result of the continuously fluctuating source voltage. In the case of an inductor such as the coils we have been discussing, the reactance (symbol X) is identified as ***inductive reactance*** (symbol X_L).

Inductance and Phase Relationships

In the case of an AC circuit containing only resistive loads, voltage and current remain constantly in phase (Figure 2-16). Both voltage and cur-

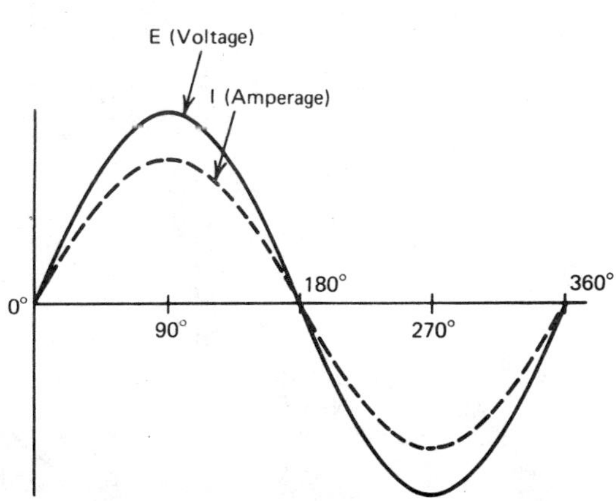

Figure 2-16
Phase relations of voltage and current in a resistive loaded circuit.

rent rise together, fall back to 0, reverse to maximum, and return to 0 again still together. The relative amplitudes of the voltage and amperage curves as illustrated are meaningless. They differ only for clarity of illustration.

With a circuit containing induction the induced counter voltage always opposes change in the source current so that as the current is rising, the counter emf tends to hold its value below what would be attained by the source voltage alone. Then as the current decreases the counter, emf again opposes the decrease. The result is that both the rise and fall of the current lag behind the rise and fall of the source voltage through the entire cycle. If circuit contained only pure inductance the phase relationships between voltage, counter voltage, and current would appear as in Figure 2-17.

The induced voltage is greatest when the current is ***changing*** most rapidly. This is when the current passes through 0 at 90° and 270°. Since the source voltage is opposite to the induced voltage, it too peaks at the same time, but in the opposite direction. The induced voltage drops to 0 when the current is changing most slowly which is when it peaks at 0° and 180°. Since the source voltage is opposite to the induced counter voltage it too drops to 0 at the same times. Thus in a hypothetical circuit that contains only inductance the current would lag behind the voltage by 90°. In practice it is impossible to build a circuit containing only inductance, and no resistance whatsoever. Consequently, in practice due to the inevitable presence of resistance to a greater or lesser degree, the 90° phase lag between current and voltage never does occur.

Capacitive Reactance

The capacitor is another component that acts significantly differently in an AC circuit than it does in the presence of DC. A capacitor is a device that can store an electrical charge. The size of the charge it can store is the

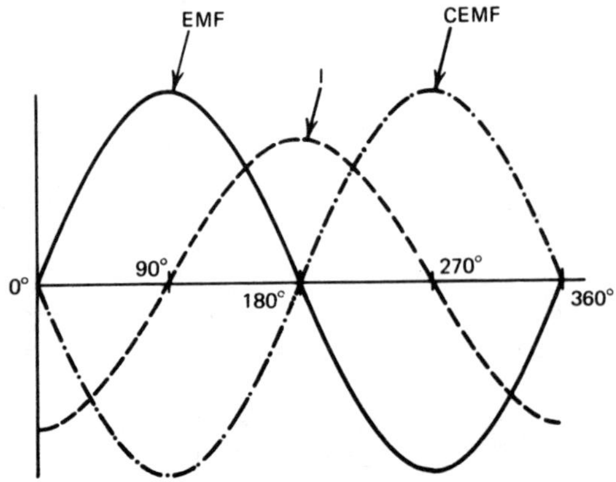

Figure 2-17
Voltage and current phase relationships—purely inductive circuit.

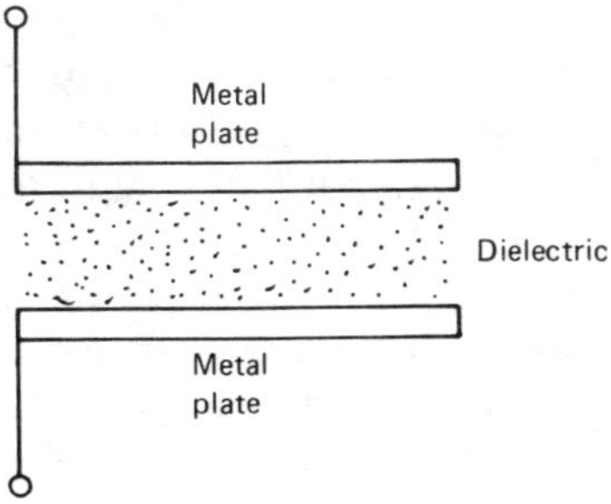

Figure 2-18
Capacitor.

measure of its ***capacitance.*** It is basically an extremely simple construction (Figure 2-18) consisting of two plates of conducting material separated by a layer of dielectric or nonconducting material. The dielectric can be paper, mica, ceramic, even air. Simply placing a dielectric between two plates creates the capability for storing an electrical charge. The only other requirement is a source of voltage. A battery will do quite well. When connected to DC the capacitor is charged by the movement of electrons from plate A to the positive terminal of the battery, and the movement other electrons from the negative terminal of the battery to plate B (Figure 2-19). When the potential across plates A and B is both equal and opposite to that of the battery, the capacitor is fully charged, or as fully charged as it can be from that power source. The capacitor, once charged, will hold its charge for varying lengths of time and can be discharged at will for various purposes as in an automotive ignition system.

When the DC power source is replaced by AC the picture changes considerably. As the AC voltage increases from 0 to its maximum value, current flows from the source to one side of the capacitor, building up a counter voltage in the capacitor just as the battery did, but as the source voltage moves on to its decreasing curve, the counter voltage drives a

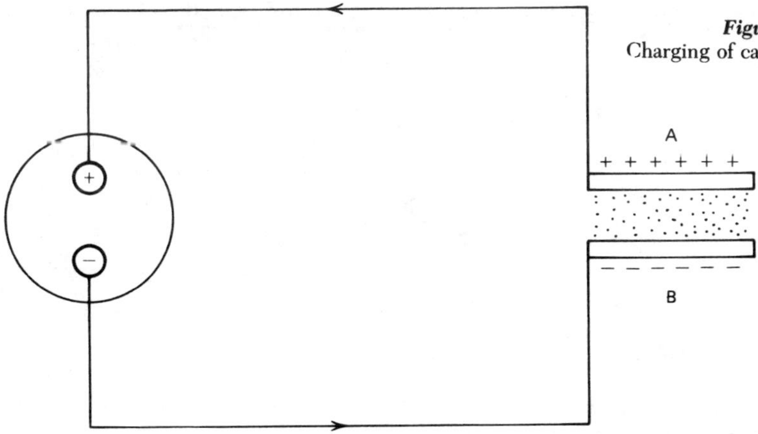

Figure 2-19
Charging of capacitor.

current back from the capacitor to the source. The source voltage now reverses, rising to its opposite maximum. At this point the current driven by the counter voltage in the capacitor and the current driven by the source are both moving in the same direction which is toward the other side of the capacitor. A new counter voltage now builds up on the opposite side of the capacitor. As the source voltage fades from its second maximum, the counter voltage in the capacitor drives a current back from the capacitor to the source to start the cycle all over again.

During this entire cycle current has been flowing throughout the circuit except through the dielectric of the capacitor. If a device such as a light bulb had been connected to this circuit it would light, indicating a steady flow of current in the circuit. For this reason, although nothing actually passes through the dielectric layer of a capacitor, it is said that AC can flow through a series circuit containing a capacitor.

The plates of a capacitor will readily store a charge; thus current can flow onto the plate of an uncharged capacitor essentially unopposed. With no resistance initially there is no voltage drop which means current will lead voltage in a capacitive circuit. In the inductive circuit the inductor opposed changes in current. In a capacitive circuit the capacitor opposes changes in voltage. This opposition is called ***capacitive reactance*** (symbol X_c). Like inductive reactance, it is measured in ohms.

Capacitance and Phase Relationships

In a circuit containing pure capacitance (Figure 2-20), current will flow from the source toward the capacitor only when the source voltage is rising. The maximum current will flow when the voltage is rising most rapidly. This will be at 0° and 180°. At 0° (with the voltage crossing OV) the current flow is at its maximum positive value. From 0° to 90° the voltage continues to rise, but at a rate that steadily slows. As the voltage has been rising, the current has been dropping. At 90° the voltage has

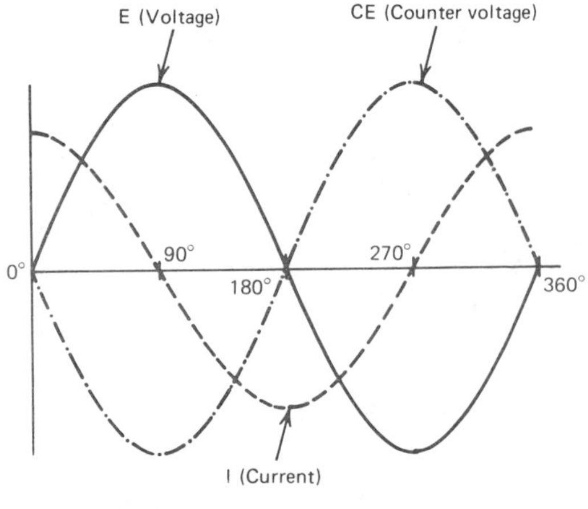

Figure 2-20
Voltage and current phase relationships—purely capacitive circuit.

peaked, the capacitor is fully charged, the counter voltage has also peaked, and the current has reached 0.

Source voltage now starts to fall, as it does so current starts to flow from capacitor to source, and counter voltage decreases as well. At 180° the source voltage has reached OV, current has reached maximum negative value while with the capacitor now discharged the counter voltage has also reached OV.

Source voltage now reverses and steadily increases but at a diminishing rate. Meanwhile current begins to decrease from its maximum negative value toward 0 while the capacitor is becoming recharged in the opposite direction from its first charge. At 270° the source voltage is again at maximum, and current again at 0. As the source voltage falls, the counter voltage drives the current back up to start the cycle again at the 360° mark.

The effect of capacitance, then, is to cause the current to lead the voltage, an effect opposite to that of reactance. In this hypothetical circuit containing only capacitance the current leads the voltage by 90°. For a circuit to contain only pure capacitance is just as impossible as for one to contain only inductance. Either one is necessarily combined with resistance, thus the phase difference will not ever be a full 90°.

Power in an AC Circuit

As we saw earlier in Watt's Law, the power consumed by a circuit is the product of the voltage times the amperage

$$P = E \times I$$

This law holds true as long as voltage and amperage are in phase. The relationship between voltage, current, and power are graphed in Figure 2-21 for such a situation. Where both voltage and current are positive,

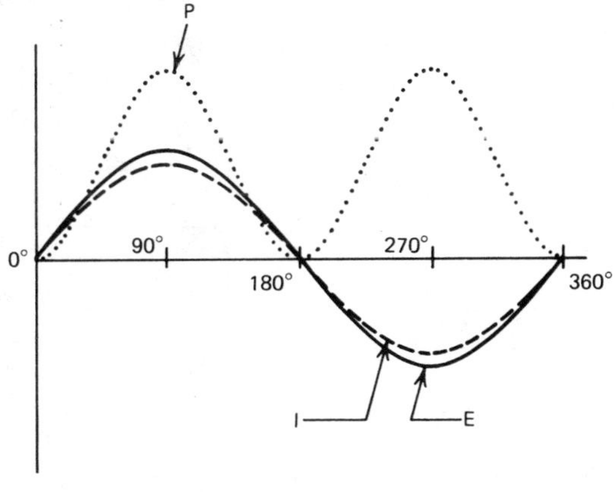

Figure 2-21
Power, current, and voltage graph with current and voltage in phase.

power is positive as well. During the second half of the cycle when both voltage and current are negative, power is still positive since the product of two negative numbers is a positive one.

In a purely resistive circuit, voltage and current will be exactly in phase as shown. However, as we have just seen in a circuit containing inductance current will lag behind voltage, and in a circuit containing capacitance the reverse occurs with current leading voltage. The phase differences in such circuits have a significant effect on the power those circuits consume.

To demonstrate this fact, let us look at the purely inductive circuit with current lagging behind voltage by 90°, and examine its power graph (Figure 2-22). For the first 90° of the cycle, although the voltage is positive, the current is negative. The multiplication of a positive and a negative gives a negative result showing that power is flowing from the circuit back to the source. During the second 90° both voltage and current are positive; thus the multiplication of the two produces a positive result. During the third 90° current is positive, but voltage is negative. Again positive times negative results in a negative product. During the final 90° of the cycle both current and voltage are negative, giving a positive result. During two quarters of the cycle power flowed from the circuit back to the source, and for the other two quarters power flowed out from the source to the circuit. At the end they cancel each other out with the result that the circuit finally consumes no power. Power merely flows from the source out into the circuit, and back.

Of course this pattern will only hold when current and voltage are 90° out of phase with each other, that is only in purely inductive or a purely capacitive circuits. We have also noted that the hardware just does not exist that will allow construction of such completely pure circuits. It is impossible to eliminate resistance. Therefore, in the real world current and voltage get out of phase in circuits containing inductance or capacitance, or even combinations of both, by something a good deal less than

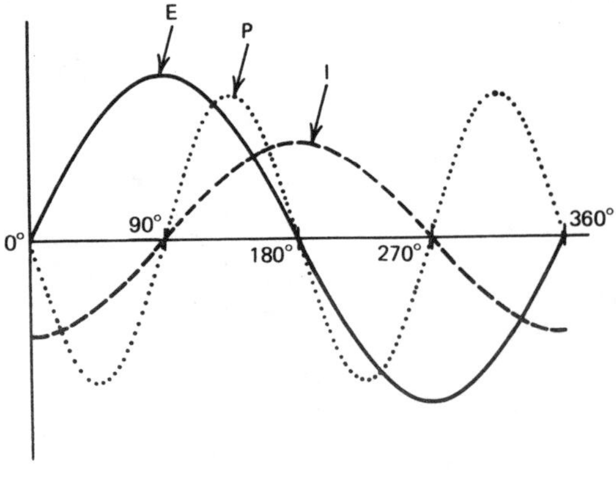

Figure 2-22
Power, current, and voltage graph with E and I 90° out of phase.

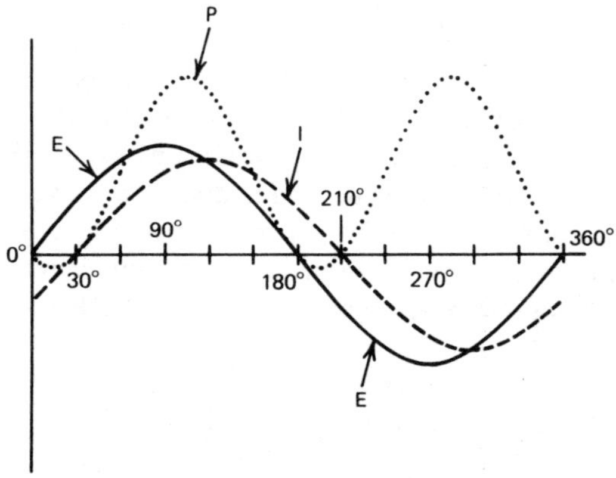

Figure 2-23
Power, current, and voltage graph with I lagging E by 30°.

90°. As an example let us examine what happens when forces combine to cause current to lag behind voltage by 30° instead of 90° (Figure 2-23). Between 0° and 30° and between 180° and 210°, power is negative, flowing back to the source. However, between 30° and 180° as well as between 210° and 360°, power is positive; thus the net power consumed is by no means zero, but rather a considerable positive quantity. This positive quantity, however, is considerably lower than a comparable circuit in which current and voltage are perfectly in phase (Figure 2-21) due to the periods of negative flow.

Power Factor

When current and voltage are out of phase, power is flowing back to the source rather than from it during parts of the cycle. This means that the ***true power*** actually consumed is necessarily less than the ***apparent power*** mathematically derived by multiplying voltage by amperage. The ratio of true power to apparent power is called the ***power factor.*** The formula for finding the power factor is:

$$\text{Power factor} = \frac{\text{True Power}}{\text{Apparent Power}}$$

The power factor may be expressed as a fraction, a decimal, or a percentage. A power factor could be given as three quarters, 0.75 or 75 percent. At 1.0 or 100 percent, power factor, wattage, and volt-amperage are equal—the ideal condition.

The practical importance of the power factor is that it directly measures the efficiency with which the energy fed into a circuit is utilized. A system with a 75 percent power factor is getting 75 watts of useful power out of every 100 volt-amperes being delivered (and billed) by the utility company. The other 25 watts are used fighting counter-voltage generated in the circuit by power using devices connected to it.

Apparent power is easily calculated by measuring voltage with the voltmeter, and amperage with any of several types of ammeters and then multiplying the values thus obtained. True power, however, can be determined only by measuring with a wattmeter. Since true power can only be found by wattmeter measurement, it is referred to in watts to clearly distinguish it from apparent power which is given in volt-amperes.

Commercial Generation and Distribution of AC Power

The AC power supplied by utility companies is generated continuously in very large 3 phase alternators at a voltage of 13,800 volts alternating at 60 Hz. This 13,800V, 60 Hz power passes through a bank of step-up transformers that greatly increase the voltage for transmission purposes. The magnitude of the step-up in any particular instance depends on how much power is required, and how far it must be transmitted. The step-up could be to as low a value as 23,000 volts, or could go as high as 765,000 volts. As the power requirement at the destination increases, or the distance to the destination increases, or both, the voltage increases.

The reason for stepping the voltage up to very high values for transmission purposes is that current causes power loss during transmission not voltage. Since power (P) is the product of voltage and amperage, any given power value can be transmitted at a very, very low amperage as long as the voltage is made sufficiently high, and extremely high voltage levels can easily be obtained with step-up transformers.

Wherever power is to be taken from the main transmission line for use in a local area, it passes through a sub-station containing banks of step-down transformers. Here it is reduced in voltage to either 2,300 or 4,100 volts for distribution through that area. At this point most of it is split into single phase. At the final delivery point, it passes through another step-down transformer. In residential use, and much commercial use as well, it is cut to 120/240 V and enters the building as will be discussed in Chapter 9. Industrial and commercial uses requiring 440 V supplies are beyond the scope of this book.

CHAPTER 3

TOOLS AND SAFETY

A good many of the tools used by the electrician are everyday, and commonplace, used by carpenters and plumbers as well. The fact that they are commonplace means that they are often taken for granted, and as a result they are frequently misused and improperly cared for. A craftsman is any field cannot do first class work with second or third class tools. You have undoubtedly heard, over and over again, the old saying: "a good craftsman does not blame his tools" for his difficulties. Unfortunately that saying generally is not properly completed. The reason a good craftsman does not blame his tools is because one of the first things a good ***craftsman*** will insist on is ***good*** tools. He will not attempt to do the job with poor ones.

Conventional Hand Tools

Hammers are available in a wide variety of sizes ranging from a light 7 oz. tack hammer to the 20 oz. framing hammer. For the electrician's purposes a hammer no heavier than the 13 oz. size is adequate. The curved claw (Figure 3-1) is generally more useful than the rip claw type for electrical work. Wood handles in hammers will eventually work loose, but they can readily be tightened by simply driving the steel wedges in deeper. If one gets really loose a new wedge may be needed. To promote straight driving and avoid marring finished surfaces, keep the hammer face clean.

The electrician will need several screwdrivers. The three types of tips are shown in Figure 3-2. Tip "c" is sometimes called an ***electrician's tip,*** and sometimes a ***cabinet tip.*** Obviously the name depends on who

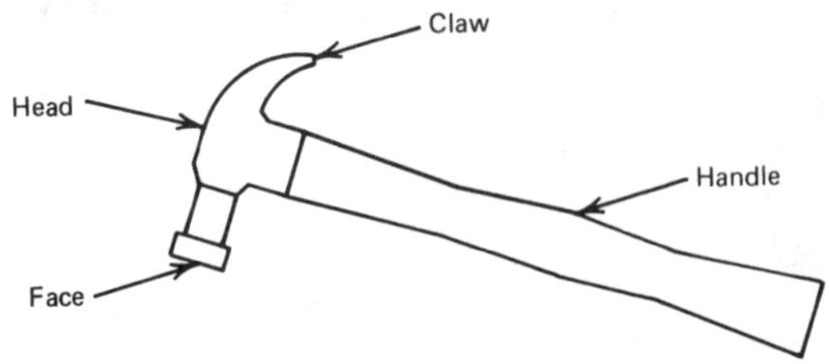

Figure 3-1
Claw hammer.

uses it. In any event the electrician should have a #3 and a #6 screwdriver with this type tip whatever the supplier calls it. A medium and a large standard tip plus a Phillips will be useful along with a small stubby type (Figure 3-3). Since the screwdriver is not a cutting tool one does not automatically think of it getting dull, but it most certainly does. Both flat and Phillips tips with use become worn and rounded and no longer fit snugly into the screw heads. Worn tips will severely mar screw heads, often making it impossible to withdraw a properly tightened screw. Flat screwdriver tips can easily be sharpened by filing, but once a Phillips tip is worn discard it.

A couple wood chisels kept viciously sharp are helpful additions to the electrician's tool kit. A 1″ and a 1½″ will be used to make notches for cable, greenfield, or conduit in wood framing. If allowed to become dull a wood chisel is worse than useless, it is dangerous. A chisel that is too dull to cut wood may very well bounce back out of the wood and cut you instead.

The hacksaw (Figure 3-4) will be used to cut both flexible conduit (greenfield), and armored cable (BX) as well as pipe conduit in circumstances when a standard pipe cutter is not available. Hacksaw blades vary from 18 teeth to the inch up to 32 teeth to the inch. For electrical work the general rule is stay with the finest blade available. A coarse blade cuts faster, but will bind badly in any of the various types of tubing. The finer the blade the more smoothly it will cut through thin materials.

The ***keyhole saw*** (Figure 3-5) is the tool for cutting rectangular, or irregular, holes in walls, floors, or ceilings. Generally two types of blades are available. There is a general purpose blade with 11 teeth to the inch, and a hacksaw blade at 32 teeth per inch. To cut for a wall box, or a recessed ceiling light, first mark the outline of the desired cutout on the

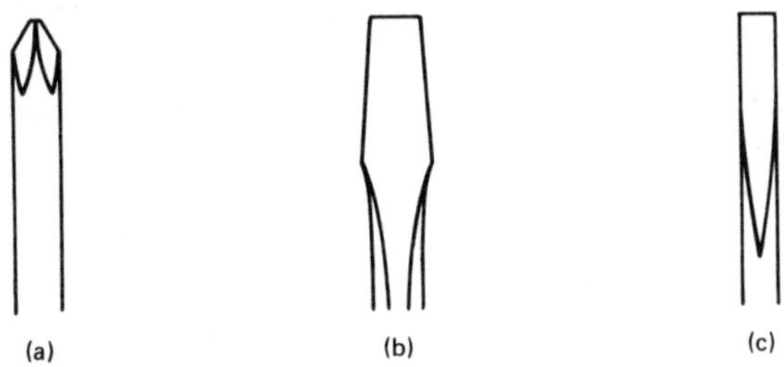

Figure 3-2
Screwdriver tips:
(a) phillips head;
(b) standard;
(c) electrician/cabinet.

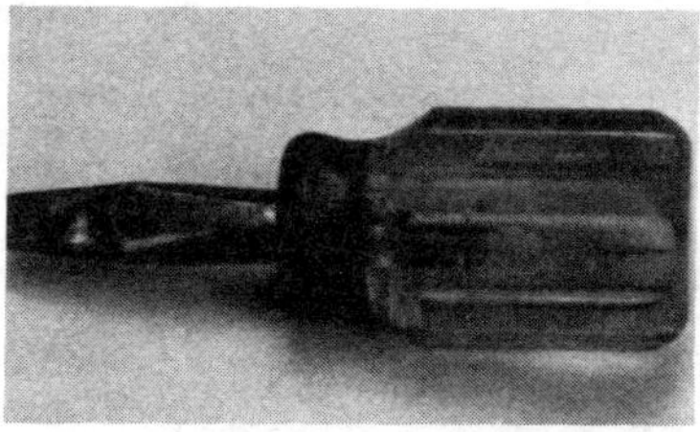

Figure 3-3 Stubby screwdriver.

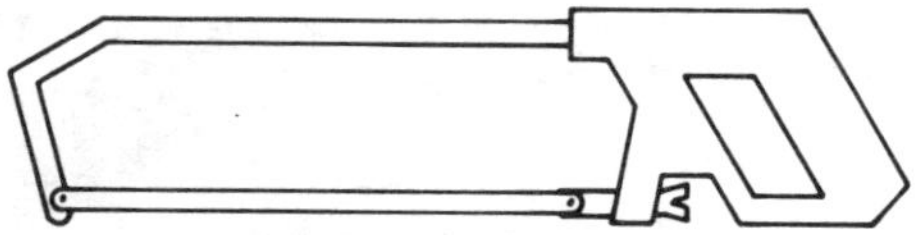

Figure 3-4 Hack saw.

wall or ceiling. Then within that outline drill a starting hole big enough to get the end of the saw through. Then cut over to the outline, and on around it. Use ***very light pressure*** on a keyhole saw. They bind and the blade bends very easily. Don't be in a hurry. Light pressure and patience are the watchwords for success with a keyhole saw. If you bend a blade (which is very likely), it can be more or less straightened by hand, but once bent a blade can never be fully straightened again and it never again cuts quite true.

The adjustable crescent wrench (Figure 3-6) is primarily of use to the electrician when installing or removing compression couplings on thinwall conduit. This is shown in the section describing conduit work in Chapter 10. For ½″ and ¾″ conduit compression couplings and box connectors, a 6″ crescent wrench is adequate. As the conduit size increases so should the wrench size. Someone doing a lot of work with conduit and using a great many compression couplings should have two wrenches. One is used to hold the center of the coupling while the ends are tightened with the other. A word of caution to those who sometimes forget—a crescent wrench is ***not*** a hammer; it is not made for pounding.

The electrician fondly hopes he will not often need to use a carpenter's bit and brace (Figure 3-7). He would certainly rather drill whatever holes he needs with an electric drill; however, if no power is available the bit and brace will have to do. The overall length of the bit and brace combined, as well as the turning radius of the brace, limit its usefulness. When there is no other way, however, it will somehow make most necessary holes.

The same reamer plumbers use to ream galvanized pipe (Figure 3-8) is used by the electrician to ream conduit after it is cut. As we shall see in Chapter 10, the reaming of conduit after cutting is of vital importance to prevent the burr left by the pipe cutter, or hacksaw, from damaging the insulation on the wires as it is pulled through.

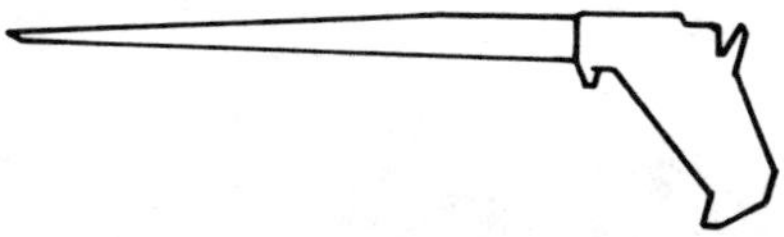

Figure 3-5
Keyhole saw.

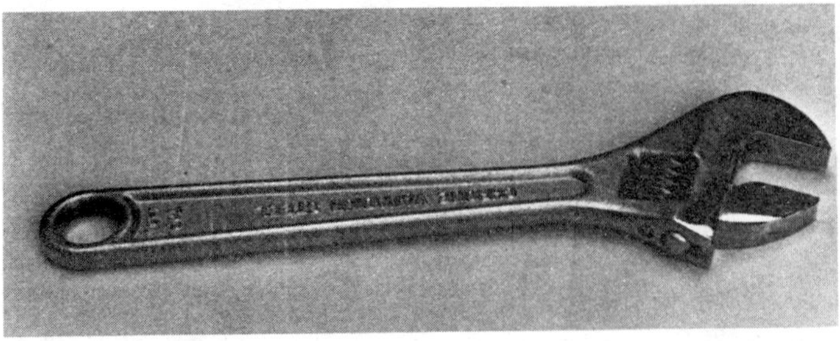

Figure 3-6
Crescent wrench.

Flexible conduit, Liquidtight, and BX are normally cut with a hacksaw which will also cut thinwall, rigid, or nonmetallic (PVC) conduits. But a plumber's pipe cutter (Figure 3-9) does a much better job on the latter three. The most important and vulnerable part of the pipe cutter is the cutter wheel itself. In normal use, they will eventually become dull and need replacement. If used improperly (or abused), the wheel may be nicked or warped which will produce ragged or crooked cuts, or it will not cut at all. If this happens, replace the cutter wheel, and be more careful.

The small pocket tubing cutter (Figure 3-9b) that plumbers use to cut small diameter copper pipe will cut ½″ steel thinwall conduit, but go easy on the cutter wheel. Feed it slowly—remember it was made to cut copper and it is being used to cut steel, a very different material. Note also that the small reamer on the side of the tubing cutter will satisfactorily ream copper, but will not adequately ream ½″ steel conduit. A full size pipe reamer will be necessary.

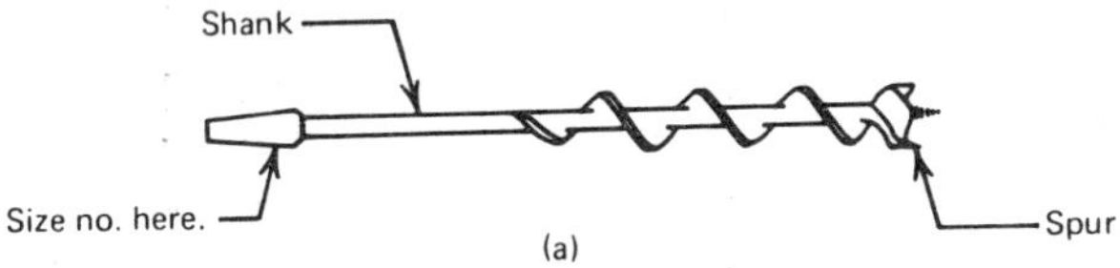

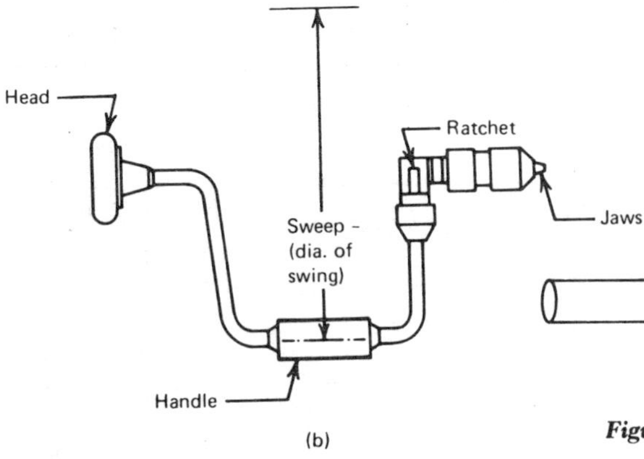

Figure 3-7
Bit and brace: (a) wood bit; (b) hand drill brace.

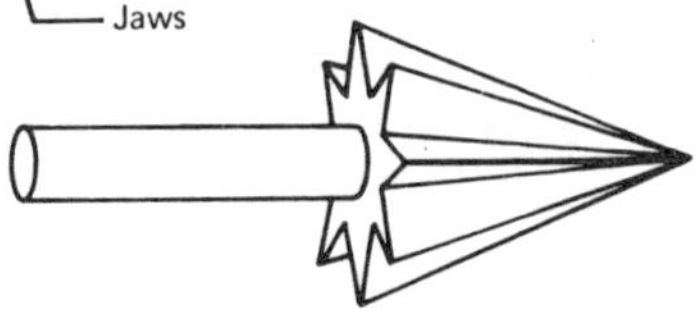

Figure 3-8 Pipe reamer.

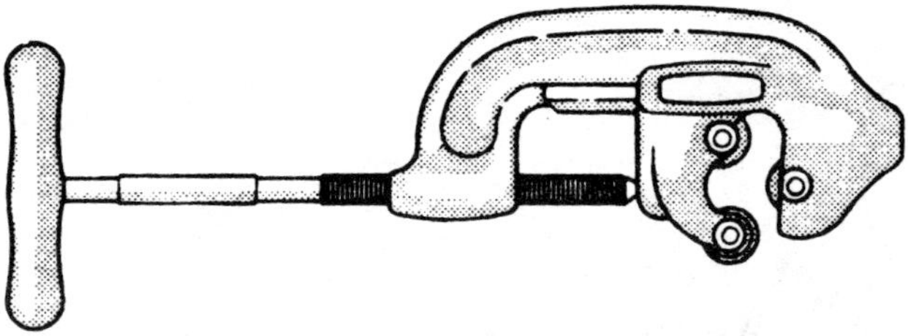

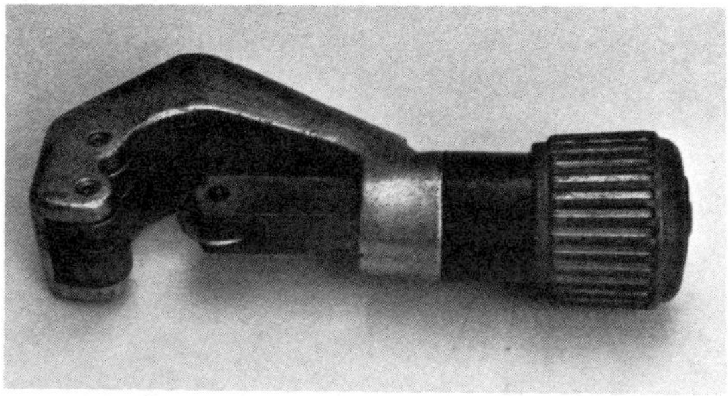

Figure 3-9
(a) Pipe cutter; *Reprinted from* Electrical Wiring Fundamentals, *McGraw Hill Publishing Company. (b) tubing cutter.*

Electrician's Hand Tools

The majority of residential wiring requires nonmetallic sheathed cable (Figure 5-1). In order to make connections and splices the inside insulated conductors must be stripped of the exterior plastic, protective sheathing. This is done with a ***cable stripper*** (Figure 3-10). The stripper is placed over the cable, and the sharpened cutter pressed through the sheathing. As the stripper is pulled off the cable, the cutter makes a slit through the sheathing. The author has found these strippers easier to use than a pocket knife, but of uniformly poor quality of manufacture. They are very cheap sheet metal stampings and consequently have a very short service life. But they are cheap in price as well. Buy them by the dozen, and as soon as one gets dull throw it away and use another. Why no manufacturer has elected to produce a decent quality cable stripper is mysterious.

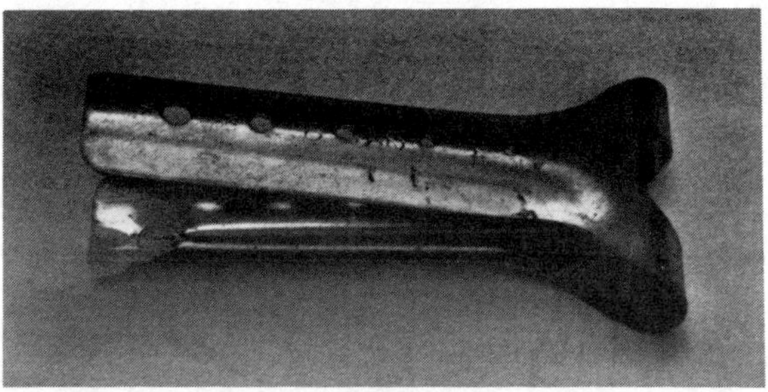

Figure 3-10
Cable stripper.

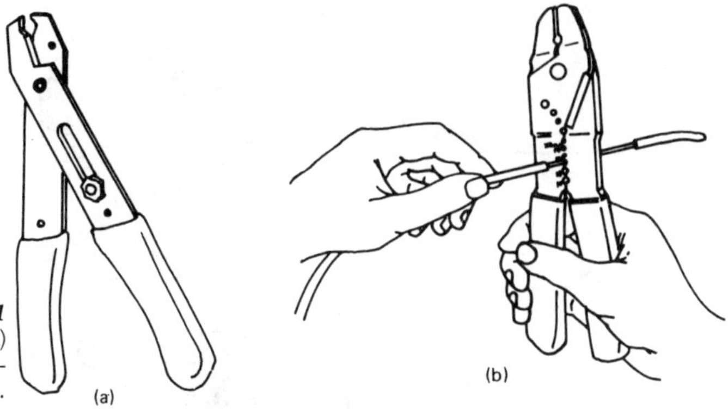

Figure 3-11
Wire stripper: (a) scissors type; (b) multipurpose tool.

Once the cable sheathing has been stripped the insulation on the ends of the individual wires must still be stripped before any connections or splices can be made. This is done with a ***wire stripper*** (Figure 3-11). Wire strippers, fortunately, are vastly better made than cable strippers. The scissor type (Figure 3-11a) is compact, light, and will handle the bulk of the average electrician's stripping which lies in the size range from #14 AWG to #10 AWG. This tool does require a bit of practice to enable the user to cut through the plastic insulation, and stop there before damaging and weakening the wire.

The ***multipurpose tool*** (Figure 3-11b) is a combination wire stripper, crimper, and threading die, which will cut small machine screws to desired length, and repair the end thread after the cut. The writer has found this tool inferior to the scissor type as a wire stripper while admitting it does have other useful capabilities.

The ***needlenose plier*** (Figure 3-12) is used as a wire cutter, but its primary use is to bend an end loop in a wire to go around a screw terminal. Like all pliers to be used for electrical work get a pair with insulated handles.

Individual wires can be trimmed with the needlenose pliers, or the scissor type wire stripper, but to cut off an NM (Non-Metallic) cable it is much easier to use a substantial pair of diagonal cutters (Figure 3-13). In

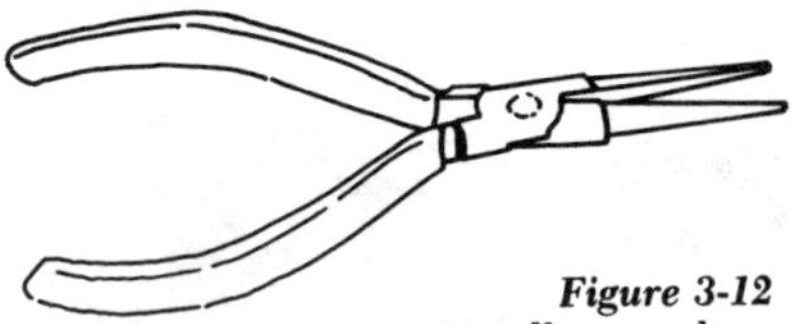

Figure 3-12
Needlenose plier.

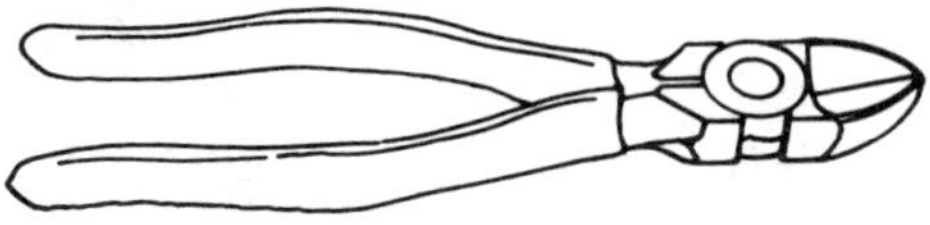

Figure 3-13
Diagonal cutter.

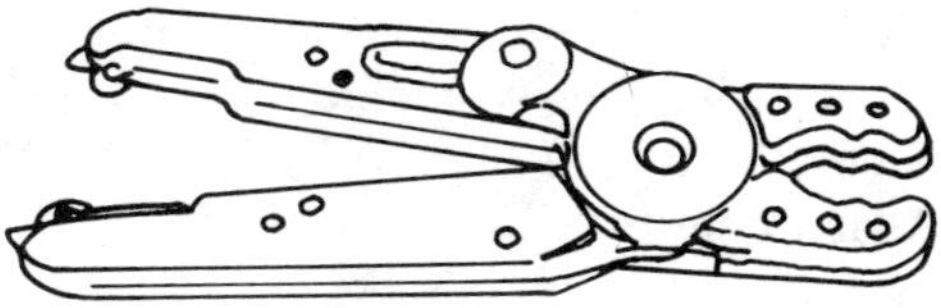

Figure 3-14
Fuse puller.

some areas these are familiarly known as ***dyks.*** These too should have insulated handles.

Among the overcurrent protective devices discussed in Chapter 9 you will find cartridge fuses. These fuses are most easily inserted and removed with a ***fuse puller*** (Figure 3-14). A fuse puller is simply a curved jaw pliers made entirely of non-conductive material. They come in several sizes to match the diameters of the various cartridge fuse cylinders. In a pinch a pair of metal pliers with insulated handles can be used, but a fuse puller is preferable from the point of view of safety of the fused equipment and the electrician who is changing fuses.

Thinwall conduit is discussed among some other materials in Chapter 5, and again in connection with rough wiring in Chapter 10. The process of bending it to the required shapes and angles calls for a ***conduit bender*** (Figure 3-15). This tool will bend conduit to any required angle and will also make saddle bends and offsets as well as the various complicated shapes sometimes necessary to follow architectural irregularities. Different conduit sizes require different benders because the code specifies different minimum bending radii for each conduit size in accordance with Appendix Table 346-10.

A conduit bender is a curved channel into which the conduit fits as it is bent. For each conduit size, the curve of the bender is the arc of a circle having the code specified minimum radius for that size. A small conduit size can be bent with a bender intended for a larger size, but the reverse cannot be done. For example, ½″ conduit can be bent with a ¾″ bender.

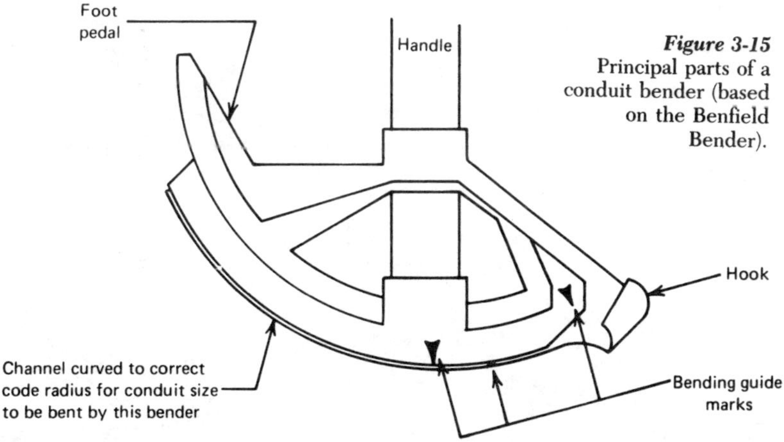

Figure 3-15
Principal parts of a conduit bender (based on the Benfield Bender).

The bend will be to a 5″ radius rather than the 4″ minimum allowed for ½″. This will be acceptable to any inspection department. However, ¾″ conduit cannot be bent with a ½″ bender because it will not fit in the tool. As it would be unacceptable if it fit, it is just as well that the tool itself prevents any such mistake.

An experienced electrician who works with conduit regularly can bend and fit it with amazing accuracy purely from practice and experience. Those who lack practice or experience or both will find a bender such as the Benfield (Figure 3-15) a great help. This maker supplies benders furnished with various calibrating marks cast into the tool as well as an instruction manual detailing how to use them. By following the instructions carefully a rank beginner can do satisfactory work almost at once.

Electrical metallic tubing (thinwall conduit), rigid metal conduit, rigid non-metallic conduit, flexible conduit (greenfield), liquidtight flexible conduit, and surface raceways (Wiremold) will be described in detail in Chapter 5. They all have one characteristic in common which is that they start out as empty protective tubes through which wires are pulled after the protective "plumbing" is in place. Pulling the wiring through any of these materials calls for another specialized electrician's tool called a ***fish tape*** (Figure 3-16).

A fish tape is a steel tape about ⅛″ wide and about 1⁄16″ thick wound on a reel. They vary in length from 25′ to 100′. At the far end, the tape is turned back on itself to form a small loop. After a run of conduit has been bent as necessary and installed in place in the building, a fishtape is passed through it from one box to another where the required wires are attached to the loop in the end of the tape so they can be pulled back through the conduit. Pulling wires through conduit is described in detail in Chapter 10, but it cannot be successfully done without the proper tool which is the fish tape.

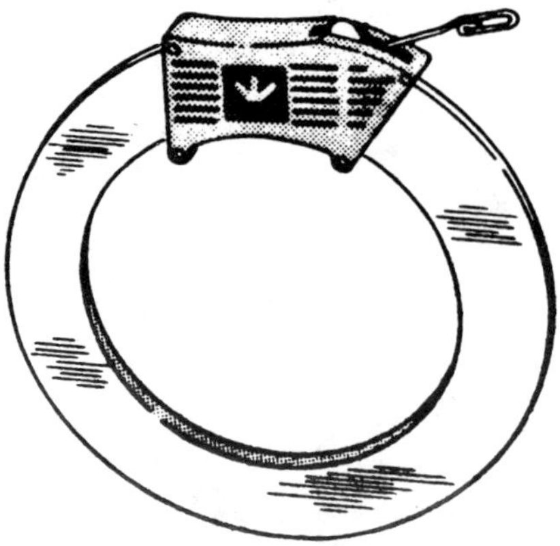

Figure 3-16
Fishtape. *Reprinted from* Electrical Wiring Fundamentals, *McGraw Hill Publishing Company.*

Power Tools

The power tool most frequently needed in electrical work is an electric drill. The usual ¼″ drill turning at 1750 rpm is satisfactory, if there is not a great deal of work required. It will handle the drilling needed for modest repair and alteration jobs, but when there is an entire house or other small building to be wired, it should be supplemented by a ⅜″ slow speed, and perhaps a ½″ slow speed as well, depending on the size of the total job.

Most of the holes drilled for electrical wiring are moderately large in diameter: ¾″ or 1″, and some larger. Let's look for a moment at what happens to drill bits as diameter increases. A ¼″ drill has a circumference of a little over ¾″ (.7854″). Turning at the standard 1750 rpm its rim speed is approximately 114.5 ***feet*** per minute. In wood at that speed the ¼″ drill gets rather hot, but it can handle it. Now when the drill diameter increases to ¾″ its circumference becomes more than 2¼″ (2.3562″). That drill turning at 1750 rpm develops a rim speed of 343.66 feet per minute. At that speed the tip of a drill bit gets extremely hot very quickly. The heat very quickly dulls the bit. When there are a few dozen holes to be drilled through 1½″ thick studs plus a lot more holes through 3″ of partition top plates, a standard speed drill will destroy drill bits at an unacceptably rapid rate.

That 343.66 feet per minute was the rim speed for only a ¾″ drill. Most of the holes an electrician drills will be 1″. The 1″ bit has a circumference of 3.14″ which at 1750 rpm means a rim speed of 458 feet per minute. A slow variable speed drill with a top rate of 600 rpm brings the rim speeds of ¾″ and 1″ drills down to 117.75 and 157 feet per minute respectively. At those speeds the bit life increases to acceptable levels. A fixed speed 450 rpm drill motor is much better yet. Using the 1" bit, rim speed decreases to 117.75" per minute.

The bit used for drilling wood framing members is called alternately a ***spade bit*** or a ***speed bit*** (Figure 3-17). The electrician should keep a fine metal file for touching up the edges of his bits. Drill bits kept sharp will reward the user many times over by producing clean holes quickly. A dull bit, by contrast, will produce ragged holes and take unconscionably long to do so in addition.

When sharpening spade bits (Figure 3-18), be careful to maintain angle A sloping down from the outside tip toward the center of the drill. At the same time angle B, the angle of the cutting edge, must be maintained as well. If either of these angles is allowed to vary significantly from what they were when the drill was new it will cut poorly, or not at all. Rather than waiting for a drill to get really dull before sharpening, touch it up while it is still cutting well, and never have to fight a dull drill.

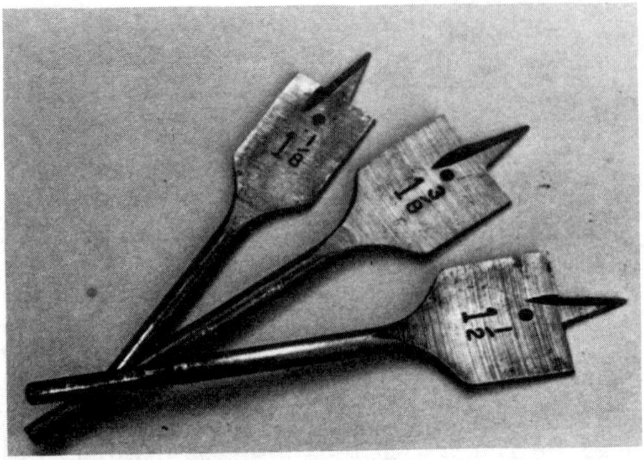

Figure 3-17
Spade bits.

In addition to making wiring holes in the building framework, the electrician's other major use for the electric drill is to make holes in masonry. Here holes are needed either for the insertion of fasteners or to pass wiring through a masonry wall. In either case a slow speed drill is an absolute must.

Carbide tipped masonry bits (Figure 3-19) must turn slowly or they will overheat and dull in ***seconds.*** Turn them slowly, and lean on them hard. They need plenty of pressure to cut well. Press hard for no more than 20 or 30 seconds at a time, then give them a few seconds to cool before leaning on them again. If you are drilling in concrete and dust stops coming out of the hole while you are drilling and pressing as hard as

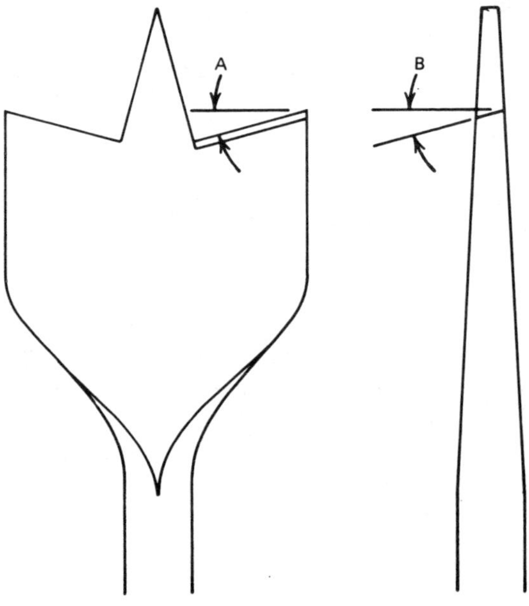

Figure 3-18
Sharpening angles for spade bits.

Figure 3-19 Carbide tip masonry drills.

possible, you've hit a piece of rock aggregate. Stop before ruining the drill. Get a ***star drill*** (Figure 3-20) and hammer to break up whatever is in the way until the masonry bit will cut again.

The other power tool an electrician will find useful is a ***sabre saw*** (Figure 3-21). Most of the rectangular or irregular shaped holes for which the keyhole saw is used can be more rapidly and accurately cut with a sabre saw. Be aware that the sabre saw cuts primarily on the up-stroke; therefore when the preservation of a smooth top surface is important, use a fine tooth blade. It will not cut as fast as a coarse tooth, but it will leave a much smoother top surface. Padding the foot of the saw with masking tape will also help to avoid marring an expensive finished surface as the saw passes over it.

Test Instruments

Whether installing a new electrical system, adding to an existing one, or troubleshooting a malfunction, a few simple test instruments will be necessary. A basic kit of three inexpensive instruments will handle most

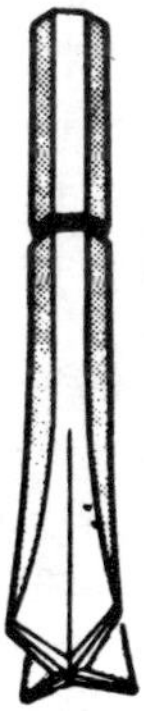

Figure 3-20 Star drill. *Reprinted from* Electrical Wiring Fundamentals, *McGraw Hill Publishing Company.*

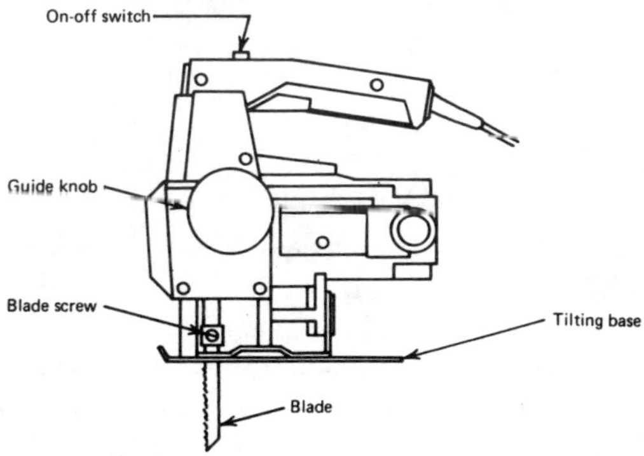

Figure 3-21 Saber saw.

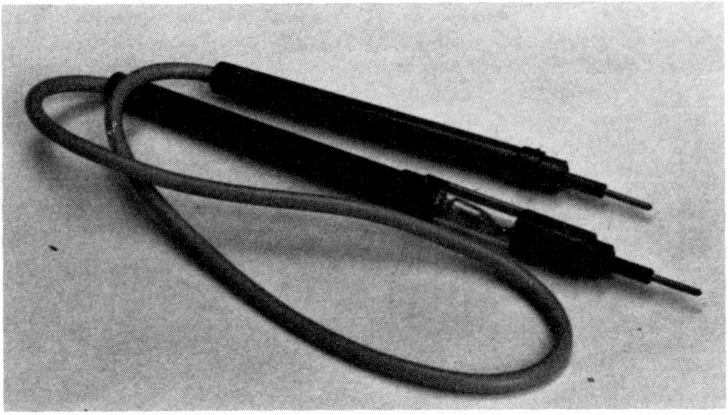

Figure 3-22
Test light.

problems. These and a fourth more expensive one will handle testing needs of the average electrician. The three basic test instruments are a test light, a multimeter (mentioned in Chapter 1), and an outlet analyzer. The fourth, a clamp-on ammeter, is not used as often as the others, but is particularly useful in tracking down circuit loading problems which will be discussed in Chapter 13.

The ***test light*** (Figure 3-22) is an inexpensive, compact, lightweight, and easy to handle device for tracing power through a circuit as well as testing switches, outlets, fuses or breakers. A word of caution—do not buy a small, cheap test light. Get the best you can find. It will still be inexpensive, and it is worth the cost difference.

A test light consists of two probes connected to each other by a wire while between a neon light is placed. With one probe on ground and the other on a wire or terminal that should be hot, the test light illuminates showing that the power is there. No light, no power. If the light shows with one probe on each end of a cartridge fuse, the fuse is blown. When the light does ***not*** show, the fuse is good, providing it does light with one probe at one end of the fuse and the other on ground.

To use a test light to test a 120V power receptacle (Figure 3-23) first push the probes in the two slots. The light should come on. With probes in the short slot and the ground hole it should also light, but with probes in the long slot and ground it should not light. If it lights with probes in long slot and ground, but does not with them in the short slot and ground, polarity is reversed.

To check a single pole switch put it in the "ON" position, then with one probe on ground place the other alternately on each of the two terminals. It should light both times. If it does not, the switch is bad. With a breaker in the "ON" position, probes on the breaker output screw and the ground buss in the breaker box will light the test light if the breaker is OK.

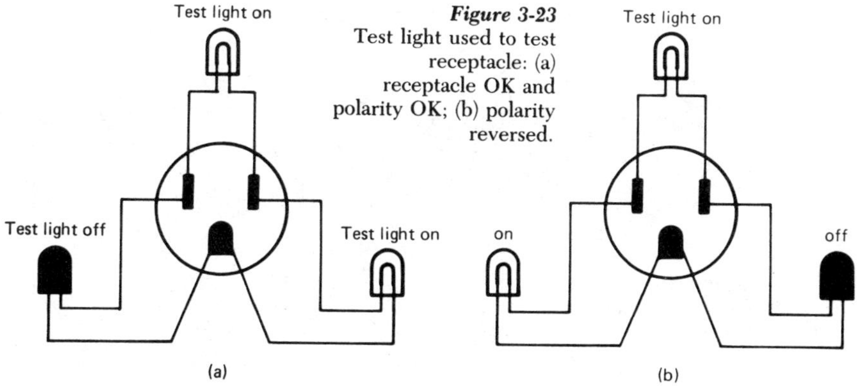

Figure 3-23 Test light used to test receptacle: (a) receptacle OK and polarity OK; (b) polarity reversed.

The test light will show where there is power and where there is not, but it does not indicate whether the voltage is correct. Most of the time the voltage for which you will be looking will be 120 volts AC nominal. 120V nominal means between 110V and 120V. Line voltage from the utility company does occasionally rise above or drop below these limits, but only under unusual or emergency conditions.

The instrument to use to read the actual voltage is the *multimeter* (Figure 1-7). As mentioned in Chapter 1 be sure when looking for AC voltages to place the function selector switch on the AC side, and on a scale higher than the voltage you expect to find. When looking for nominal 120V readings use the 250V scale. When looking for 240V start on the 500V scale, then switch back to the 250V scale. This is to protect the meter movement in case the line voltage should happen to be running high for any reason.

In addition to reading voltages the multimeter will read resistance, or the lack thereof as the case may be. Consequently it is used to check the continuity of wiring, and also to trace shorts. The part of the multimeter that reads resistance is called an ohmmeter. To use the ohmmeter first switch the selector to one of the ohm scales—usually there are three: × 1, × 10, and × 100 meaning that the reading on the scale is to be taken as is (× 1), multiplied by 10 (× 10), or multiplied by 100 (× 100). Note also that the ohms scale (the one at the top of the dial Figure 3-24) reads from right to left. 0 is to the right and infinity is at the left. All the voltage scales read the other way with 0 at the left and the high end at the right.

Always check to be sure the meter is functioning properly before taking any readings. This is done by shorting the test probes to each other after turning the selector switch to an ohms scale. The meter should then read 0 ohms, meaning the needle will be on 0 at the right hand end of the upper scale. If it is close, but not right on, adjust it with the ***ohms adjust*** dial. If it cannot be adjusted to read 0, probably the battery is low and must be replaced.

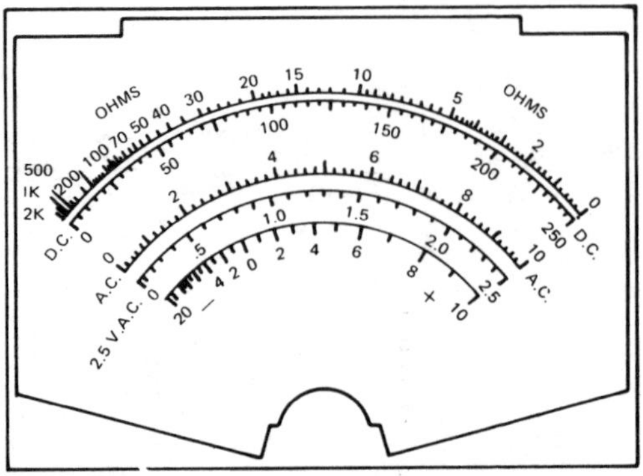

Figure 3-24
Multimeter dial plate. Ohms scale reads right to left; all others left to right.

Resistance measurements cannot be taken while the power is on. Any attempt to do this will damage the meter. Use the voltage scales of the multimeter first to make certain the power is off before switching to the ohms scales for resistance readings. A reading of 0 on the meter means that there is no resistance between the two points on which probes have been placed. For example, in any duplex receptacle box, if one probe is placed on the white wire (common) and the other on the bare or ground wire when the circuit is wired properly the meter reads 0. With one probe on the black wire the meter should stay on infinity (all the way to the left) with the other on either white common or ground. A 0 reading this time means a short.

Multimeters vary in cost from around \$20.00 to \$1,000.00. For building wiring a very small, very inexpensive one is adequate. The cheaper models are about 4″ × 2½″ by a bit over an inch thick and have two great advantages over the bigger expensive ones for our purposes. They are small enough to fit in a shirt pocket leaving both hands free to help you climb or crawl into some unlikely place.

The plug-in ***outlet analyzer*** (Figure 3-25) will quickly and easily indicate whether an outlet has been wired properly. When plugged into a live outlet, the way they light up, or do not light up, will show whether polarity is correct, and will also immediately reveal an open ground or an open common. There are a couple different versions of this tester, but they all have instructions on them as to how they should be read. Since these units are very inexpensive, cost is no reason for being without one.

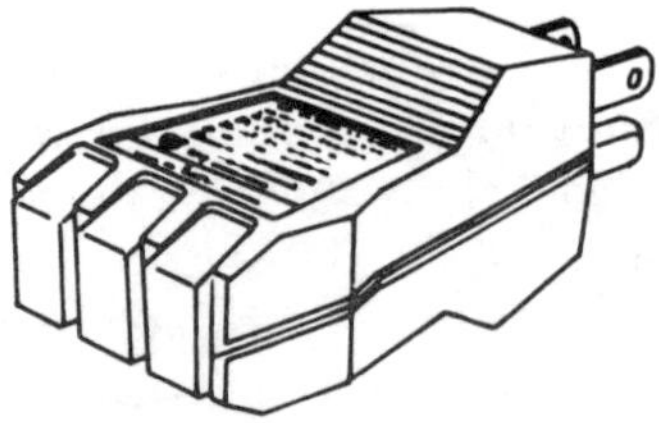

Figure 3-25
Plug-in outlet analyzer.

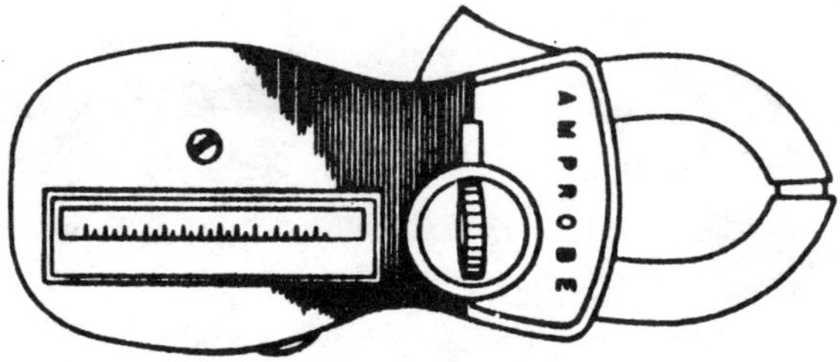

Figure 3-26
Clamp-on ammeter.
Reprinted from
Electrical Wiring
Fundamentals,
McGraw Hill
Publishing Company.

The fourth handy test instrument is a ***clamp-on ammeter*** (Figure 3-26). This instrument will read the amperage flowing in a circuit without interfering in any way with the flow. The hot wire only is placed between the jaws of the meter. The magnetic field produced by the current passing through that wire is translated by the meter into amperage and shown on the vertical dial of the particular instrument illustrated. Other manufacturers of similar instruments have different dials, but all operate by reading the magnetic field created by a current passing through a single conductor. Most clamp-on ammeters are equipped with test probes and can be used as AC voltmeters as well. Some have an additional attachment allowing them to be used also as ohmmeters.

Basic Wiring Techniques

Since electrical wiring involves the use of some tools that may be unfamiliar it also involves some techniques and procedures requiring the use of those tools that may also be unfamiliar.

The cable stripper (Figure 3-10), as described earlier, is used to slit the sheathing on the ends of NM cable. That sheathing should be slit for a distance of 6″ to 8″ from the end, and the slit sheathing cut away along with any paper or fiber filler inside the cable leaving only the insulated conductors and the bare ground wire. After cutting away the sheathing, observe the insulated conductors carefully to make certain the cable stripper did not cut through the insulation on the insulated conductors. If it did the damaged areas will have to be taped, or cut the wires all off and start over again. When its end has been stripped of sheathing the cable is mounted in the electrical box so that the sheathing ends in the box clamp (Figure 3-27). The wires inside the box are free of cable sheathing.

After the cable has been stripped of sheathing the ends of the individual conductors must be stripped of insulation in order to make connections. Using the wire stripper (Figure 3-11) the insulated conductors are stripped back ½″ to ⅝″ from their ends. A common mistake made by beginners is to strip the cable sheathing too short, and the wire insulation too long. To strip the insulation too long is to invite trouble. Strip an

(a)

(b)

Figure 3-27
(a) NM cable correctly clamped in connector; (b) connector in box.

insulated conductor no more than absolutely necessary to make the required connection as described below.

One commonly required connection is the splicing of two or more wires to each other. Splices in the building power wiring will be done with solderless connectors. Most of your splices will be done with small size wires (#14 up to #10) and will be secured with wire nuts (Figure 3-28). To make a splice the insulation is stripped as noted above, the wire ends to be spliced placed inside the wire nut, and the wire nut twisted clockwise until the splice is tight and firm. When the splice is completed no uninsulated wire should extend beyond the end of the wire nut. There are several sizes of wire nuts so as the number, or size, or both of the wires to be spliced increases so does the size of the wire nut.

When the number of wires to be spliced, or the size of the wires has increased beyond the capacity of the largest available wire nuts, the splices must then be secured with a split bolt connector (Figure 3-29).

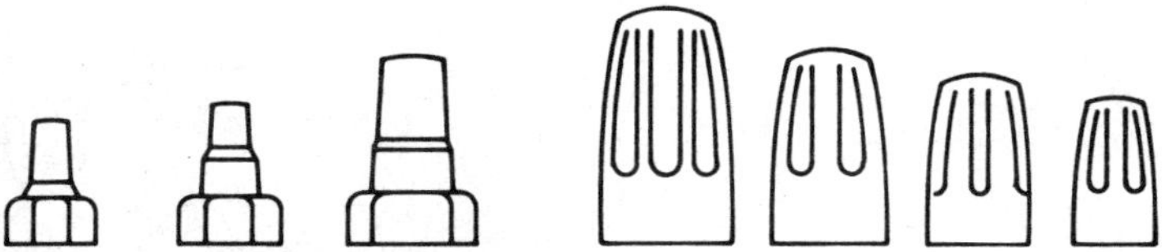

Figure 3-28
Wire nuts.

Figure 3-29
Split bolt connector.

These too come in a variety of sizes to handle the various larger wire diameters. Note that the wire nut having an outside shell of insulating material needs no further insulation. Thus a splice made with a wire nut is already insulated while one made with a split bolt connector is not. A split bolt connector splice must be well and thoroughly taped with an approved electrical tape. Medical adhesive tape or masking tape will not suffice.

Ultimately wires must terminate at some device or other be it a switch, a receptacle, a fixture, or at the supply end in the breaker box. Where the wire ends on a terminal screw (Figure 3-30), it shall be wrapped around the screw clockwise in the same direction the screw will turn when being tightened. Unless the screw terminal is clearly marked otherwise, no more than one wire may be attached to a single screw per NEC 110-14(a). When two or more wires require a common attachment they shall be spliced together in a wire nut along with a *pigtail* (Figure 3-31) which is used to make the common attachment to the screw terminal.

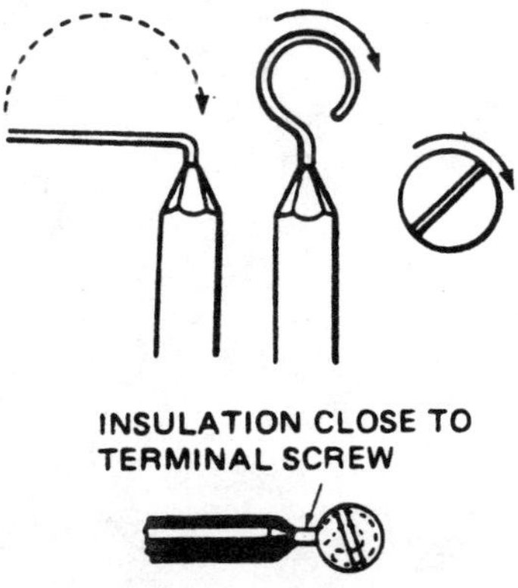

Figure 3-30
Wire properly connected to screw terminal

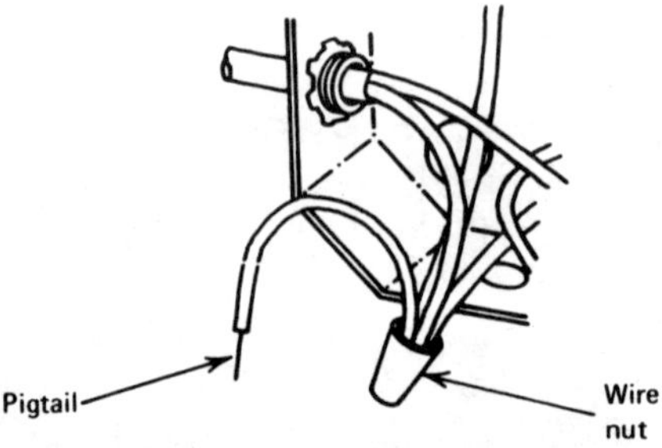

Figure 3-31
Splice with pigtail for terminal attachment.

A fairly recent development for making connections to switches and receptacles is the ***push-in terminal*** (Figure 3-32). Here the wire is simply pushed into a hole in the back of the device where it is secured by an internal spring clamp type contact. This writer's enthusiasm for the push-in terminal is a good deal less than overwhelming because the area of metal to metal contact is very slight. The screw terminal provides a vastly larger area of metal to metal contact between the wire and the terminal than is possible with the push-in. The primary advantage of the push-in is that it is a time and labor saver. Many connection failures and burn-outs have been observed at push-in terminals that would not have occurred at screw terminals.

Safety

In doing electrical work the primary safety concern is the avoidance of electrical shock. Many people who have experienced mild electrical shocks from time to time, or who are inclined to boast their machismo have a tendency to view shocks rather lightly. This view is unwise.

As we shall see a bit further on, the smallest circuits used in building wiring have a current carrying capacity of 15 amperes at 120 volts. The amount of current that will cause serious physical damage varies in each

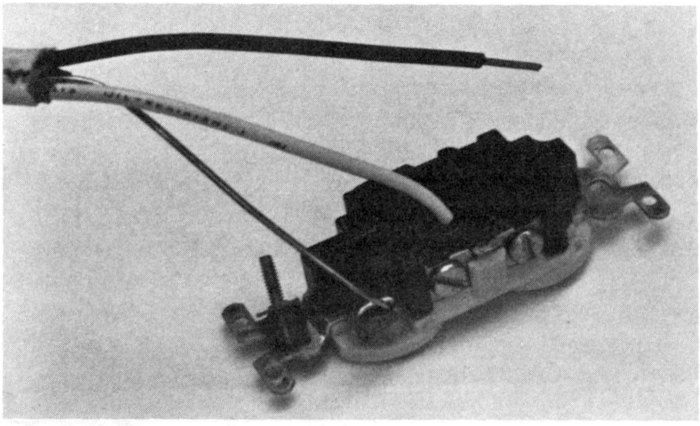

Figure 3-32
White wire in push-in terminal.

individual case depending on age, general physical condition, and the conditions under which the body makes contact with the electrical supply. However, in general a current of .01 amperes will be unpleasant enough to get a person's attention. The current available in the ***smallest*** circuit in the building is ***1500 times that amount.*** Less than .1 ampere is quite likely to be fatal. In other words the smallest circuit is capable of killing the average person 150 times over. While electricity is not to be feared, this fact should be a sufficiently sobering thought to produce a most desirable wholehearted respect.

Fatalities due to electrical shock result from stoppage of breathing, or interruption of heart function, or both, caused by paralysis of the muscles. Severity of a shock will depend on the resistance of the body to the passage of a current, the voltage pushing that current, the path of the current through the body, and the length of time the current flows.

The resistance of a person will be least when wet and contact is made to bare skin. Therefore, when working on electrical apparatus keep dry. Use tools with insulated handles. Wear gloves when working around live conductors. Wear thick, preferably rubber soled shoes, and do not stand in damp areas to work on live circuits.

Since the major danger from electrical shock is stoppage of heart or breathing, the most dangerous path for an electrical current though the body is up one arm, across the chest (through lungs and heart), and down the other arm. Thus when working be sure that if one hand could touch a live conductor, there is no way the other one could contact ground—put it in your pocket, or put a glove on. A path up one arm, down through the torso, and out through one leg or the other is not as bad as arm to arm, but since it goes through so many vital organs it is by no means good.

The next best thing to avoiding an accident is to know what to do in case one occurs—and do it quickly. A most common result of electrical shock is paralysis of the voluntary muscles; a person touches a live conductor, is paralyzed, and cannot let go. Should this happen to a colleague, shut off the power immediately. If this cannot be done, separate the victim from the live wire or terminal as quickly as possible. ***Do not touch him with your bare hands in the process.*** If you do, the current passing through him might go through you and paralyze you as well. Throw a dry piece of rope over him and pull him off, or throw a dry shirt or sweater over and pull him off with the sleeves. If he has a heavy jacket on, pull him off by the jacket.

Once the victim is safely loose from the power supply, check at once to see whether he is still breathing, and whether the heart is still functioning. If not, cardiopulmonary recuscitation (CPR) should be administered at once. American Heart Association, Red Cross, and many hospitals give CPR lessons to anyone who wants to take them. The lessons take approximately 6 hours. While in some places there is a nominal charge for the course, in many places it is free.

It cannot be emphasized too strongly that in a case of failure of heart or lung function, a matter of a few seconds' delay in administering CPR may mean the difference between life and death. Consequently there should always be someone on every jobsite familiar with this technique.

Electrical shock can easily be avoided, if a few simple safety rules are followed:

Before working on wiring shut off the power.

Then test with voltmeter or test light to make sure power is really off.

Leave a note at fuse or breaker box warning others not to turn circuit on.

Use only screwdrivers, wrenches, or pliers with insulated handles.

Stand on dry material and preferably on insulating material (wood, concrete, etc.).

Replace worn or damaged cables on electrical power tools.

Short any capacitor to ground before touching terminals.

In addition to electrical shock the electrician is exposed to all other normal hazards around a construction jobsite and must therefore observe the same safety precautions as everyone else. An electrician can fall or have things fall on him just the same as anyone else. Be alert, be careful, be safe. To emend an old saying among aircraft pilots: "there are old electricians and there are bold electricians, but there are very few old, bold electricians."

CHAPTER 4

CONDUCTORS

While ***conductors*** might seem a pompous and technical way of saying ***wires***, actually it is not. A wire is bare metal; a conductor is metal *with* or *without* a covering of some kind. The meanings of the two words are not the same. Article 310 is the part of the NEC dealing with conductors. Section 2 (a) of that article states clearly that "conductors shall be insulated." This is immediately followed by the confusing statement "Exception: where covered or bare conductors are specifically permitted elsewhere in this Code."

They have said, then, that a conductor is insulated except when it is covered or when it is bare. To put it another way, a conductor is insulated except when it is not! Obviously, this needs some explanation.

In accordance with the code, to be considered "insulated" a wire must be fully encased in a material that has been tested and found acceptably resistant to the passage of electrical current at the thicknesses specified for that material in Table 310-13 of the code (Appendix A). By contrast a covered conductor is a wire sheathed in such a way that either the material used, or the thickness of that material, or both, do not result in a coating of sufficient resistance to the passage of electricity to be considered an electrical insulation. Covering simply acts as a protection for the metal inside, not the people out side.

Wire Materials

Wires are manufactured for a variety of purposes and from many different metals: brass, copper, silver, aluminum, tungsten, and various steel alloys. However, for use as power wiring in buildings only copper, aluminum, or copper-clad aluminum are approved [NEC 310-2 (b)].

While aluminum is considerably less expensive than copper, the cost differential is absolutely its only advantage. In many other respects aluminum is considerably less desirable. Most importantly it has considerably higher resistance to electrical current than copper. This means that a circuit wired in aluminum will always require a larger gauge wire than the same circuit wired in copper.

All metals when exposed to air and moisture form oxides. (A familiar metal oxide is the rust that forms on ferrous metals, such as iron and steel.) For some metals this oxidation process is self-limiting as with copper or brass. Either of these metals will form a thin surface oxide coating that effectively seals the metal beneath from further oxidation until that surface film is removed. In the case of aluminum, as with steel, oxidation of the metal is not self-limiting. Once aluminum starts to oxidize it will continue indefinitely converting the metal to aluminum oxide, a fine white powder. As long as aluminum is covered by insulation it is also protected from direct contact with the air, and thus from oxidation. However, the end of an aluminum wire that has been stripped of its insulation in order to make a connection is now exposed and subject to oxidation. Once stripped, aluminum wire should be treated with an anti-oxidant, and even then should be periodically inspected for signs of the formation of white aluminum oxide powder. When any is found, clean the aluminum with a fine abrasive, and retreat with the antioxidant.

In the case of the copper-clad aluminum the copper surface film protects the aluminum beneath from oxidation as long as that copper film remains intact, which is precisely the problem with it. The copper film is too easily scraped. For example, stripping insulation leaves areas of bare aluminum subject to oxidation.

In addition to the wire sizing and oxidation difficulties, aluminum has other disadvantages. Its coefficient of expansion is considerably greater than that of copper, meaning that it expands and contracts more with changes in temperature. This expansion and contraction of the metal may result in loose connections at odd and out of the way points in the circuit. At ***best*** these loose connections fail electrically, meaning the circuit, or some part of it, goes dead; at ***worst*** a loose connection becomes a high resistance point that causes the wire to heat and either arc to a terminal, or melt off insulation, allowing arcing between wires. The arcing, in turn, can and has caused many fires.

A further disadvantage of aluminum is its brittleness. Copper is far more plastic than aluminum. It withstands bending and rebending without fatiguing, crystalizing, and breaking longer than aluminum. All in all, the general consensus among people in the electrical field is that the best thing to do with aluminum wire is ***nothing.*** Avoid using it if possible.

Terminal and Device Markings

Due to the important differences in the properties of the two approved metals, copper and aluminum, the terminals and devices to which wires may be attached will not necessarily accommodate both. Some will take copper only, some will take copper or copper-clad aluminum, but not bare aluminum, and some will take any of the three, copper, copper-clad, or bare aluminum. In order for the electrician to avoid connecting aluminum to terminals that will not accept it, a system of markings has been adopted to forewarn him.

Receptacles and switches rated for 20 amperes or less must be marked ***Co/Alr,*** if aluminum may be connected to them. If not so marked, only copper or copper-clad aluminum may be connected unless they are marked ***copper only.*** In any case only copper or copper-clad aluminum may be used with the push-in type pressure terminals (Figure 4-1) commonly found on switches and duplex receptacles. Bare aluminum, where permitted, must always be connected to a screw type terminal.

Receptacles rated at 30 amperes and up must be marked ***Al-Cu,*** if aluminum may be connected to them. If not so marked, only copper or copper-clad aluminum may be connected.

It is imperative that aluminum be connected only to terminals or screws made of metals compatible with it. Look carefully for the marking ***Co/Alr*** on devices rated at 20A or less, and the mark ***Al-Cu*** on those rated at 30A or over. If these markings cannot be found on a particular device, do ***not*** under any circumstances connect aluminum conductors to it.

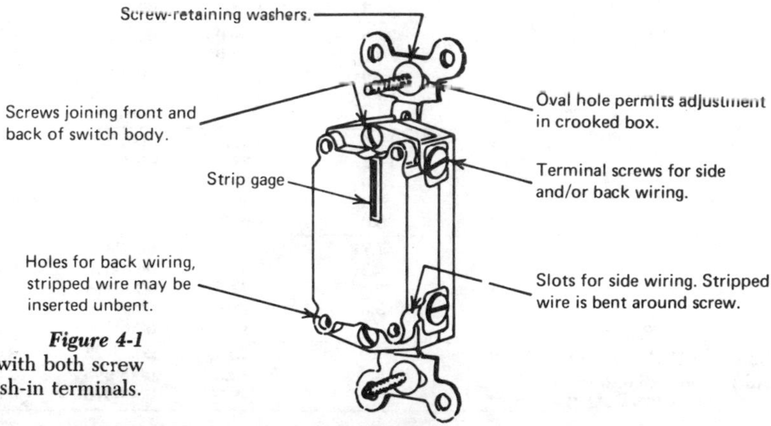

Figure 4-1
Switch with both screw and push-in terminals.

Wire Size

In order to provide a uniform standard of reference nearly all wire manufactured in the U.S. is sized to conform to the American Wire Gauge (AWG). The standard sizes manufactured in accordance with this gauge are identified by gauge numbers that range from #46 up to #4/0 (0000). Yes, you read it correctly. The smallest size, #46, is the largest number. The numbers decrease (Figure 4-2) as the wire thickness increases up to #0, then #2/0 (00), #3/0 (000), and finally the largest AWG size, #4/0 (0000).

In building power wiring we are not concerned with any sizes smaller than #14 because NEC Table 310-5 limits such wiring to #14 copper or #12 aluminum as the smallest sizes permissible. Smaller size wires are permitted for low voltage signal, alarm, and control circuits such as

Gauge number	Diameter (mils)	Cross section Circular mils	Cross section Square inches	Pounds per 1000 feet
0000	460.0	212,000.0	0.166	641.0
000	410.0	168,000.0	.132	508.0
00	365.0	133,000.0	.105	403.0
0	325.0	106,000.0	.0829	319.0
1	289.0	83,700.0	.0657	253.0
2	258.0	66,400.0	.0521	201.0
3	229.0	52,600.0	.0413	159.0
4	204.0	41,700.0	.0328	126.0
5	182.0	33,100.0	.0260	100.0
6	162.0	26,300.0	.0206	79.5
7	144.0	20,800.0	.0164	63.0
8	128.0	16,500.0	.0130	50.0
9	114.0	13,100.0	.0103	39.6
10	102.0	10,400.0	.00815	31.4
11	91.0	8,230.0	.00647	24.9
12	81.0	6,530.0	.00513	19.8
13	72.0	5,180.0	.00407	15.7
14	64.0	4,110.0	.00323	12.4
15	57.0	3,260.0	.00256	9.86
16	51.0	2,580.0	.00203	7.82
17	45.0	2,050.0	.00161	6.20
18	40.0	1,620.0	.00128	4.92
19	36.0	1,290.0	.00101	3.90
20	32.0	1,020.0	.000802	3.09
21	28.5	810.0	.000636	2.45
22	25.3	642.0	.000505	1.94
23	22.6	509.0	.000400	1.54
24	20.1	404.0	.000317	1.22
25	17.9	320.0	.000252	0.970
26	15.9	254.0	.000200	0.769
27	14.2	202.0	.000158	0.610
28	12.6	160.0	.000126	0.484
29	11.3	127.0	.0000995	0.384
30	10.0	101.0	.0000789	0.304
31	8.9	79.7	.0000626	0.241
32	8.0	63.2	.0000496	0.191
33	7.1	50.1	.0000394	0.152
34	6.3	39.8	.0000312	0.120
35	5.6	31.5	.0000248	0.0954
36	5.0	25.0	.0000196	0.0757
37	4.5	19.8	.0000156	0.0600
38	4.0	15.7	.0000123	0.0476
39	3.5	12.5	.0000098	0.0377
40	3.1	9.9	.0000078	0.0299

Figure 4-2
Table of AWG numbers and nominal diameters: #46 to #4/0.

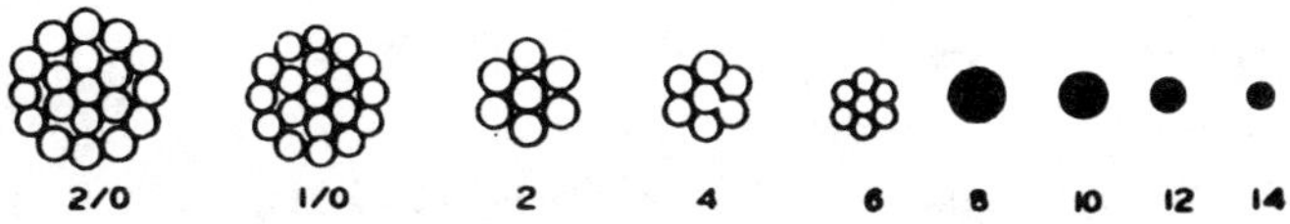

Figure 4-3
AWG wire gauges, actual size: 14, 12, 10, 8, 6, 4, 1, 0, 2/0, 3/0, and 4/0.

intercoms, thermostats, bells, buzzers, sprinkler timers, or burglar alarms. Also, smaller sizes are permitted for various exposed flexible cords. For fixture wire ampacity, operating temperature, minimum wire size, etc., see Article 402.

The standard AWG wire sizes most commonly used in building power wiring are shown actual size in Figure 4-3. Note that from #14 all the way up to #4 only even numbers appear. The odd numbered sizes are not made for electrical wiring.

Wire size 4/0 AWG is by no means the largest wire made. 4/0 is merely the large end of the American Wire Gauge sizes. Wires vastly bigger than that are commonly made, and they are identified by their cross-sectional areas as measured in thousands of circular mils. This designation is abbreviated MCM. The next electrical wire size larger than 4/0 is 250 MCM. From there on the numbers in the designation increase as the wire size increases because the numbers now equal the cross-sectional area of the wire.

This change in numbering systems is clearly evident in Appendix A, Table 310-16. Wire sizes are given in the far left and far right columns. From the top of the column numbers decrease as wire size increases, then halfway down the page the sequence changes. The number 0000 is followed by 250 then 300 and so on. At that point AWG numbers changed to MCM.

Wire Construction

As used in building power wiring all wire sizes from #14 up to #8 are solid wires meaning that a single round strand of metal makes up the full thickness of the wire (Figure 4-3). All wires larger than #8 are stranded, meaning that a large wire is made up of several smaller wires twisted together. Size #8 AWG marks the dividing line. It can be obtained either solid or stranded, and is used in building wiring both ways.

Sizes #14, #12, and #10 are also available in stranded form. However, in the permanent parts of a building electrical system, they are not customarily used in this form. Solid wire in these sizes is easier to strip without damaging the metal, and it makes more positive connections when spliced and when connected to terminals.

The reason for changing to stranded wire at #6 is that it and larger sizes in a solid rod would be stiff and unwieldy to handle. Splices in these larger wire sizes are made using pressure clamps such as the split bolt connectors shown in Figure 3-29. Connections at switches, breakers, or outlets are made in clamp or set screw terminals.

Insulations

Within buildings, as well as on the way to and from them, power wiring is exposed to a wide variety of environmental conditions. Temperatures may vary from sub-freezing cold to extreme heat. While one area is absolutely dry another may be moderately damp, while yet another is soaking wet. These differences in temperature and humidity might vary from one place to another in and around a building, or they may vary in a particular spot with the seasons of the year, or from day to day, or even from hour to hour.

In addition to the variations in temperature and humidity, power wiring may be exposed to a wide variety of oils, greases, chemicals, tars, gases, fumes, or liquids, some of which might have a corrosive effect on the metal in the conductors while others might attack the insulations or sheathings used to protect the conductors. In order to deal safely with this variety of conditions, a wide selection of different insulations has been developed. Each provides a high degree of protection from one or more of the unfavorable kinds of conditions that might be encountered. No one is suitable for all conditions.

Since the purpose of a residential building is to provide an environment suitable for human habitation, extreme conditions of temperature, humidity, or exposure to corrosive chemicals are not normally encountered in such buildings. The extremes, particularly in terms of corrosive materials, occur in industrial structures and occasionally in commercial buildings. However, a really qualified electrician, regardless of his area of specialization, should be acquainted with the general range of available materials and have an overall understanding of their properties and uses.

In Table 310-13 (Appendix A) the code recognizes a total of 26 different insulation types for use in general wiring. Although the average electrician will seldom, if ever, use more than, perhaps, a half a dozen of these, for reference purposes all are included here. Many, as will quickly become apparent, are for very specialized applications.

All of the types starting with the code letter "R", such as RH, RHH, and RHW are rubber. Rubber, once the primary insulation material, is now largely supplanted by plastics. "R" insulations are currently found on main building feeders or on some major sub-feeders, but not used on general branch circuit wiring. Seldom, if ever, will the electrician encounter any of these insulations in residential work.

The ones he will use consistently will be from the "T" or thermoplastic group. These are THHN, THW, THWN, and TW. Of these, TW and THW are those used most frequently.

TW is the standard insulation type used in NM cables which will be discussed in detail in Chapter 5. Since the vast majority of residential wiring is done with NM cables it follows that TW is the insulation with which the residential electrician will work daily. Socrates would like this material. It's moderate in all things. It is moderately heat resistant—up to 140°F. It is moderately moisture resistant, and, not unimportant, it is moderate in cost by comparison with some of the more highly resistant materials whose properties are not needed in residential applications.

When wiring is done in either rigid or flexible conduits, it is likely to be done with THW rather than TW. Due to its higher operating temperature combined with moisture resistance, the THW fits a wider range of situations; hence it is routinely being used to replace TW in conduit runs.

When yet a higher temperature resistance is necessary, switch to THHN. THW is good up to 90°C (194°F) only for special applications, while THHN is good to that temperature for general purpose wiring. Another reason for employing THHN might be the insulation thickness. On a #14 wire, as an example, using THW the insulation has to be 30 mils thick. With THHN it need only be 15 mils thick–only ½ as thick.

As we shall see in Chapter 10, the matter of insulation thickness becomes a consideration of primary importance when planning to pull wiring through any of the various types of conduit. Per NEC Chapter 9 Table 1, you are permitted to fill no more than 40 percent of the cross sectional area of a conduit with conductors. As we well know by now, a ***conductor*** includes both the wire and its associated insulation. Therefore, the thinner the insulation the more space there will be left for wires.

To define the number of conductors permitted in various types of conduits the code repeatedly refers to the percentages of fill specified in Table 1, Chapter 9. In residential work we seldom pull either very many or very large wires, hence this limitation is seldom a problem. If you intend to do considerable conduit work, get the complete 1999 NEC and refer to Tables C1 through C13A for full details.

UF as the cable sheathing for cables consisting of two or more conductors plus ground is used for underground feeds to decorative lighting and for supplying power to outdoor utility outlets in the vicinity of buildings and on the outside of buildings as well where exposed wiring is needed. However, as will be discussed in the next chapter, in these applications the UF type insulation forms the cable sheathing which protects individual conductors insulated with TW.

When the electrical power supply reaches a building via an underground service entrance (Chapter 9) the buried wires connecting the transformer to the building meter box will probably be USE. This material is similar to UF in terms of moisture and temperature resistance; however, a thinner layer of USE will do the same insulating job as a good deal thicker layer of UF. NEC specifically permits the use of UF on underground feeders, but just as specifically forbids its use on underground service entrances for which USE is clearly approved.

Figure 4-4 Typical conductor markings as required per Code.

The extensive group of high temperature rated plastic insulations in Table 310-13 are for use in special applications mostly in dry locations. In any event none of them have become at all popular for use in general building power wiring.

Markings

In order that there shall be no doubt, question, or confusion concerning wire size, insulation type, approved maximum voltage rating, or responsible manufacturer, all individual conductors and all factory assembled cables shall be properly marked (NEC 310-11). Properly marked means that at intervals of not more than 24" all individual conductors and all assembled cables (Chapter 5) shall be durably marked showing wire size in AWG or MCM whichever applies (Fig. 4-4).

Marked at intervals of not more than 40":

a. Proper letter designation for insulation type being used as TW, THW, THHN.

b. Maximum voltage for which that insulation type and thickness is approved.

c. Name or trademark of manufacturer.

Color Code

In addition to the identifying markings described above, NEC also requires that the insulation on individual conductors be color coded in order that conductors being used for certain specific purposes may instantly be identified.

NEC 250-119 states that the grounding conductor may be bare, covered, or insulated. When covered or insulated, it shall be colored green or green with one or more yellow stripes.

NEC Article 200-6 requires that all grounded conductors (called commons or neutrals) shall be colored either white or grey. No other colors may be used for this purpose. Occasionally it may be absolutely impossible to obtain a piece of wire of the length, size, and insulation type needed for a particular run that is at the same time colored either white or grey. In such cases, it is permissible to use another available color. However, when this is done, in absolutely every box where that

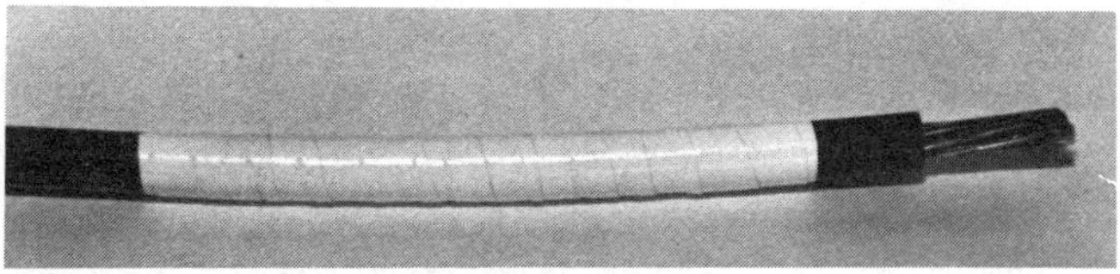

Figure 4-5 Black conductor reidentified with white tape as required per Code.

wire becomes accessible, it must be reidentified (Figure 4-5) with either white paint or white tape so that there can be no doubt that regardless of its original color the wire is in fact being used as a common (grounded) wire.

Ungrounded (hot or power) conductors [NEC 310-12 (c)] must be any color other than white, grey, or green since those colors have been specifically assigned to common and ground respectively. The colors most frequently used for the ungrounded leads are black, red, and blue in that order. As we have just seen, when white or grey are not available it is permissible to reidentify another color with white paint or tape, and use the wire as a common. The reverse is also permissible. Perhaps black, red, or any other power color is not available, but white or grey is. In that case white or grey may be reidentified with black paint or tape ***everywhere*** the wire is accessible after which it may be used as a power wire.

There are specifically two situations in which it is permissible to use a white insulated wire as an ungrounded power wire ***without*** reidentifying it. These cases are covered in NEC 200-7(c) and deal with the use of white as a power color in factory assembled cables only. Two conductor cables are made up with black, white, and a bare ground. Three conductor cables have black, red, white, and bare ground. In a ***switch loop*** the white lead in either two or three wire cables may be used to bring power into a switch as long as a proper power color (black or red) is used to return to the switched load. In this case the white wire need not be taped with black.

The other case when white may be a power wire without reidentification is when it is used as a ***traveler*** in a multiple switch circuit. Both the switch loop and the traveler exceptions are discussed in detail in Chapter 7. They are noted here in connection with the basic NEC color code primarily to emphasize that while there ***are*** instances when white may be used as a power color without reidentification there are ***no cases*** of the reverse. A power color may absolutely never be used as common without proper reidentification.

Ampacities

Amperage, as was explained in Chapter 1, is the measure of the rate of flow of an electrical current. The NEC imposes strict amperage limitations on conductors used in building wiring based on both the wire sizes, measured in AWG or MCM, and the insulation type with which that wire is covered. These limitations are clearly shown in Table 310-16 (Appendix A).

In this table the wire sizes are listed twice, in both the extreme right and the extreme left columns. Temperature ratings and insulation types are also listed twice across the top of the table. Four columns of each will be found above the heading ***Copper.*** Four more appear above the heading ***Aluminum or copper-clad aluminum.*** In the columns below those headings are listed the permissible amperages for the various wire sizes and insulation types listed.

As an example let us look up the allowable amperage for a #12 AWG wire in copper and insulated with TW insulation. Go down the left column of wire sizes to the number 12. The column to the right of the wire sizes is headed ***Types TW & UF.*** In line with wire size 12 in that column appears the number **25+.** At the bottom of the table there is a note after the symbol **+.** It states that #12 copper marked with the obelisk shall be rated at 20 amperes. Any number in the table not followed by the obelisk mark (+) is the actual amperage allowed for that wire size and insulation. A #8 copper wire with the same TW insulation is rated at 40 amperes. The same #8 wire with THW insulation—next column to the right—is good for 50 amperes. The same #8 size but in aluminum with the THW insulation is only good for 40 amperes (6th column to the right of the wire sizes).

Remember, NEC lists minimum requirements. It is always permissible to exceed these code minimums, but never to fall short of them. #12 copper with TW insulation is good for 20 amperes; thus it is permissible to use it for a 15 amp circuit, if desired. On the other hand #14 copper with the same TW is only good for 15 amps so do not try to get away with using it for a 20 amp circuit.

CHAPTER 5

GROUPED CONDUCTORS

With the one exception of the lead connecting the ground bus in the breaker box to the grounding electrode system, all wire runs in building power wiring must contain at least two conductors plus a ground path. That ground path might be an interconnected system of grounding wires such as is used in NM type cable, or it might be provided through a metallic protective casing such as electrical metallic tubing, usually called simply conduit.

Since all power wiring runs must consist of at least two conductors plus ground, a number of different methods for grouping wires have been approved in order to provide for the varying degrees and types of protection that are required under different operating and environmental conditions. The code clearly defines the uses and limitations of each approved method.

Nonmetallic Sheathed Cable—Type NM, NMC, & NMS (NEC Art. 336)

For a number of years the majority of residential power wiring has been done with nonmetallic sheathed cable. It is properly designated as NM cable and defined in the code as a factory assembly of two or more insulated conductors having an outer sheath of moisture resistant, flame-retardant, non-metallic material.

A typical example of this cable is shown in Figure 5-1. The insulation used on the individual conductors is type TW. The outer sheathing is a flame-retardant and moisture-resistant plastic. The grounding conductor in this cable is uninsulated although wrapped with paper. Note that the sheathing carries markings indicating wire size (14), number of wires (2), the presence of the ground wire (G), the cable type (NM), the maximum voltage for which the cable is approved (600V), and the name of the manufacturer (Anaconda). It is required that all of this information appear

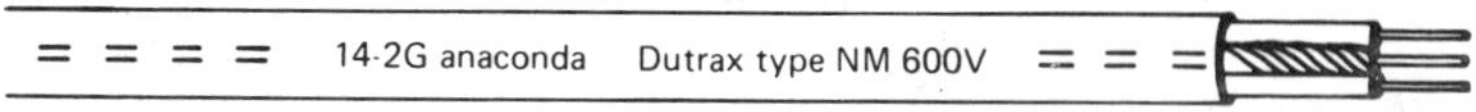

Figure 5-1
NM cable.

on the sheathing of NM cable repeated at intervals not exceeding 2 feet. NM and NMC cables come in both two wire and three wire with ground. Wire sizes run from #14 to #2 AWG.

Type NMC differs from type NM in that the outer sheathing of NMC, in addition to being flame retardant and moisture resistant, is also fungus and corrosion resistant. NMS also contains signal or control wires.

Types NM and NMC cables are permitted by the NEC to be used in single or multi-family dwellings and other structures not over three floors above grade. While the national code permits the use of NM and NMC for stores, offices, or other commercial or industrial buildings as long as they are not over three stories, it is good to ascertain what local codes have to say on this subject. Many local codes provide that all wiring in structures other than residential shall be metal protected.

Type NM is permissible for both exposed and concealed runs in normally dry interior locations. This means that it can be passed through the voids in concrete block walls, as well, as long as such walls are normally dry. It may not be used where it will be exposed to corrosive fumes, nor may it be embedded in concrete or plaster. It is not permissible, for example, to cut a channel in a masonry wall or a concrete floor, place NM cable in the channel, and then seal it in place by covering with plaster or concrete.

NMC may be used anywhere that NM is permitted. It may also be used in damp areas, or in the presence of corrosive fumes. Unlike NM, it may be embedded in a shallow chase in masonry, concrete, or adobe. Neither NM nor NMC may be used for service entrance cables.

Non-metallic cables must be secured to the structure within 12″ of the point where they emerge from a box, and every 4½′ thereafter until within 12″ of another box. This may be accomplished with staples or straps—most commonly staples are used (Figure 5-2). Both NM and NMC cables must be secured to metal boxes with cable clamps (Figure 5-3). These cable clamps may be metal (a), plastic (b), or they may be an integral part of the box itself (c). Clamps are not required when plastic boxes are used.

Where a cable could be exposed to physical damage it must be protected by guard strips or conduit. Where it passes through a floor, conduit is required, and that conduit shall extend at least 6″ above the floor. When cables pass through studs, joists, or rafters (Figure 5-4), the hole for the cable must be at least 1¼″ from the nearest edge of the wood,

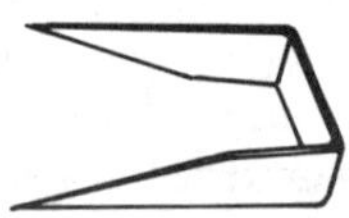

Figure 5-2
Staple for NM cable.

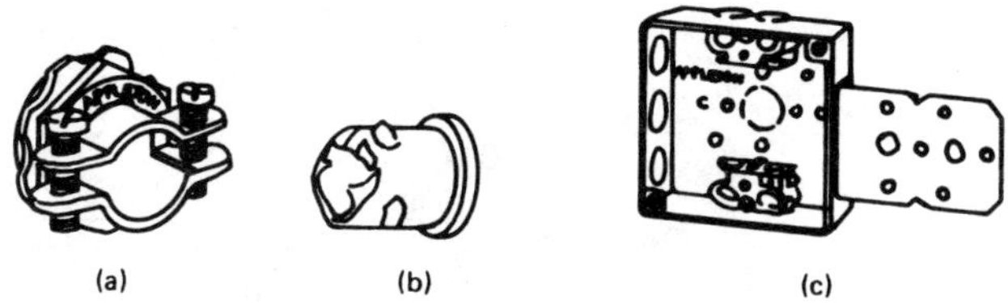

Figure 5-3 NM cable box connectors: (a) metal; (b) plastic; (c) internal within box.

or, as in the case of a notch, there must be a steel cover plate 1/16" thick to protect the cable from drywall or panelling nails. Where cables are run through attics, a distinction is made between what is termed an accessible and an inaccessible attic. If there is a permanent ladder or stairway leading to the attic, it is considered accessible. If there is merely a scuttle hole in the ceiling, making it necessary to bring a ladder to gain access, it is classified as inaccessible. In an inaccessible attic it is only necessary to protect cables on top of the joists within 6' of the scuttle hole by flanking them with guard strips. In an accessible attic, all cables running across the joists plus any cables within 7' above the joists will require guard strips (Figure 5-5).

Service Entrance Cable—Type SE & USE (NEC Art. 338)

Type SE has a flame retardant and moisture resistant cover. While USE may be used underground, it is not flame retardant. In SE and USE cables of two or more conductors, one conductor may be uninsulated. This type is not to be used for interior branch circuits.

When all conductors are insulated, Type SE may be used in interior wiring. When so used it shall be installed to the same requirements as Type NM cable.

Underground Feeder Cable—Type UF (NEC Art. 339)

UF cable is similar to NM cable in that it consists of insulated conductors plus an uninsulated ground wire bound together inside a plastic sheathing (Figure 5-6). The insulation on the conductors is again TW. The exterior covering is flame retardant; moisture, fungus, and corrosion resistant; and is suitable for direct burial in the ground. A sunlight resistant type is also available, making it suitable for general outdoor use whether buried or exposed. The sunlight resistant type is specifically marked. The other markings on UF cables also resemble those required on NM. Wire size is given along with cable type, maximum voltage, and manufacturer name. Available wire sizes again run from #14 to #2.

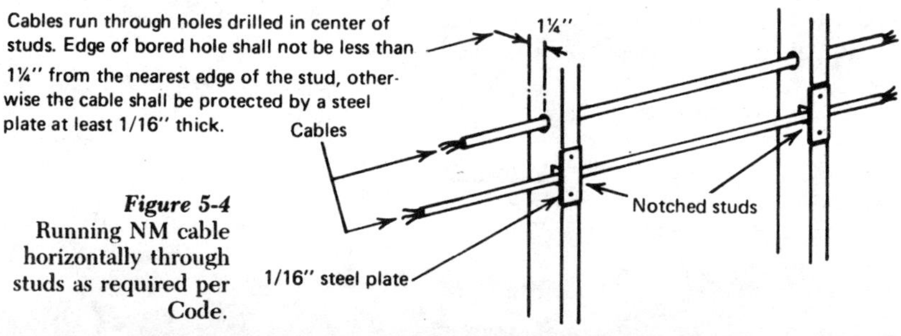

Figure 5-4 Running NM cable horizontally through studs as required per Code.

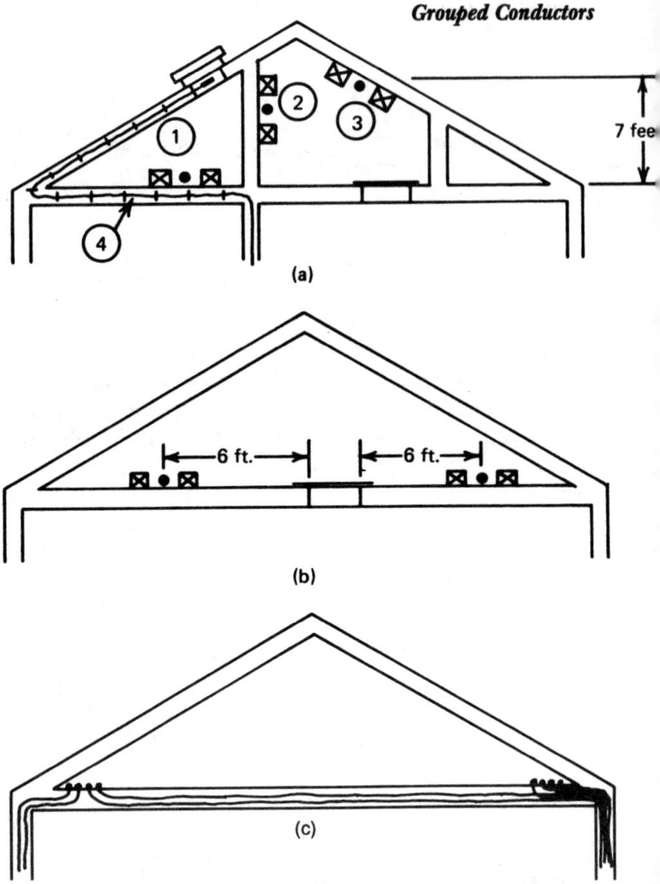

Figure 5-5
Protection of NM cable in attics: (a) accessible; (b) inaccessible; (c) no protection required.

The major visible difference in the cable itself is that while the outside covering of NM cable is a comparatively thin sheathing, UF is actually a solid plastic extrusion in which the insulated conductors and the uninsulated ground wire are completely embedded.

UF cable is approved for direct burial in the ground provided that it be buried at least 24" deep. If a 2" thick concrete layer is placed above the cable, the depth of burial may be reduced by 6". If there is a building slab above the cable, that slab is considered adequate protection and no additional burial depth is necessary. Appendix A, Table 300-5 gives necessary depth requirements for buried wiring.

In addition to underground use, UF cable may be used in any of the interior applications for which NM or NMC cables are approved. When used for such purposes, the UF installation shall conform to the same requirements as NM or NMC, particularly as regards protection from physical damage. UF may not be used as a service entrance cable,

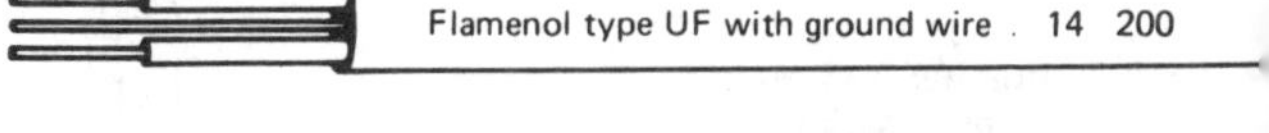

Figure 5-6
Type UF underground cable.

Figure 5-7
AC or BX cable.

and it may not be embedded in concrete. If UF is to pass under a 4" slab, for example, the slab is adequate protection, a mentioned above. Prior to pouring the slab, however, the cable should be placed under a thin layer of sand so it will not become embedded in the finished slab.

Armored Cable—Type AC (NEC Art. 333)

AC cable is more commonly known as ***BX*** cable. It consists of a fabricated assembly of insulated conductors cased in a flexible spiral metal sheathing (Figure 5-7). The flexible metal armor of AC cable is accepted as a grounding conductor; thus the separate grounding conductor included in NM and UF cables is omitted. However, an internal bonding strip is included in AC cables to supplement the effectiveness of the spiral armor as this grounding conductor. Where AC cables are attached to boxes, approved bushings and connectors must be used (Figure 5-8).

AC cable displays the usual marking information. Wire size, number of wires, cable type, maximum approved voltage, and maker's name are found on a tag attached to the coil in the case of AC cable. The wire sizes available in AC cable range from #14 to #1. Both two and three conductor cables are available. The designation ACT indicates that the conductors are insulated with a thermoplastic insulation, and ACL indicates that a lead covering is applied over the conductor assembly greatly improving its moisture resistance.

AC cable may be used anywhere that NM is permitted; it may also be used for underplaster extensions and may be embedded in plaster finishes on masonry. It must be secured with staples or straps within 12" of each box, and at intervals not exceeding 4½′ in runs between boxes. Bends must be made so that the cable armor is not damaged, and the inside radius of any bend must be at least five times the diameter of the cable being bent. AC cable, like NM, may be fished through the voids in

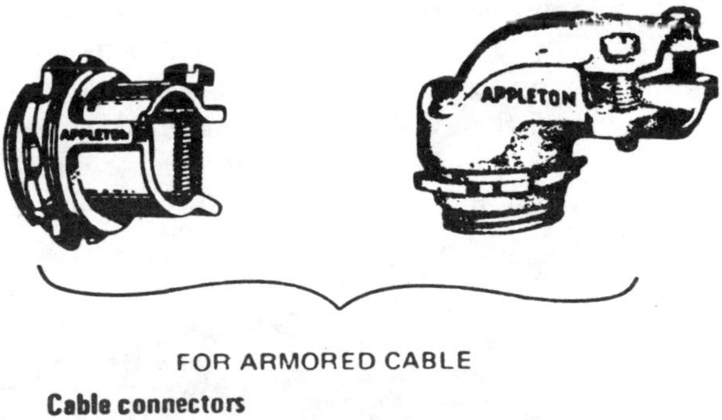

Figure 5-8
Box connectors for BX cable. *Courtesy of Appleton Electric Company.*

Figure 5-9
Flexible metal conduit—greenfield.

concrete block walls when those walls are normally dry. When such walls are subject to considerable dampness, or are below grade, type ACL may be used.

When used in attics, curiously enough, the limitations regarding protection for AC cable are the same as for NM. In inaccessible attics protection is only required within 6′ of the scuttle hole. However, in accessible attics all cables passing over the ceiling/floor joists must have guard strips. Any above the floor level up to 7′ will also need guard strips (Figure 5-5).

Flexible Metal Conduit (NEC Art. 350)

Flexible metal conduit is also a spiral metal armor (Figure 5-9) similar in exterior appearance to AC type cable. It differs from AC cable in that it contains no wires until the electrician puts them in. Use of this type of material any smaller than ½″ electrical trade diameter is not permitted as a permanent part of the building wiring, although ⅜″ trade size may be used in some cases to carry power to motors, and to connect an outlet box to a light fixture.

There are, of course, no markings stating the number of wires, type of insulation, voltage rating, etc., because this material is merely an empty armor through which the necessary wires must be pulled. In accordance with 350-12 the number of conductors permitted in flexible metal conduit shall not exceed the percentage of fill specified in Appendix A, Table 1, Chapter 9. Incidentally, this material is often referred to in the trade as *greenfield.*

It is commonly used concealed in stud or concrete block walls. It may not be used in wet locations unless the wiring pulled inside is of a type approved for the conditions, and water is not likely to enter enclosures to which it is connected.

Although it is a type of metal conduit, this flexible type is acceptable as a grounding means, only if both the conduit and the fittings being used are approved for grounding. Otherwise it is permitted as a grounding means only up to lengths of 6′. If the total length of the ground return path is longer than that, a separate ground wire must be added.

The box connectors used with flexible metal conduit thread inside the spiral of the material (Figure 5-10). It is also possible to connect two

Figure 5-10
Greenfield box connectors. *Courtesy of Appleton Electric Company.*

a b

pieces of flex with a coupling that threads into the inside of both as shown.

Like both NM and AC cables flexible metal conduit must be secured within 12″ of each outlet box, and every 4½′ from there on until within 12″ of the next box. Between boxes a run of flexible metal conduit shall not contain more than 4 quarter bends. That means no more than 360° in bends; a quarter bend equals 90°. Also such bends shall have sufficient sweep so that the armor is not damaged, meaning an inside radius of at least 5 times the diameter of the armor being bent.

Liquidtight Flexible Conduit (NEC Art. 351)

Liquidtight is another type of spiral flexible metal armor similar to greenfield except that the spiral turns are rather tighter, and an outer plastic jacket has been added, that plastic being both liquid tight and sunlight resistant (Figure 5-11). Like greenfield, liquidtight carries no markings for wire size, insulation, or voltage rating because wires must be pulled as needed. The number of wires of different sizes and insulation types that may be pulled shall agree with Appendix A, Table 1, Chapter 9.

While NEC permits the use of liquidtight in a wide range of applications, it is expensive. Consequently, its use in residences is quite limited. It is often used in short runs for wiring outdoor air conditioning equipment or for wiring disposals under kitchen sinks.

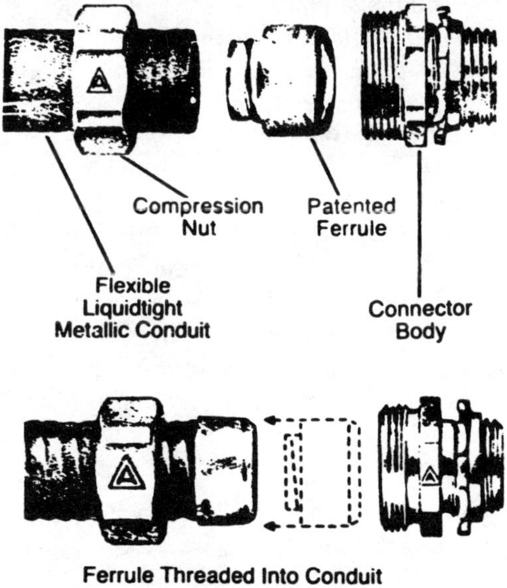

Figure 5-11
(a) Liquidtight flexible metal conduit; (b) box connector.

Liquidtight flexible conduit is always attached to boxes by means of special liquid tight compression connectors (Figure 5-11). In the rare case when liquidtight is installed on or in walls it must be secured with staples or straps every 4½′ and within 12″ of boxes the same as any of the other flexible wiring materials.

NEC also recognizes *non-metallic* liquidtight flexible conduit. In residential work it may be used in the same places as the metallic type except that extreme cold may make it brittle, limiting its use outdoors to moderate climates. Lengths are limited to no more than 6'. Grounding conductor may be installed either inside or outside the conduit, but when outside may be no more than 6' in length.

Rigid Metal Conduit (NEC Art. 346)

While NEC refers to, and permits, the use of both ferrous and nonferrous metals as rigid metal conduit, in practice galvanized pipe is generally used—the same type galvanized pipe commonly found in water supply and gas lines. As always the use of pipe smaller than ½″ trade size is not allowed.

Since this material is plain, everyday steel pipe, it can be cut and threaded the same as water or gas pipe. When it is cut, all cut ends must be reamed to remove any rough edges. If any rough edges are left, they are likely to tear the insulation when insulated conductors are pulled through. Threading is done with the same type threader and dies that a plumber would use on the same pipe. Standard American Pipe threads are to be used throughout.

The number of wires of specific sizes and insulation types that may be pulled through pipe conduit of the various trade pipe sizes shall conform to Appendix A, Table 1, Chapter 9.

Between boxes or fittings the number of bends allowed are again limited to no more than the equivalent of four quarter bends or 360°. This is to facilitate the pulling, removal, or repulling of conductors. The minimum radius of bends for various conduit sizes are listed in Appendix Table 346-10. When space does not allow for bends of the radii required in the table, fittings called pulling ells may be used (Figure 5-12). Whenever a pulling ell is used, the count on bends can be started again.

Pulling ells (Figure 5-12) are made with either side, or the long end, or both sides removable. This makes it possible to get into them even when located in all sorts of unlikely and difficult nooks and crannies. In addition to ells there are tees and straight conduit bodies. Any of these fittings may also be used with flexible metal conduit, which has already

Figure 5-12
Conduit fittings used with rigid, EMT, and flexible metal conduits.

been discussed, or with electrical metallic tubing (EMT), which will be discussed next.

Rigid metal conduit must be secured to supporting material within 3′ of each box or fitting, and every 10′ thereafter until within 3′ of another box or fitting. In accordance with Appendix Table 346-12 with conduit sizes of 1″ and greater the spacing between supports increases with the conduit size.

Rigid metal conduit formerly was used far more commonly than it is at present. In a great many small residential and commercial buildings, the only rigid metal conduit used, if any, will be for the mast supporting the overhead service entrance. Far more commonly used, because it is much easier to handle, is electrical metallic tubing.

Electrical Metallic Tubing—EMT (NEC Art. 348)

EMT or ***thinwall*** conduit is by far the most widely used non-flexible conduit. Except for situations where it will be subject to severe physical damage, it may be and is used in virtually all places where rigid conduit may be used. It may be used exposed or concealed in or on walls, floors or ceilings. With proper protection and fittings, it can be embedded in concrete or buried in the ground. Also properly protected it may be used in areas subject to severe corrosion.

It is preferable to rigid conduit as it is much lighter, thus easier to handle. It is also easier to cut and ream. Bending to follow fairly complex forms is quite easy with the smaller diameters. Threading is not only not necessary, it cannot be done. Connections are made between lengths of EMT and boxes, or between successive lengths of conduit using either compression type connectors and couplings, or setscrews (Figure 5-13).

For the number of wires of different sizes and insulation types that may be used in EMT refer again to the familiar Appendix A, Table 1, Chapter 9. There may be no more than a total of four quarter bends in a single run between boxes, to facilitate pulling the required wiring through. The minimum radius for bends in EMT in the various available

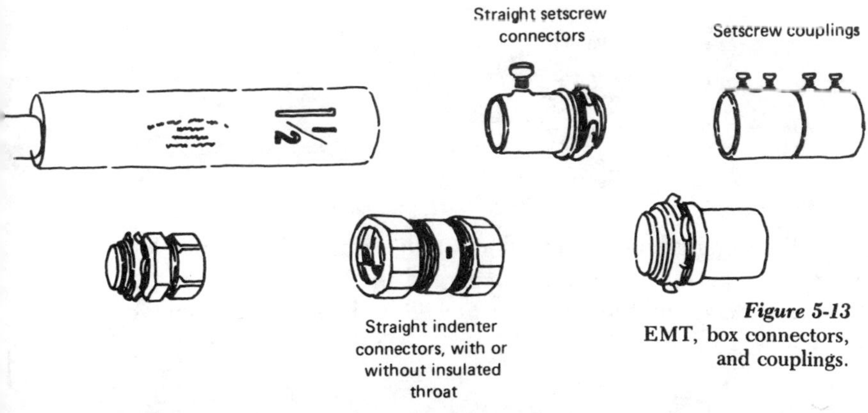

Figure 5-13
EMT, box connectors, and couplings.

trade sizes is the same for rigid and will be found in Appendix A, Table 346-10.

The same fittings—pullings ells, tees, and straights—used with rigid are also used with EMT, and as with rigid the counting of bends starts again at any fitting, or box. EMT must be securely attached to supporting structures within 3′ of every 10′ thereafter. The material is supplied in 10′ lengths. This insures that there can be no full length of it supported only by the couplings at both ends, as this would be likely to be too weak for safety.

Since EMT is a continuous metal armor made up of parts in solid, firm electrical contact with each other it also constitutes an acceptable continuous ground return path making the inclusion of a separate ground wire unnecessary.

Rigid Non-Metallic Conduit (NEC Art. 347)

Although other materials are approved for this purpose, most of the rigid non-metallic electrical conduit presently used is polyvinyl chloride (PVC). Other approved materials are asbestos cement, fiberglass epoxy, or high-density polyethylene.

Non-metallic conduit may be used above ground or underground and may be concealed in walls, ceilings, or floors. It may be used in wet locations and has excellent resistance to many highly corrosive chemicals. Since PVC is designed for connection to couplings, fittings and boxes using a solvent type cement, it is absolutely waterproof, thus ideal for many really wet areas in industrial and commercial structures.

NEC Article 347 does not limit the conductors allowed in a single rigid conduit by number but instead limits the percentage of the cross section that may be filled to that given in Appendix Table 1.

As always, no conduit smaller than ½" trade size is permitted. Supports for rigid non-metallic conduit shall be attached within 3' of each box, and from there as follows:

Conduit Size	*Support Spacing*
½″–1″	3′
1¼″–2″	5′
2½″–3″	6′
3½″–5″	7′
6″	8′

Every length of non-metallic conduit must be marked at least every 10′ giving the manufacturer's name and the type of material permitting the electrician, and the inspector as well, to be sure that approved material is being used.

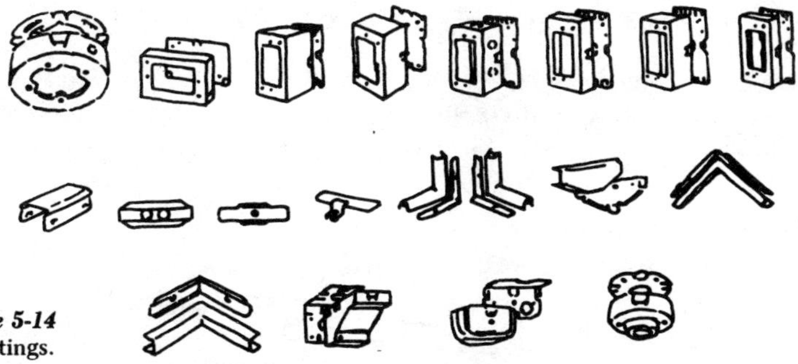

Figure 5-14
Wiremold fittings.

Surface Raceways–(NEC Art. 352)

Occasionally in residential applications and quite often in offices and commercial applications it is either necessary or desirable to surface mount electrical wiring that must be visually acceptable to a degree that exposed NM cable or exposed EMT simply is not. For this purpose complete systems of raceways, boxes, and fittings, both metallic and nonmetallic, designed for surface mounting are available under the trade name "Wiremold."

Surface raceways are permitted only in dry areas, and a variety of fittings, couplings, connectors and boxes are available (Figure 5-14). These allow considerable flexibility in the design and installation of surface mounted wiring. The joints between fittings in the metal Wiremold systems are a type of compression connection. NEC accepts these metal to metal contacts as an approved ground path and does not require an additional ground wire inside the raceway.

There are several different sizes of Wiremold metal raceways intended to handle different numbers of wires of various sizes and insulations. These are illustrated and their conductor capacities shown in Figure 10-14, on page 230.

Surface nonmetallic raceways, like their metallic counterparts, are permitted only in dry areas. They may not be concealed and may not be exposed to physical damage. They may pass transversely through dry walls or partitions as long as the part inside is unbroken. The number of conductors permitted in this raceway are:

Wire Size	*THHN*	*TW*
14 AWG	10	6
12 AWG	7	5
10 AWG	5	4

Inclusion of a grounding conductor is, of course, required when non-metallic raceways are used.

*Wiremold is a registered trade name.

Multioutlet Assembly (NEC Art. 353)

The multioutlet assembly is a surface mounted raceway (Figure 5-15) in which outlets spaced are placed at intervals of either 6" or 12". The raceway contains the necessary wiring to serve the installed outlets.

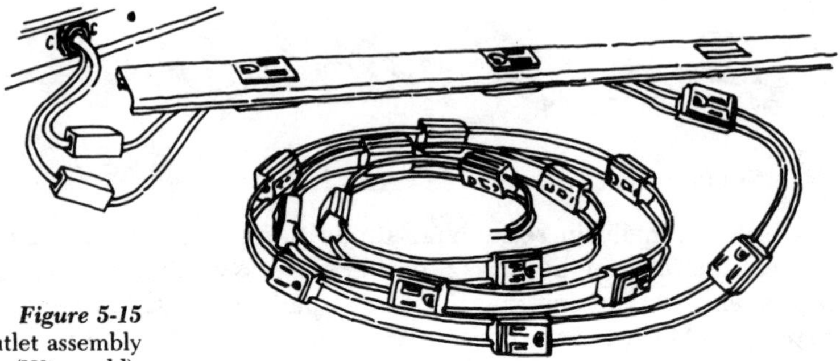

Figure 5-15
A multioutlet assembly (Wiremold).

These assemblies are available in lengths of 3′, 5′, and 6′. Suitable connectors are also available to provide for attachment of multioutlet assemblies to existing wall outlet boxes, conduit, flexible conduit or Wiremold.

CHAPTER 6

ELECTRICAL BOXES

All receptacles, switches, and fixtures must be installed in appropriate electrical boxes in order to conform to NEC 300-15, which requires that all connections of conductors to devices be made in boxes. Some light fixtures are supplied with junction boxes as an integral part of the fixture. Those not so supplied must be attached to boxes that are parts of the power wiring system. NEC 300-15 further states that all connections, or splices, of conductors to each other must be made in boxes or ***fittings.*** Fittings, also called conduit bodies, were mentioned in Chapter 5 and shown in Figure 5-11.

The number of conductors that must meet and connect in, or pass through, a particular box varies greatly as does the number of devices mounted there. Also how a box can be installed in a wall and the amount of space available for it vary considerably as well. In addition, as we shall see, the NEC places a number of restrictions on how many conductors may be brought into a box of given size and how they shall be brought in, as well as how the box shall be installed. As a result of the broad range of both requirements and restrictions relating to the use of electrical boxes a considerable variety is now available in both metal and plastic.

Metal Boxes

Metal boxes may be used with all of the various wiring systems: NM cable, rigid or EMT conduit, flexible conduit, BX cable, or liquidtight flexible conduit. There are two basic shapes. Rectangular boxes are designed to conform to the standard sizes of switches and receptacles. Octal or round boxes are for mounting ceiling or wall light fixtures. When a box is needed solely as a joining or junction point for wiring, either a rectangular or an octal box may be used.

Single Gang Boxes

The most commonly used single gang type is called a ***handy box.*** The ordinary use for this box is as the mounting place for a single switch or a duplex outlet. As we shall see in Chapter 7, certain other devices may be mounted in it as well.

Figure 6-1 shows some of the different handy boxes available. While at first glance they may look very similar each is significantly different from the others. All of the boxes illustrated are fabricated from stamped sheet steel. All of the circles on sides, bottom and ends are partially cut out so as to be easily removable. They are called ***knockouts*** because wherever a cable or conduit is to be connected to the box one of these disks is literally knocked out to provide an opening for mounting the appropriate connector (Figure 6-2). As we shall see later, knockouts may

Figure 6-1
2⅛″ × 4″ handy boxes—metal.

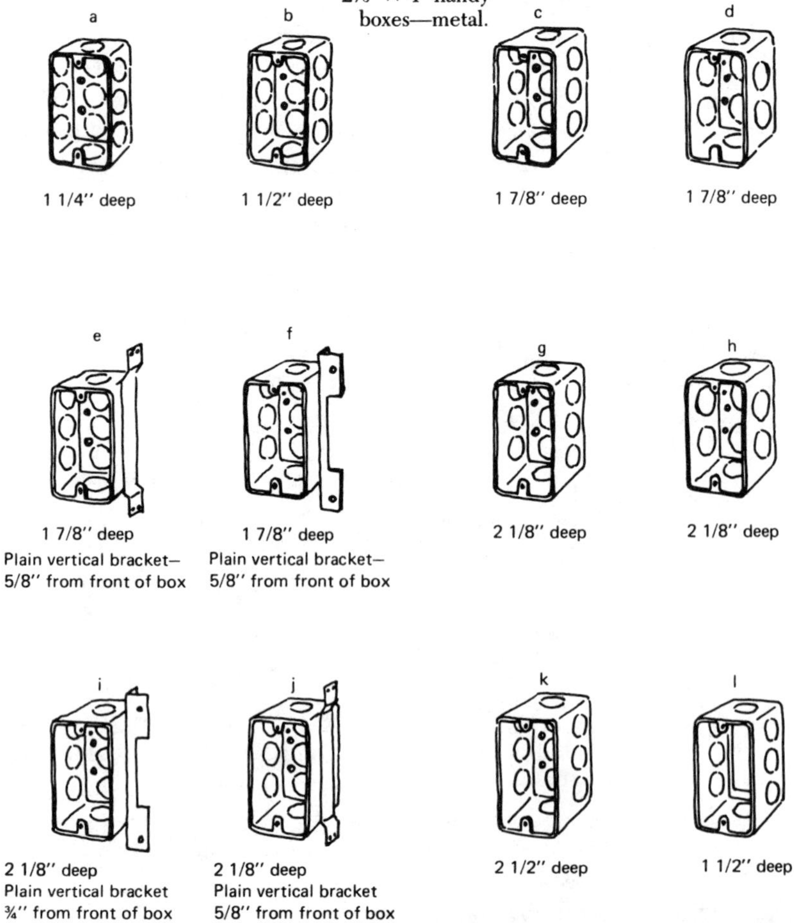

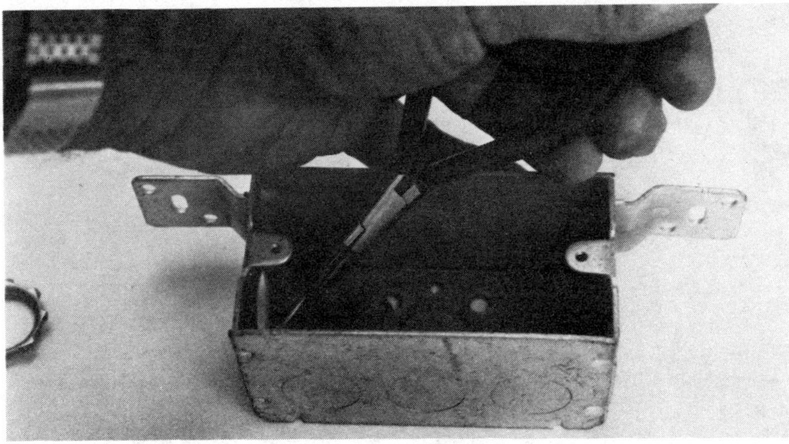

Figure 6-2
Removing knockout.

be removed only from openings that are to be used. The small front flanges top and bottom are correctly spaced and threaded to take the mounting screws for standard switches or outlets.

The boxes shown in Figure 6-1 a, b, and c all have the same number of knockouts, and all of them are of the same ½″ trade size. The difference between these boxes then is only in depth. The reason for this variation is that the number of conductors that may be brought into a box according to code is limited by its cubic volume. Since the length and width of a single gang box are limited by the standard dimensions of switches and outlets, cubic volume, and therefore permissible number of conductors, can only be increased by adding to the depth as shown here.

Box d has all the same dimensions as box c, but the knockouts have been changed. Box d has fewer knockouts but they are made for ¾″ trade size connectors.

Boxes e and f are still as deep as c and d, but they are provided with side brackets for easy attachment to wall studs. The bracket on e allows it to be mounted with the box front at any desired distance forward of the stud, while f assures a constant measurement. The choice will depend on whether the need is for flexibility or uniformity.

The boxes in Figure 6-1 g and h are deeper than the previous ones, meaning they are approved for more conductors. Box g is stamped for ½″ trade size knockouts while h has ¾″ holes. The two types of mounting brackets that were used on e and f are used again on i and j, which are deeper boxes, both with ½″ knockouts throughout.

Box k is the deepest of all, but is otherwise similar to a, b, c, or g. However, box l is a new type. This is an extension box. Its back has been removed and a pair of screw holes made that will line up with those on the front flanges of any of the other handy boxes allowing it to be mounted on top of one of them greatly increasing the internal cubic volume available for wires.

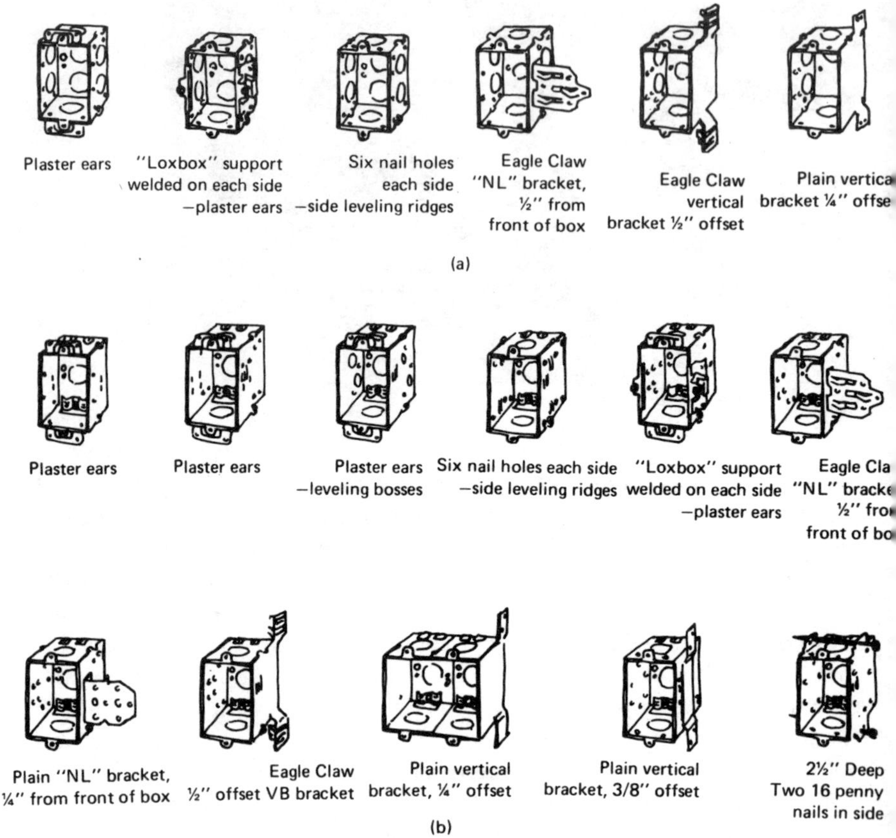

Figure 6-3 3″ × 2″ square corner switch boxes: (a) with conduit KO's; (b) with nonmetallic sheathed cable clamps.

In wood frame construction the majority of handy boxes are mounted on wall studs using brackets similar to those shown in Figure 6-1. Other methods of installing handy boxes will be discussed in Chapter 10 in connection with rough wiring.

Another type of single gang box is the square corner device box (Figure 6-3). These are also made of sheet steel, many having removable sides. This allows two or more to be joined together to accommodate combinations of switches, or switches and outlets. Note that with these boxes the top and bottom front flanges threaded for switch or receptacle mounting are outside not inside. This is because the box is 3″ high instead of the handy box dimension of 4″.

Square corner boxes are available in a wide variety of configurations in terms of knockouts and mounting brackets. Some are provided with

Figure 6-4
Bevel corner device boxes—metal.

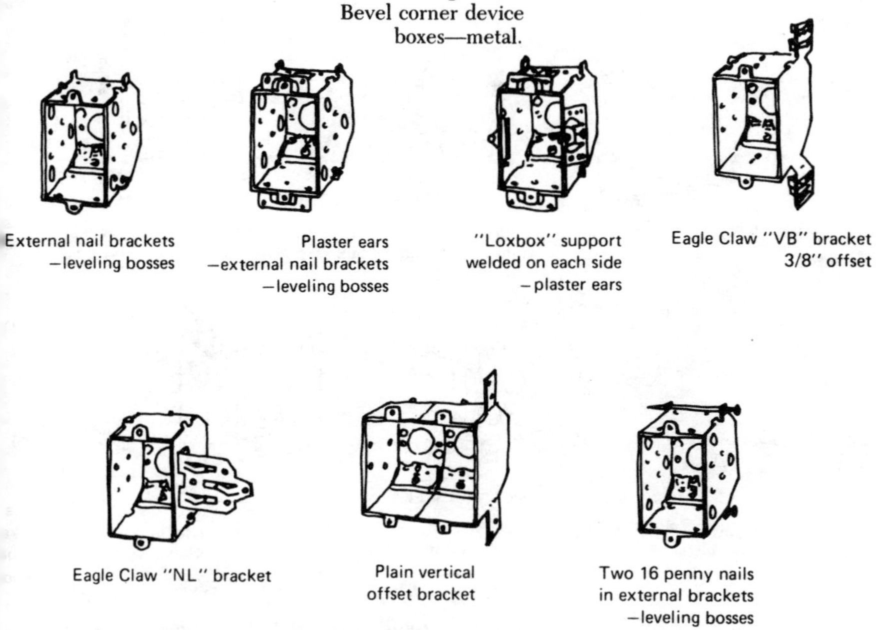

knockouts only for round conduit or NM connectors. Others have round knockouts as well as internal NM cable clamps. Due to the reduction in the height, square corner boxes are generally deeper than handy boxes in order to maintain adequate internal volume for conductors. The internal cable clamps readily facilitate installations using NM cable, since the separate cable clamps required in all round knockout holes can be omitted.

Another single gang box that can be connected to others to form various multiple combinations is the bevel corner type (Figure 6-4). These are also supplied with internal cable clamps for NM cable, and a selection of mounting methods is possible.

4 Inch Square Boxes

When two switches, or two outlets, or a switch combined with an outlet must be placed in a single box the 4″ square, or 4 by 4 box is used (Figure 6-5). This box may also be used in an instance where only a single device is required, but the box is also a junction for more conductors than the single gang boxes will take.

4″ square boxes are supplied with many different arrangements of knockouts and mounting brackets as well as in several depths ranging from 1¼″ up to 2⅛″ to accommodate various numbers and sizes of conductors. Some have all ½″ knockouts, others are all ¾″. Many have differing

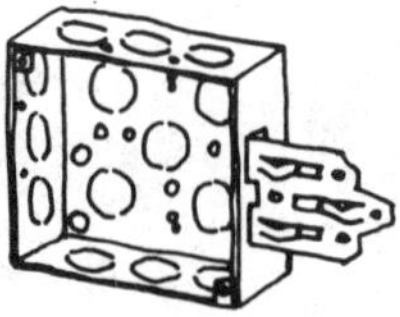

Eagle Claw NL bracket flush with front of box

Eagle Claw vertical bracket flush with front of box

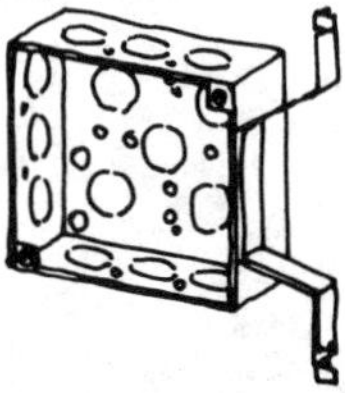

Offset vertical bracket Flush with front of box

Plain vertical bracket with barbed hooks

Figure 6-5
4″ square boxes—metal.

combinations of ½″ and ¾″ knockouts. Boxes with all 1″ knockouts can be obtained as well as ones with combinations of all three.

Many 4″ square boxes have corner mounting flanges as shown in Figure 6-5, to accept any of the various cover plates shown in Figure 6-6. Others have two pairs of flanges exactly the same as the single pair on a handy box to permit direct attachment of two switches or two outlets to the box without the necessity of an intermediate cover plate.

Extension rings for 4″ square boxes can be added where the number of conduits, cables, or wires that must meet in a single box becomes large (Figure 6-7). Any of these rings can be mounted in front of one of the

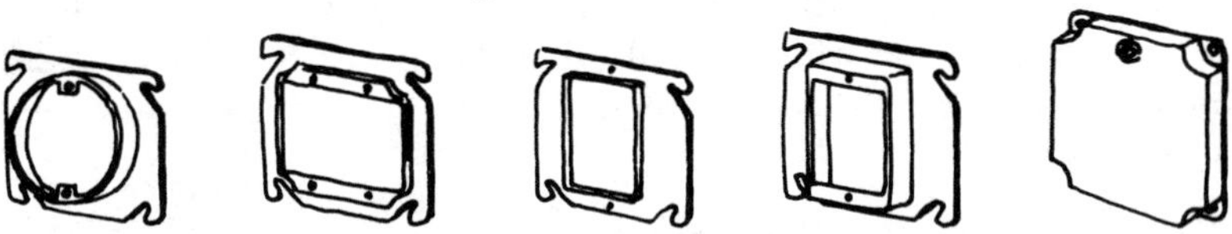

¾″ Raised ¾″ Raised ¾″ Raised ¾″ Raised ½″ Raised

½″ Raised ½″ Raised ½″ Raised ½″ Raised ½″ Raised

Figure 6-6
Various 4″ square box cover and adapter plates.

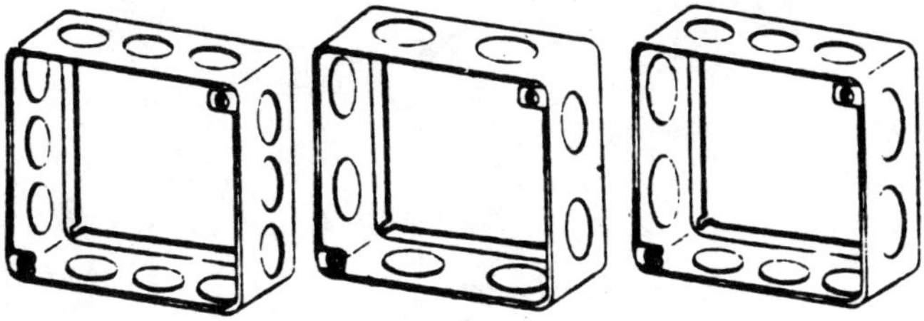

Figure 6-7
4″ square box extension rings: 1½″ deep—Two 8-32 tapped holes and two slide-on screw slots in bottom. *Courtesy of Appleton Electric Company.*

regular 4″ square boxes by screwing through the slide-on slots in the back of the ring. Knockouts on these rings are either all ½″, all ¾″, or a combination of the two.

Multiple Gang Boxes

Larger boxes with space for 3, 4, 5, or even 6 devices (switches or outlets) can be obtained when required (Figure 6-8). While a 3 gang requirement is fairly common, the 4, 5, and 6 gang are not. Consequently they may be rather difficult to find when needed. These larger boxes are stamped for a combination of ½″ and ¾″ knockouts, and appropriate coverplates for the mounting of the devices are supplied.

Cut-In Boxes

A variation of the square corner boxes shown in Figure 6-3 is one equipped with adjustable clamps on the sides (Figure 6-9). This box need not be attached to the wall studs, but is installed directly in the drywall. The plaster ears hold it on the outside while the adjustable clamps are tightened against the back side of the plasterboard. This type box is particularly useful in alteration and renovation work when switches or outlets must be added to walls that have already been finished on both sides.

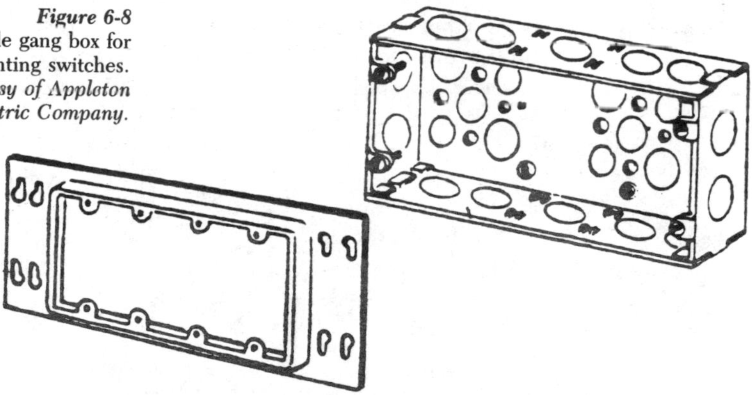

Figure 6-8
Multiple gang box for mounting switches. *Courtesy of Appleton Electric Company.*

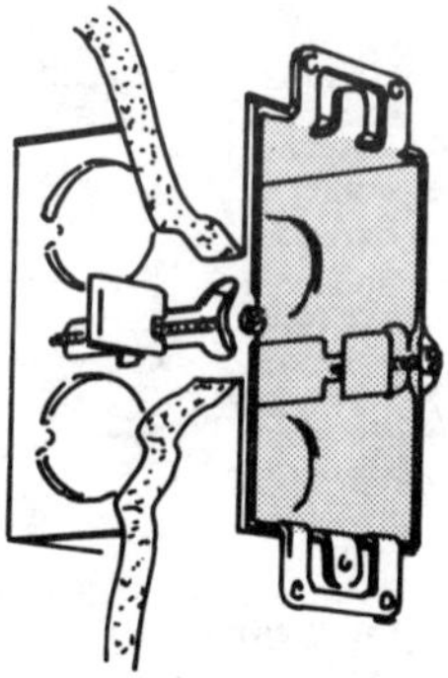

Figure 6-9
Cut-in box: install box with side brackets in drywall; side screws are tightened to bring the brackets against the wall and hold the box in place.

Another handy feature of these boxes is that they can be ganged by merely removing sides and screwing boxes to each other. In this way two, three, or more can be ganged together as needed, and then the entire composite unit installed in the wall using adjustable clamps on the ends.

Octagon Boxes

Octal boxes for the support of ceiling or wall light fixtures are either 3″ or 4″ across with two threaded flanges spaced to conform to the standard used in fixture mounting straps or on the fixture bases themselves. Octal boxes (Figure 6-10) vary both in depth, and in knockout sizing. Some have all ½″ knockouts, some ¾″, and the others are mixed.

Octals equipped with mounting brackets may be attached directly to ceiling joists. However, the location of ceiling light fixtures as dictated by visual requirements often leaves the electrician without a joist in the proper place. In that case either a block is placed between joists, as will be shown in Chapter 10, or the box is attached to a bar hanger (Figure 6-11) which in turn is attached to the joists.

Like other boxes, octals can also be deepened to provide additional space for a large number of conductors by the use of extension rings (Figure 6-12). Octal extensions are attached to the standard octal boxes by screwing them on the front in the same way as rectangular extensions are added. The knockouts vary similarly between ½″ and ¾″ as needed.

Figure 6-10
Octagon boxes—metal.

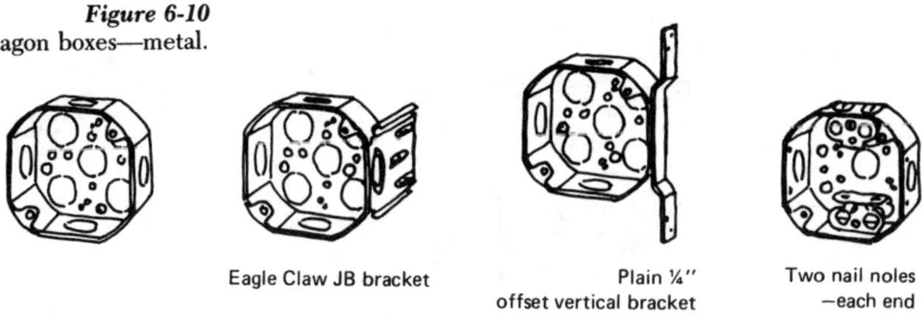

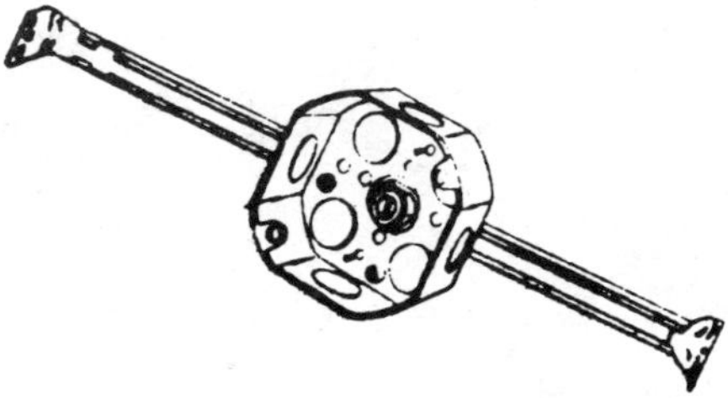

Figure 6-11
Octal box on bar hanger. *Courtesy of Appleton Electric Company.*

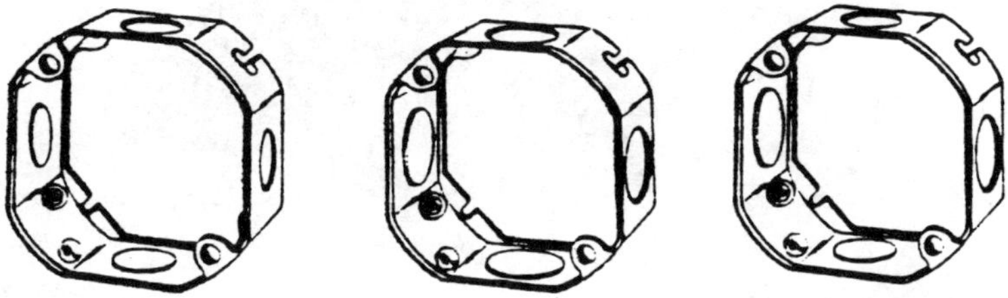

Figure 6-12
Octal box extension rings, 1½″ deep: (a) ½″; (b) ¾″; (c) ½″ and ¾″.

Weatherproof Boxes

Weatherproof, according to NEC Article 100, means "so constructed or protected that exposure to the weather will not interfere with successful operation". Naturally all boxes used outdoors for floodlights, security lights, convenience outlets, and switches must be weatherproof. In addition some boxes in wet locations indoors also need to be weatherproof type.

Weatherproof metal boxes are made of aluminum castings instead of sheet steel like other metal boxes. Instead of knockouts they contain threaded holes into which any of the standard box connectors may be directly threaded (Figure 6-13). All unused threaded connector holes must be closed with threaded plugs firmly screwed in place to maintain watertight integrity.

Weatherproof rectangular boxes are supplied in one, two, and three gang sizes. A selection of waterproof covers allows the mounting of various arrangements of waterproof switches, single, and duplex receptacles, and mixed combinations of switches and receptacles (Figure 6-14).

The receptacles used in weatherproof boxes are the same as those used for normal indoor wiring. A gasket placed between the rectangular cover plate and the box, as well as the gaskets inside the individual receptacle covers, provide the necessary weather proofing. Outdoor switches that are built into a weatherproof cover plate can be used, or switching can be done with indoor type switches placed under gasketed weatherproof lids.

Figure 6-13
Weatherproof box showing threaded connector holes.

Weatherproof boxes can be obtained either with or without mounting lugs (Figure 6-15) for wall attachment. When the box itself is not directly attached to a wall it can be supported by two or more conduits threaded into it wrenchtight [NEC 370-23(f)], after which the conduits are attached to the wall. When the box is ***not*** over 100 cubic inches in size, and does ***not*** contain receptacles or switches, and does ***not*** support a light fixture, the conduits may be attached to the wall as far as 3′ away from the box they support. However, if the box does contain switches or receptacles, or is the support for a fixture, then the conduits holding it in place must be attached to the wall within 18″ of the box (Figure 6-16). When a weatherproof box is to be supported only by conduit buried in the ground, as for example a box supplying power to outdoor shrub lights, it must then be supported (Figure 6-17) by two conduits no more than 18″ high.

Figure 6-14
Weatherproof covers: 1, 2, and 3 gang. *Courtesy of Bell Electric.*

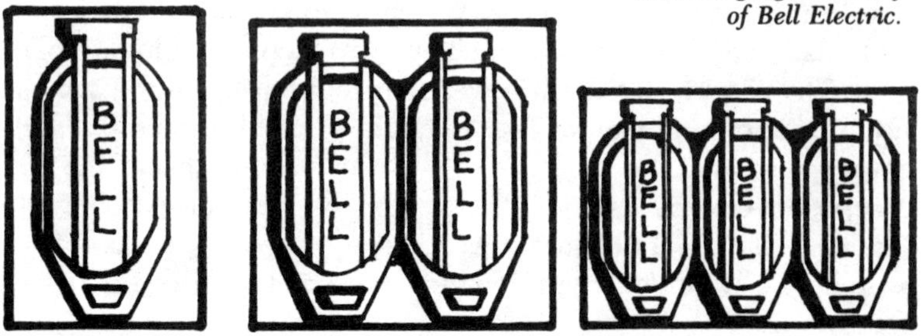

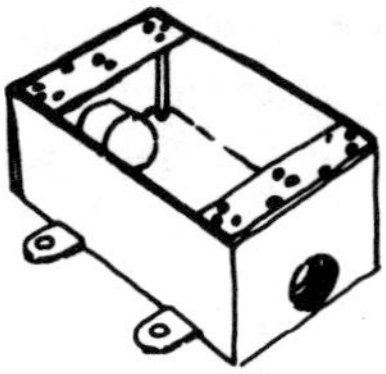	 3	1/2″	270-L
		3/4″	273-L
	 4	1/2″	271-L
		3/4″	274-L
	 5	1/2″	272-L
		3/4″	275-L

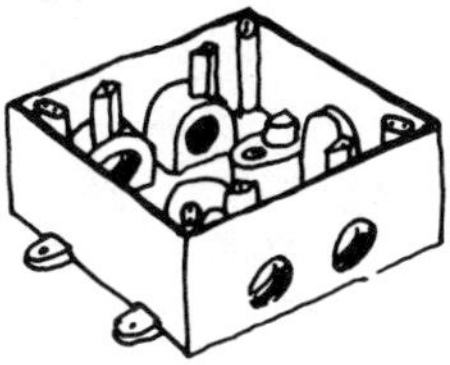	 3	1/2″	276-3L
		3/4″	277-3L
	 4	1/2″	276-4L
		3/4″	277-4L
	 5	1/2″	276-5L
		3/4″	277-5L
	 5	1/2″	276-4SL
		3/4″	277-4SL
	 6	1/2″	276-6L
		3/4″	277-6L
	 6	1/2″	276-4S2L
		3/4″	277-4S2L
	 7	1/2″	276-7L
		3/4″	277-7L

Figure 6-15
Weatherproof boxes with mounting lugs.

Figure 6-16
Weatherproof box supported by conduits on the wall.

Figure 6-17
Weatherproof box supported by two conduits in ground.

In addition to rectangulars, there are round weatherproof boxes (Figure 6-18). These are either with or without lugs for wall or soffit mounting. Like the rectangulars they may be supported solely by two or more conduits, providing the conduits are attached to the building as described above. Since round weatherproofs are most commonly used as supports for outdoor flood and security lights the 18″ maximum support spacing for the conduits will usually apply.

Plastic Boxes

Plastic boxes are referred to in the code as ***nonmetallic boxes***. They are made either of a thermosetting polyester or PVC. Code permits their use with open wiring on insulators and concealed knob and tube wiring, neither of which was mentioned in Chapter 5 on wiring systems because both methods are now obsolete. However, plastic boxes are also approved for use with nonmetallic sheathed cable (NM cable) which was discussed in Chapter 5 and which is the wiring method used for most residential work and other buildings not exceeding three stories above grade [NEC 336-5(a)(1)]. Here vast quantities of plastic boxes are used. They are light and easily installed. Since they do not require grounding, nor do they require box connectors, plastic boxes can be wired considerably faster than conventional metal ones.

Plastic boxes are definitely not approved for use with metal conduits (rigid or EMT), Greenfield, liquidtight flexible metal conduit, or armored cable. All these require metal boxes.

Figure 6-18
Round weatherproof box with floodlight mounted.

Figure 6-19
Single gang plastic boxes.

Like metal boxes, plastic boxes are supplied in two shapes: rectangular for switches and receptacles, and round for fixture mounting. Rectangulars come in 1, 2, 3, and 4 gang sizes. Singles (Figure 6-19) vary in depth from 2″ to 2½″, and all are equipped for direct stud mounting. The models fitted with nails or **V** brackets can be installed to fit any wall surfacing material: ¼″ wood paneling, ½″ drywall, ⅝″ drywall, or plaster. The right angled side bracket types are set for either ¼″ or ½″ wall surfacing materials.

Two and three gang rectangular boxes are available in either direct nailing or bracket types (Figure 6-20), and some three and four gang models are supplied with bar hangars. In the author's experience a three gang without bar hangar is more or less manageable, but a four gang really needs additional support.

The round, plastic fixture boxes are fastened with nails supplied with the box, an attached bracket, or with an adjustable bar hangar (Figure 6-21). Since boxes for ceiling fixtures require accurate placement, the bar hangar makes mounting fairly easy. In cases when the fixture is rather heavy the bar hangar also offers better support than the one sided attachments of the nail-on or side bracket boxes.

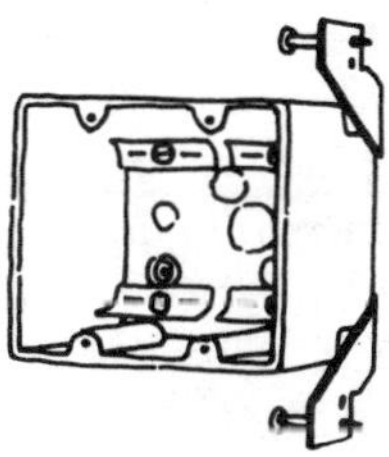

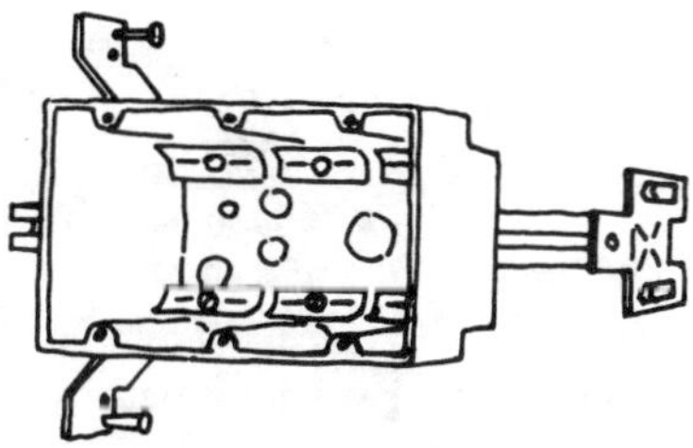

Figure 6-20
2, 3, and 4 gang plastic boxes.

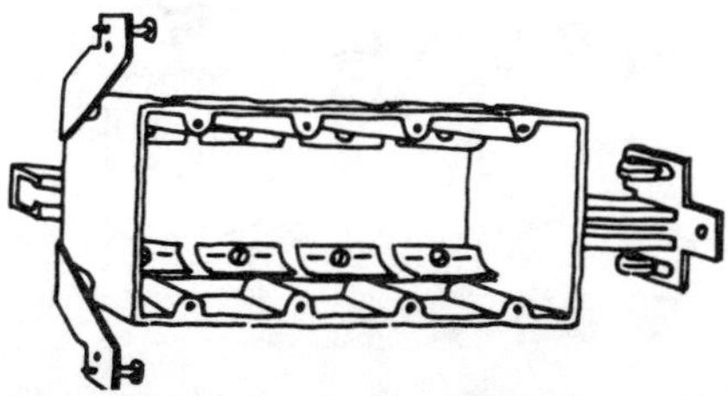

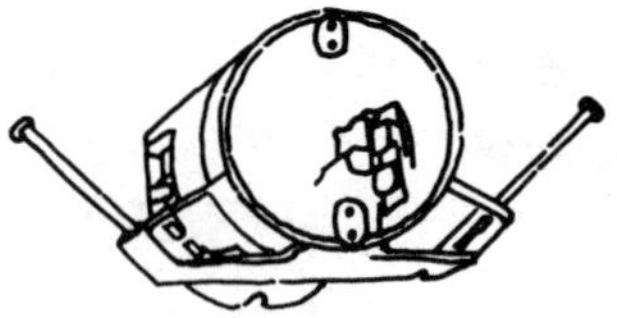

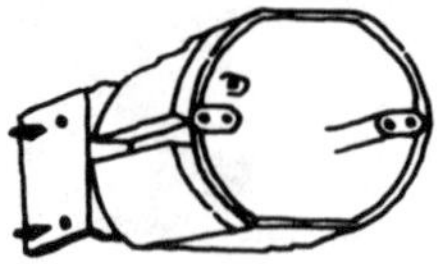

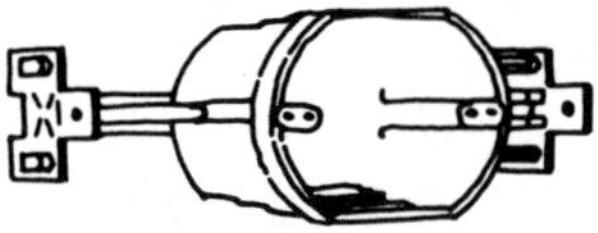

Figure 6-21
Round plastic boxes: direct nail, bracket, and bar hanger types.

Old-work or cut-in boxes (Figure 6-22) both round and single gang are available in plastic, but the singles cannot be attached to each other to make multiples in the manner of the metal cut-in box shown in Figure 6-8. The rounds can be installed directly in a ceiling without requiring any attachment to ceiling joists, but before mounting a fixture on such a box make certain the ceiling sheetrock or plaster can support its weight.

National Electrical Code Regulations Regarding Boxes

Number of Conductors Permitted in a Box

According to NEC 370-16, the number of conductors allowed varies depending on the size of a particular box as well as the wire size of the conductors. Appendix A, Table 370-16 (a) gives the conductor maximums for many of the most commonly used round or octal, square, and single gang rectangular boxes. As is immediately apparent, the permissible number of conductors is directly related to the cubic volume of the box. For example, a box 4" square by 1¼" deep (line #4) may house 9 conductors at wire size #14, but only 8 at #12. By increasing the depth to 2-1/8" the number of #14 wires permitted increases to 15.

For box sizes not included in Table 370-16 (a) the conductor maximums must be determined by computing cubic volume, and then dividing by the required volume per conductor given in Table 370-16 (b).

Figure 6-22
Plastic single gang and round cut-in boxes.

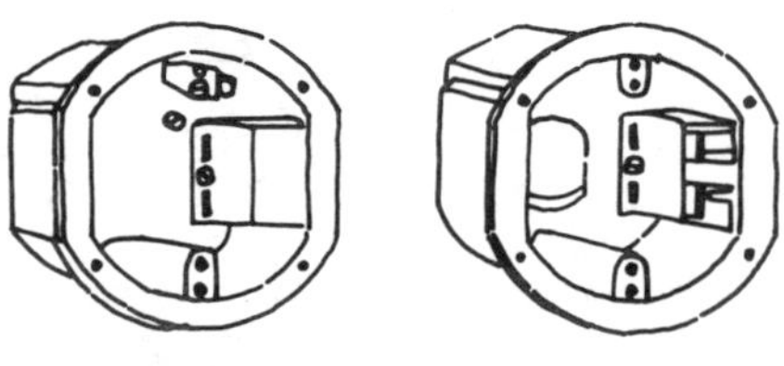

Not only does NEC 370-16 specify the number of conductors permitted in a box, it also defines how they are to be counted. At the start of Chapter 4 a conductor was defined, per NEC 310-2, as a wire that shall be insulated (except when it is not insulated). NEC 370-16 establishes a more elaborate procedure for arriving at the total of the conductors in a box:

1. An insulated wire entering a box and terminating inside the box in a splice or at a terminal on a device such as a switch or receptacle counts as 1 conductor.
2. One or more ground wires spliced together count as only 1 conductor.
3. A conductor originating and terminating inside the box (a pigtail) shall not be counted.
4. When a box contains internal cable clamps such as those shown in Figure 6-3 they are counted as 1 conductor.
5. Each switch or receptacle counts as 1 conductor
6. A conductor that originates outside a box and simply runs through it and out again without being cut or connected to anything counts as 1 conductor

In Figure 6-23 a box is shown with two three-wire and one two-wire cables entering plus three grounds, four pigtails, and a duplex receptacle. The conductor count for this box is:

a. Circuit conductors	8
b. Three grounds (count as only 1)	1
c. Receptacle	1
d. Pigtails	0
Total	10

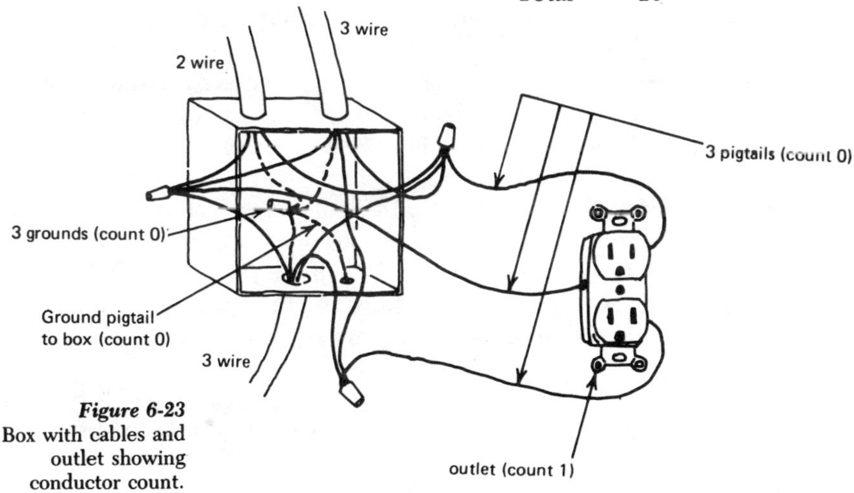

Figure 6-23
Box with cables and outlet showing conductor count.

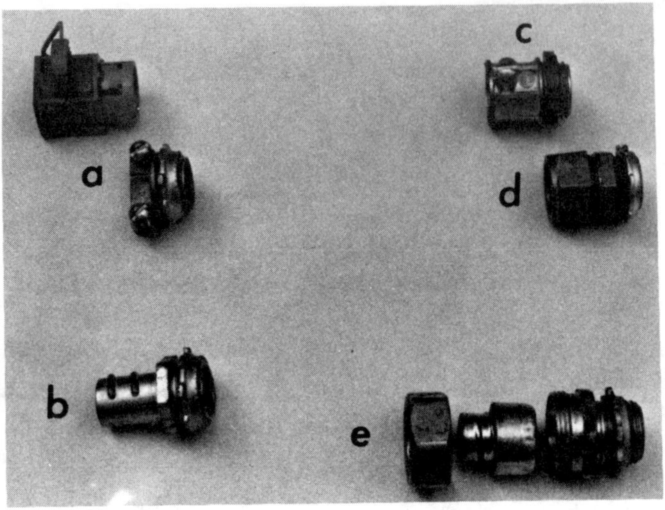

Figure 6-24
Box connectors: (a) NM cable; (b) Greenfield; (c) EMT setscrew type; (d) EMT compression; (e) liquidtight.

Entrance of Conductors into Boxes

Provisions relating to conductors entering boxes are given in NEC 370-17. Conductors entering metal boxes are protected from abrasion by the use of appropriate box connectors (Figure 6-24). Note that all of these connectors completely fill the knockout holes into which they are fitted per NEC 370-17 (a).

Greenfield and liquidtight connectors protect the wiring by screwing into the inside of the cut ends of the spiral armor. With BX cable the wiring is protected by a fiber bushing that must be slipped inside the armor end after cutting. EMT is only protected by proper reaming of the tube ends after cutting; therefore, inadequate reaming is extremely dangerous.

NM cable may be brought into plastic boxes by simply passing it through a cable knockout. Since plastic boxes do not present the same abrasion danger as metal, no connector is required. However, the sheathing must extend no less than ¼″ inside the box, and the cable must be fastened to the wall framing within 8″ of the box.

Unused Openings All unused box openings shall be closed (NEC 370-18). If by accident or error the wrong knockout in a metal box is opened, it must be plugged (Figure 6-25). In the case of a metal box, an appropriate plug is available, but if the wrong hole gets knocked out of a plastic box there is nothing to do but use another box.

Figure 6-25
Steel knockout plug.

Flush Devices Flush devices mean primarily switches and receptacles. Their boxes must be completely enclosed on back and sides (NEC 370-19), and properly support the device. A box such as Figure 6-1-1 with no back could not be used by itself as a flush device box. Any of the gangable boxes in (Figure 6-3) that have removable sides could not be used with one side missing.

Wall Box Set-back

Where boxes are set in walls or ceilings of concrete, plaster, sheetrock, or other non-combustible material (NEC 370-20) the box front shall sit back from the wall surface no more than ¼″ (Figure 6-26). However, if the wall is wood paneling or other combustible material, no set-back is allowed. The box front must be flush with the surface.

Surface Repairs

Where plaster or plasterboard wall surfaces have been broken around a box, they must be repaired so that no gaps or openings greater than ⅛" are left around the edges (NEC 370-21). In most instances this requirement is unneccessary, for the gaps around boxes that the code requires closed are unsightly and would be closed anyway.

Box Accessibility

NEC 370-29 opens, requiring that "Junction, Pull, and Outlet Boxes . . . be Accessible." What it means by accessible and what you think of as accessible are not the same. The code continues that boxes

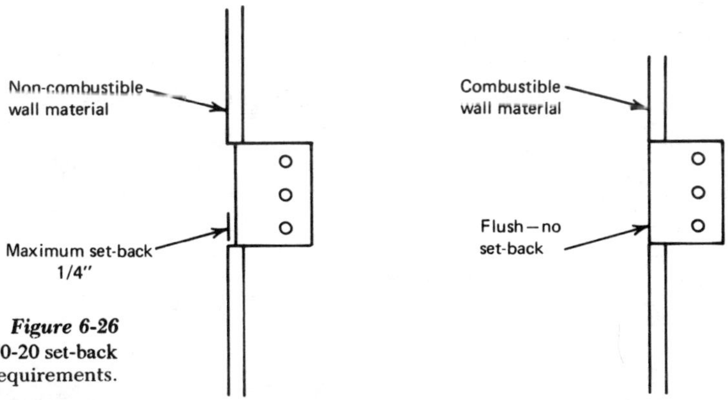

Figure 6-26
NEC 370-20 set-back requirements.

must be installed so that the wiring inside can be reached without making a hole in a wall or a ceiling, or cutting any framing out of the way.

In other words a box can be located where all manner of athletic exertion is required to reach it provided that the cover can be removed without damaging the building. However, a junction box cannot be placed behind the plasterboard facing of a living room wall unless an access door is provided so it can be reached without damaging the wall.

CHAPTER 7

WIRING SWITCH CIRCUITS AND OUTLETS

In all buildings whether residential, commercial, or industrial, nearly all lights, some appliances and machinery, and the switched half of many convenience outlets are controlled by switches or switch circuits that are integral parts of the building power wiring rather than parts of the devices being switched. This permits placing switches at considerable distances from devices and allows switches to be placed at the most convenient possible operating locations.

Switch Types

In order to accomplish various switching functions any of several types of switches may be needed (Figure 7-1).

Single Pole, Single Throw Switch

This basic ***on-off*** switch is used most commonly. Internal mechanisms vary somewhat from manufacturer to manufacturer, but the result is the same. Whether accomplished with a toggle or rocker, the two terminals on the case (Figure 7-2) are internally connected to each other, or disconnected.

Single Pole, Double Throw

The type used in building wiring (as distinct from machinery controls or electronic equipment) is more often referred to as a ***three-way switch.*** This type is an ***either-or*** not an ***on-off*** switch. It connects terminal **A** to

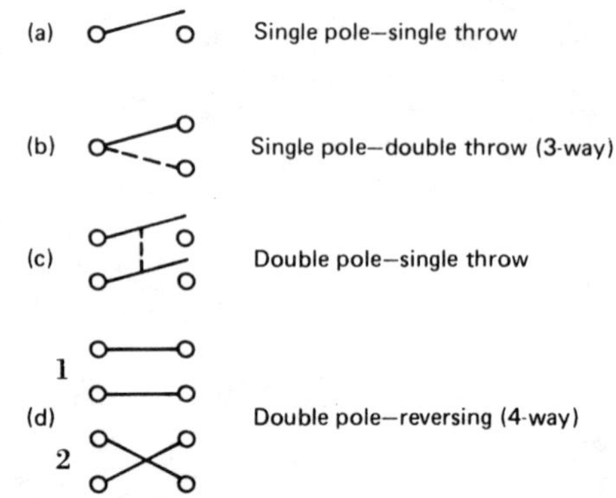

Figure 7-1
Switches: (a) single pole; (b) 3-way; (c) double pole single throw; (d) 4-way (double pole reversing).

either ***B*** or ***C***—never both, and never neither. Thus it is never ***off.*** The various ways of connecting and using this switch will be fully explained in the discussion of multiple switch circuits that will follow.

Double Pole, Single Throw

This is a second type of ***on-off*** switch. This time two pairs of terminals are simultaneously connected or disconnected. In building wiring, this switch is used on all 240V circuits to control both hot lines at once as required by code. The double breakers required for range, dryer, 240V water heaters, and split wired appliance outlets are internally double pole single throw switches. The fused disconnecting mechanisms required for small central air conditioning systems, as discussed in Chapter 11, are also double pole single throws.

In addition, the ground fault circuit interrupter discussed in connection with code requirements in Chapter 8 is also a double pole single throw since it disconnects both hot and common sides of a 120V circuit simultaneously.

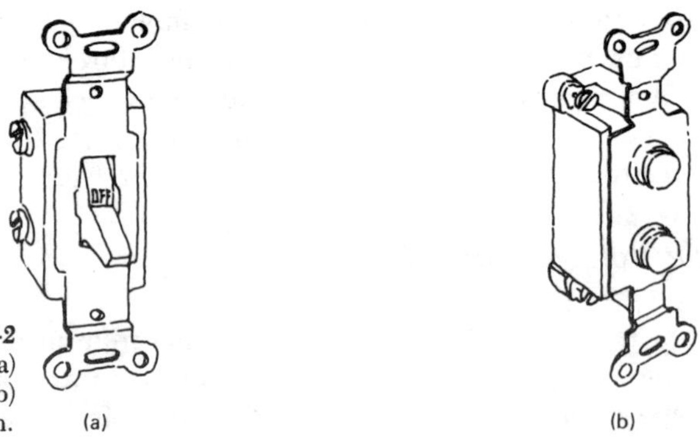

Figure 7-2
Single pole switches: (a) toggle switch; (b) pushbutton switch.

Figure 7-3
3- and 4-way switches: (a) modern 3-way switches have three terminal screws and their handles are not lettered *On* and *Off*. One screw (the dark colored one, which is the "hot" or common terminal) connects to the power.

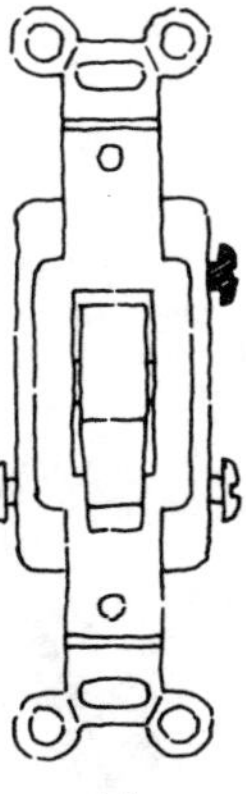

(a)

(b) A modern 4-way switch has four terminal screws: one pair for input from the 3-way or 4-way switch that feeds it, and one pair for output to the 3-way or 4-way switch that it feeds.

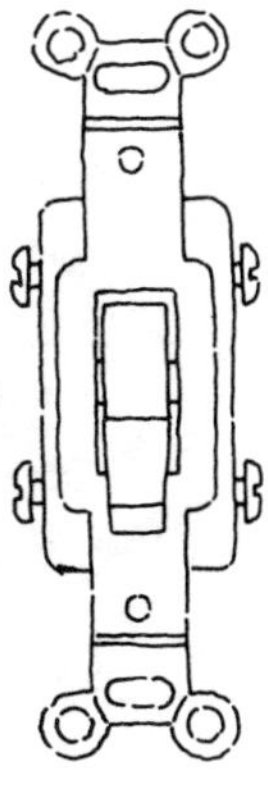

(b)

Double Pole Reversing Switch

This is another either-or switch. More commonly known as a ***four way switch,*** it is used in conjunction with two three-ways in the multiple switch circuits to be discussed shortly. Internally it connects the two terminals at one end of its case either straight through to the corresponding two at the other end (Figure 7-1 d-1), or it connects diagonally across in an **X** (Figure 7-1 d-2).

Dimmer Switches

In incandescent lighting circuits dimmer switches have become particularly popular recently in response to conservation programs although they have long been available for special effect lighting. The most commonly used dimmers (Figure 7-4) are the inexpensive rotary type, which turn clockwise to dim. Normally they are supplied with wire pigtails rather than screw or push-in terminals such as those found on single pole, three-way, or four-way switches.

Two types of dimmers are available: dimmer combined with a single pole switch (Figure 7-4a), or the dimmer combined with a three-way switch. The single pole will have two black pigtails for connections, the three-way (Figure 7-4b) will have one black and two reds. The three-way might be substituted for either one of the standard three-way switches in any of the multiple switch circuits that will be described shortly. The single pole dimmer may be used to replace any standard single pole switch. Connections to both types are made by splicing directly to the wire pigtails provided.

Remember—the dimmer switches shown are for use with incandescent lights ***only.*** Fluorescent lights can be dimmed also, but a different,

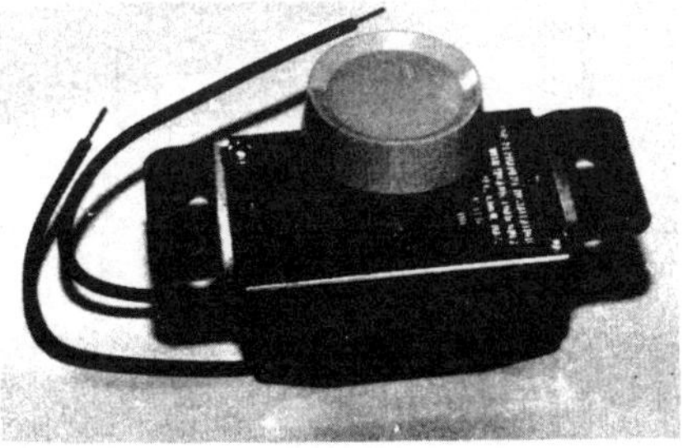

(a)

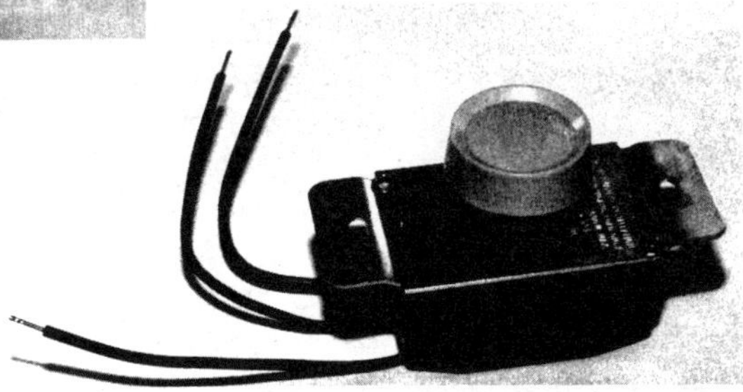

(b)

Figure 7-4
Dimmer switches: (a) single pole; (b) 3-way.

and more expensive dimmer is required. A special dimming ballast must replace the standard ballast supplied with the fixture. Consequently fluorescent dimmers are rarely used.

The average, inexpensive incandescent dimmer will handle loads of up to 600 watts. For most residential or small office applications, this wattage is generally plenty of capacity. However, make it a routine procedure to total the load that the dimmer must handle just to be sure. If the load is more than 600 watts, it will be necessary to employ a somewhat more expensive, higher capacity unit.

Pilot Light Switches

See Figure 7-5. These switches are commonly used on lighting circuits that have been set up in such a way that the light or lights controlled are not visible from the switching location. For example, one might have a light in the attic with a switch on the floor below, or conversely there could be a light in the basement with a switch on the floor above. Pilot lights are commonly used for outdoor lighting circuits such as driveway post lights, decorative shrubbery lights, or security light systems. In all cases the pilot light acts as a reminder that something not visible from the switch location is turned on.

Note well that the pilot light indicates ***only*** that a circuit has been turned on. It by ***no*** means indicates that the device at the end of the line

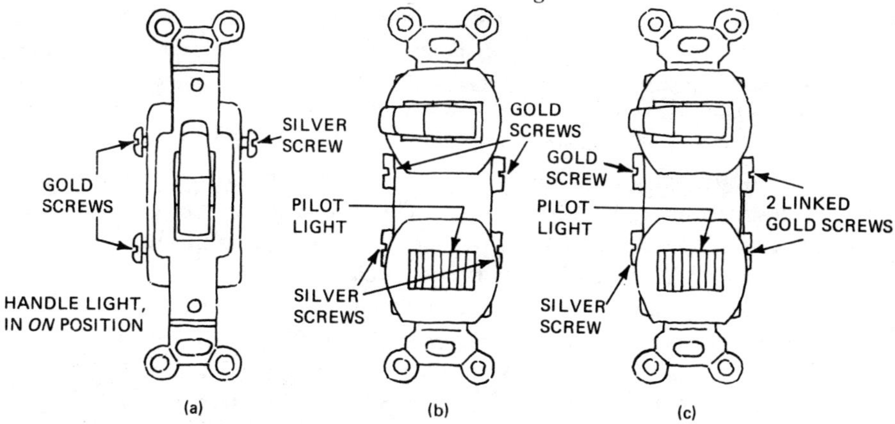

Figure 7-5
Pilot light switches with breakaway link between switch and light.

is, in fact, operating. The pilot light is wired in parallel with the load being switched. A look back at the basic series and parallel circuits illustrated in Chapter 1 (Figure 1) will remind you that devices wired in parallel operate independently of each other. Thus the pilot light at the switch could be operating perfectly while a post light at the far end of the driveway could have a dead bulb and be out of operation. The point here is that it is necessary to check the device at the far end of the line periodically to insure that it is operating.

Waterproof Switches

Often it is convenient to control outdoor devices with outdoor switches. Such switches must be waterproof (Figure 7-6). Waterproof switches are available in single pole or three-way types. This permits control of a circuit from one or two outdoor locations, or from one outdoor location plus one or more inside. These switches are made for mounting on either standard or waterproof boxes, and are supplied with a waterproofing gasket to seal the joint between the switch faceplate and the box.

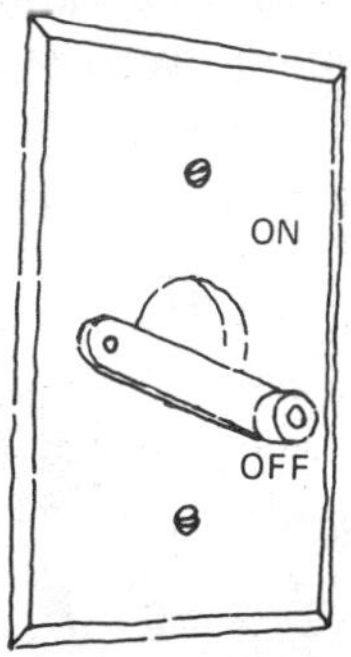

Figure 7-6
Weatherproof switch.

Photo-Control Switches

There are various requirements for lighting needed all night, but not in daylight. Such circuits can be controlled using timers, but with the change of seasons the hours of daylight and darkness change quite considerably. This means that the timers will require frequent resetting—a nuisance at best.

An alternate method for the control of such circuits is the use of photo-control switches (Figure 7-7). These devices contain a photo-electric sensor; actually it is a small photovoltaic cell, that produces a slight voltage that is sufficient to hold the switch open as long as there is light enough to activate it. When night falls and the light fails, the cell goes dead, the switch closes, and the lights go on. Naturally, the sensor unit must be adequately shielded from any customary night lighting in the area. If not, it will sense the light, activate the photovoltaic cell, and in turn open its switch shutting off its light circuit.

Photo-controls are commonly supplied with a threaded fitting that will enter a ½″ standard box knockout or will thread directly into a standard waterproof outdoor box. Connections are normally made by splicing directly to wire pigtails on the photo-control.

Key Switches

When there are children or, for various other reasons, tamperproof yet accessible switches might be useful. This need can be met with a key switch (Figure 7-8). The key switch is simply a single pole switch from which the toggle has been removed and replaced with a key slot. It may be activated only when the proper key is inserted. Electrical connections are the same as for any other single pole switch.

Pool Filter Systems, Hot Tubs, Sprinkler Systems, Night Lighting

These devices are usually controlled by timer switches (Figure 7-9). Additionally, timers are being included in an increasing number of space and water heaters as well as air conditioning in order to conserve energy.

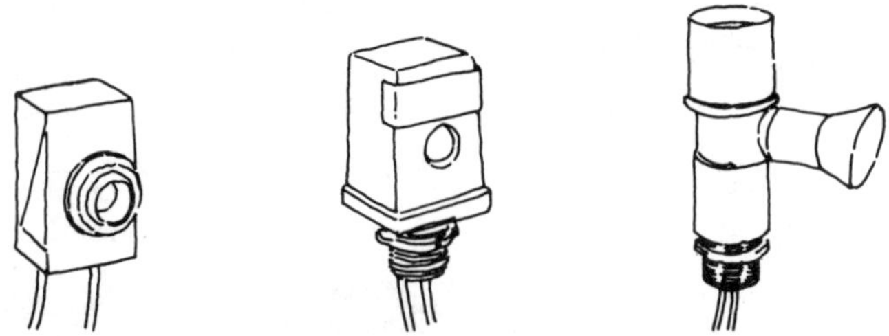

Figure 7-7
Photo-controlled switches.

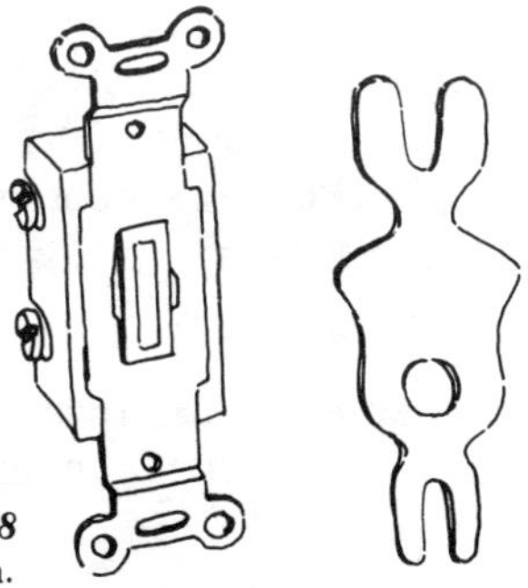

Figure 7-8
Key switch.

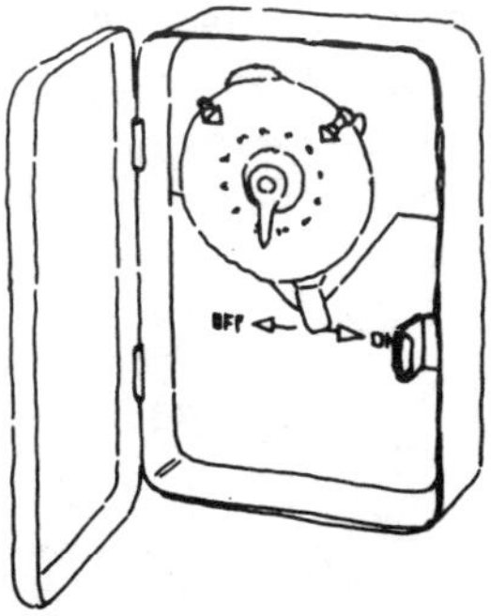

Figure 7-9
Timer switch.

Timed switching units contain a small clock operating on the 60 cycle line current. The clock turns a plate calibrated to 24 hours. Control tabs are attached to the plate that will open or close a power switch at whatever times have been set on the plate.

Wiring is connected to screw terminals. A wiring diagram illustrating the proper connections is supplied on the inside of the cover of each timer switch. Standard screwhead color coding of brass heads for power and silver heads for common will be used, but which screws are input and which output vary with different manufacturers. When installing one of these, assume that all else has already failed and read the instructions ***first.***

Wiring Switch Circuits

All building wiring runs are done as discussed in Chapter 5, with grouped conductors in one form or another: either with factory preassembled cables, or with some type of conduit through which the needed wires are pulled. In preassembled cables, the individual wires are already color coded: for power—either black or black and red; for common—white; for ground—bare. In conduit the electrician must code them correctly when he pulls his wires.

As mentioned earlier, most switches are disconnectors on the power side of the line only; the common return, or grounded side of the line, remains connected.

Wiring of Single Pole Switch Circuits

This switch (Figure 7-2) has two screw terminals, or two push-in terminals, or both. Power enters through one and leaves from the other. Internally the switch either connects or disconnects these two terminals. Consequently, it makes no difference which one is the input or output.

The basic parts of a single pole switch circuit are a switch, a load, and the wires to connect them. The incoming feeder consists of a power, or hot, wire (usually black, red or blue), a common grounded return (white or grey), and a safety grounding path which may be a grounding wire (i.e., cable or it may be a continuous metal sheathing as when EMT is used).

> ***IMPORTANT*** *The common or grounded return (white or grey) wire goes to all loads such as lights, outlets, or using devices. It does not go to switches.*

The wiring of a single pole switch circuit varies depending on where the unswitched power comes in (Figure 7-10). When the feeder enters the circuit at the switch box, in NM or other cable, the connections will be black power in to one terminal on the switch, black power out of the other switch terminal, and on to the load. White common incoming splices to white common outgoing. Neither white goes to the switch. Outgoing white connects to the load (Figure 7-10a). When the single switch controls multiple loads, the additional ones are spliced on to the switch output in whatever order happens to be convenient (Figure 7-10b).

Changing from cable to conduit, to wire the same circuit with either single or multiple loads (Figure 7-10c), only one simple change is made. In the switch box the white common merely passes through on its way to the load. Since it attaches to no device or other wires in that box, it need not be cut at all. As a standard procedure when working in conduit, any wire that simply passes through a box on its way to somewhere else should not be cut and spliced. Just leave a loop a couple inches in diameter in the bottom of the box, and continue it on to its destination. In this case the white is looped and passes on to the load.

When, as often happens, unswitched power enters the circuit at a ceiling light, that unswitched power will have to pass to a controlling switch before it gets to the light. This is a common situation because, particularly in residential applications, it is often convenient to distribute power through the building via the attic. Since code permits residential buildings, and others not over three stories above grade, to be wired in cable, right away there appears to be a problem with insulation color coding. All two conductor cables contain one black power wire, one white common wire, and a bare ground. To take only power to a switch and back, two black wires are needed since both wires are power side. This certainly seems to violate Article 200-7 of the code which says:

> A continuous white or natural grey covering on a conductor or a termination marking of white or natural grey color shall be used ***only*** for the ***grounded*** conductor.

However, read on. 200-7(c)(2) says:

Where a cable contains an insulated conductor for single-pole, 3-way, or 4-way switch loops, and the conductor with white or natural gray insulation or a marking of three continuous white stripes is used for the supply to the switch, but not as a return conductor from the switch to the switched outlet. In these applications, the conductor with white or natural gray insulation or with three continuous white stripes shall be permanently re-identified to indicate its use by painting or other effective means at its terminations and at each location where the conductor is visible and accessible.

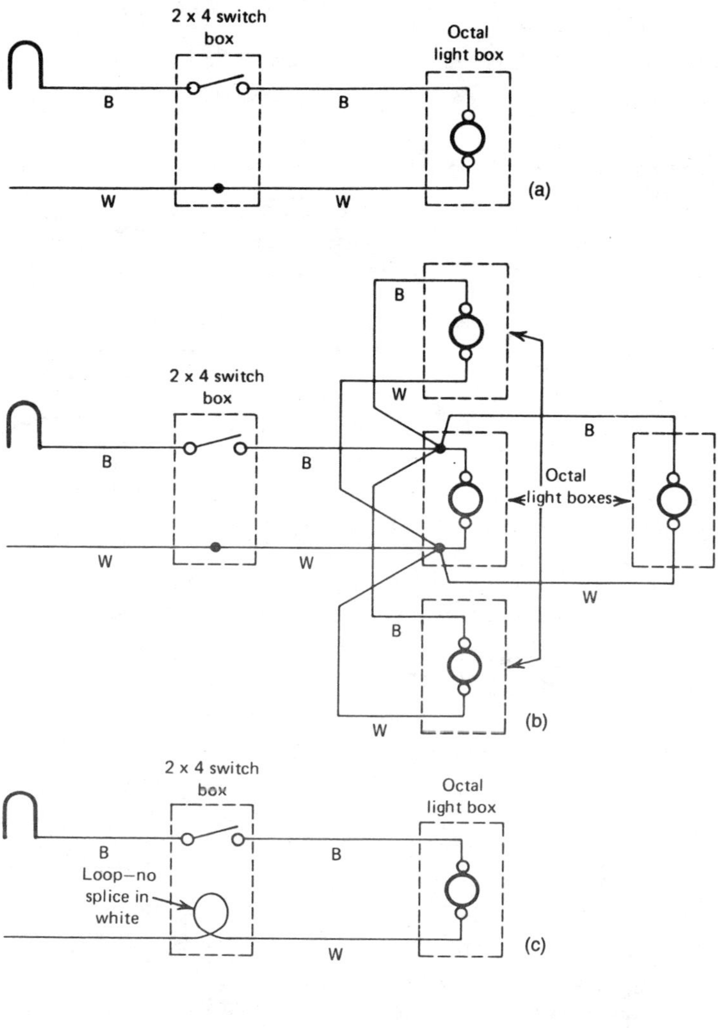

Figure 7-10 Wiring of single pole switch circuits—power in at switch box: (a) in cable; (b) in cable—multiple loads; (c) in conduit.

Thus in cable [Figure 7-11(a)], the incoming black supply line from the power source is spliced to white in the cable connecting to the switch box. There, both black and white connect to the switch with black returning power from the switch to the load. Meanwhile, white common from the power source is stopped at the load since it never goes to switches. When the switch is to control multiple loads, they may be tapped, as required, from the first one.

To wire the same circuit in conduit [Figure 7-11(b)], the color change from black to white in the power input line to the switch is not only unnecessary, it is ***not permitted***. Since the unswitched power line only passes through the light box to get to the switch, it need not be cut. It merely loops and passes through. Since the two wires into and out of the switch are both power wires, both must be black, or any other proper power color.

Wiring of Three-Way Switch Circuits

When there is a need to control a load from two different locations two three-way switches are used. Controlling a light in a stairway from either the top or the bottom is a common case.

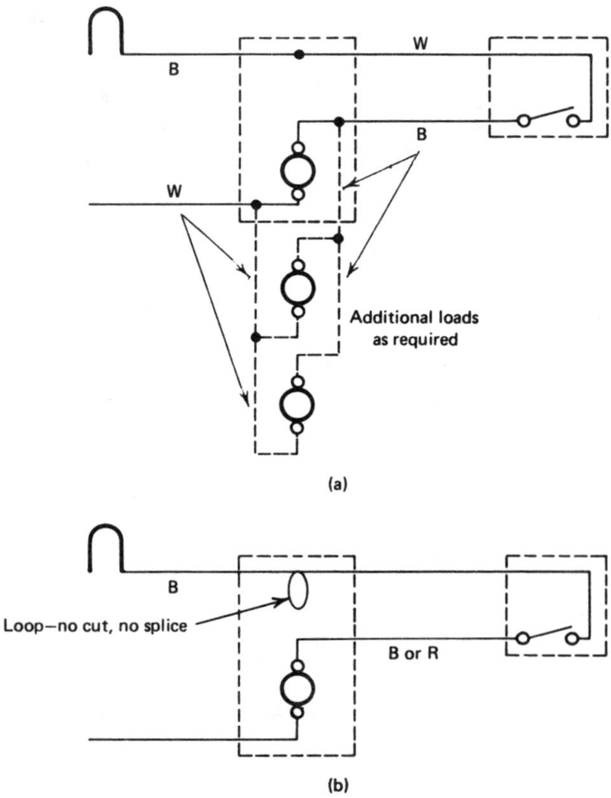

Figure 7-11
Wiring of single pole switch circuits—power in at load box: (a) in cable; (b) in conduit.

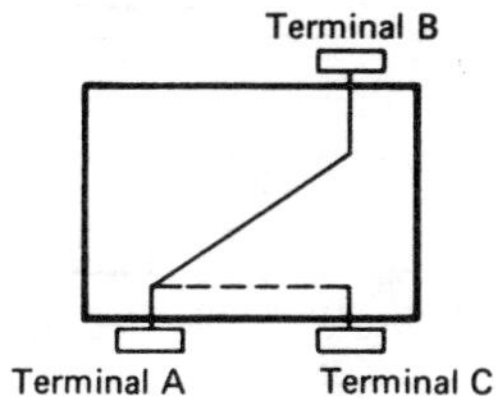

Figure 7-12
3-way switch.

Since the three-way switch is not an ***on-off*** switch, when power is supplied to terminal ***A***, at all times either terminal ***B*** or terminal ***C*** will be live (Figure 7-12). Two three-way switches are wired so that the ***B*** and ***C*** terminals on one switch are connected to the ***B*** and ***C*** terminals on the other by two wires called ***travelers.*** Power enters a three-way switch circuit (Figure 7-13) via the ***A*** terminal of one switch, and passes out through either terminal ***B*** or ***C*** to one of the travelers. At the second switch it either goes on through and out the corresponding ***A*** terminal, or it stops (Figure 7-13 b and c) because the second switch is turned to the dead rather than the live traveler.

Two three-way switches and the load they control can be wired in cable in any of five variations depending on the electrical sequence in which the parts are arranged, and the point at which power enters the sequence. Figure 7-14 shows a floor plan in which the part sequence places the two switches together, with the load at one end. Regardless of where unswitched power enters this sequence, it is wired using a three-wire cable between the switches—the other cables to and from the switches are two wire type.

Each variation illustrated differs from the others in some respects, while in others all are the same. It will be useful to note the similarities common to the three variations shown:

Figure 7-13
Power side of 3-way switch circuits.

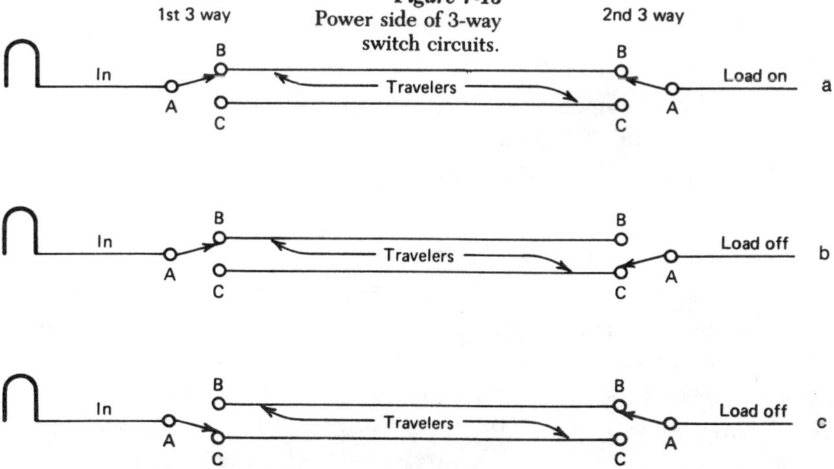

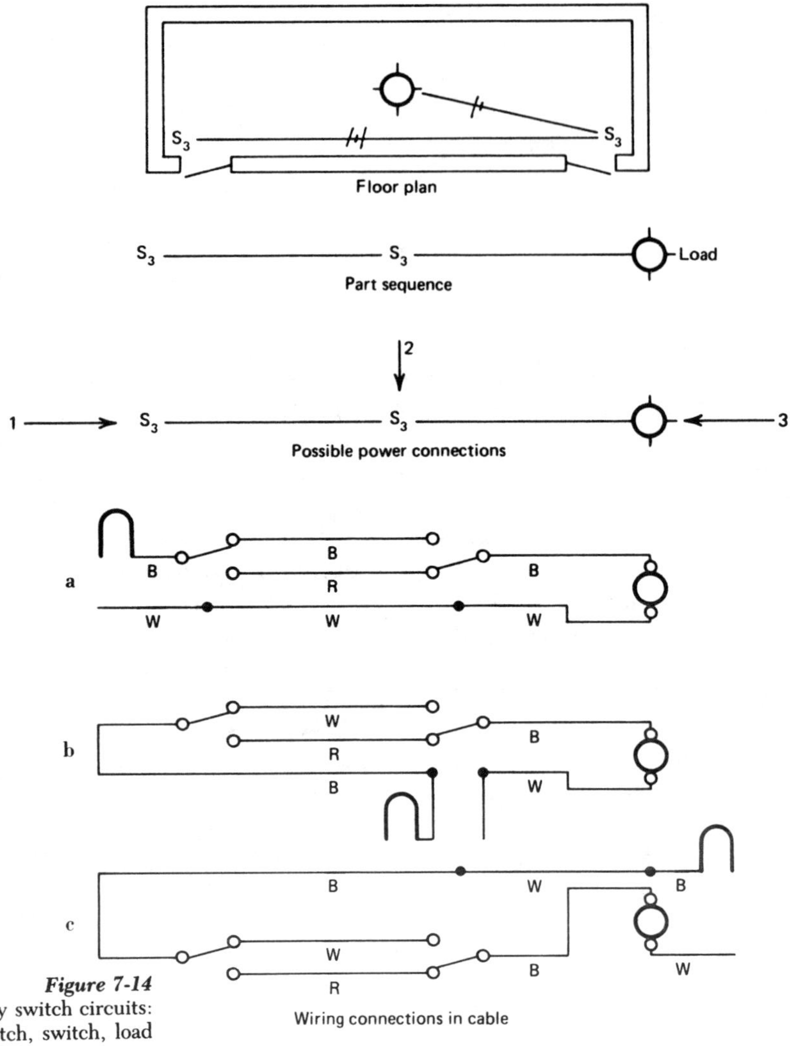

Figure 7-14
3-way switch circuits: switch, switch, load sequence.

1. White common from the power source goes directly to the load and connects to nothing else.
2. Power leads into the first switch and out of the second are both black.
3. One traveler is always red.

In variation 2, one traveler is white; and in 3 not only is one traveler white, but the incoming power wire changes from black to white and back to black again. Since all wires going to and from switches in this circuit are power wires, these applications are not code violations by Article 200-7(c), referred to earlier. Be careful to remember always that these uses of white wires on the power side of a circuit are permitted ***only in cable*** and they must be re-identified wherever visible per Article 200-7(c).

To wire the same switching circuits in conduit, both travelers will have to be power colors: two blacks, two reds, or better yet two blues to help distinguish them from whatever else happens to be sharing the same pipe. Additionally, in conduit remember not to cut and splice wires that only pass through a box without connecting to anything. Simply leave a loop and pass them on to the next box.

Figure 7-15 illustrates the other possible sequence of parts that might occur using two three-way switches and one load. When wiring this

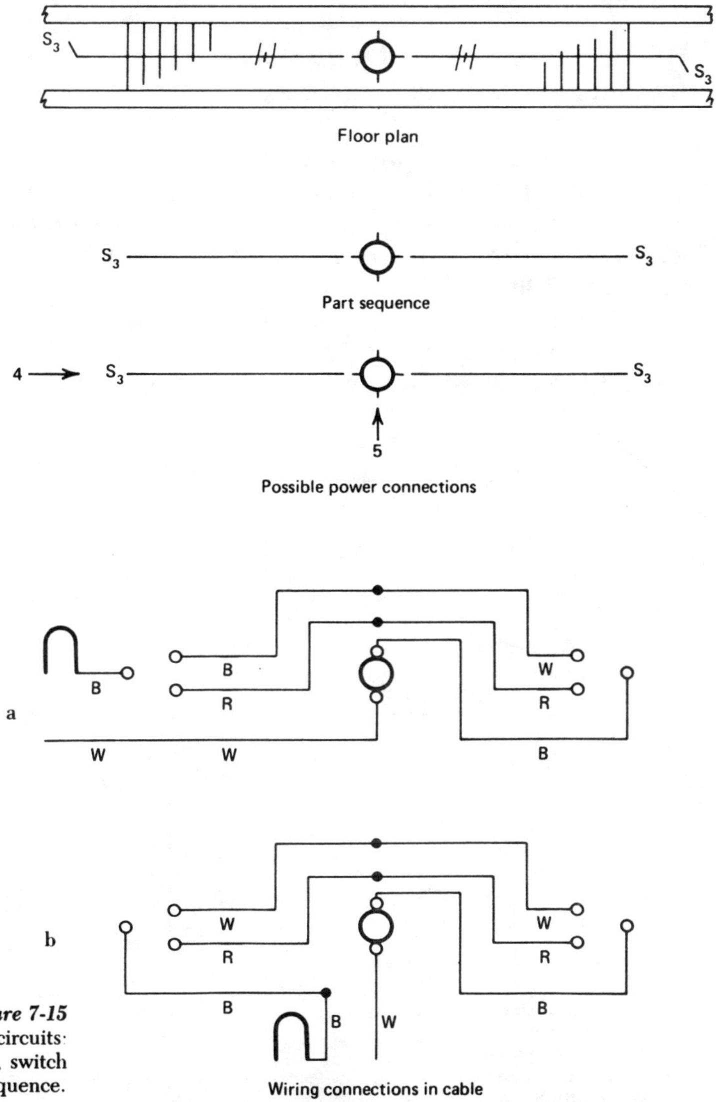

Figure 7-15
3-way switch circuits: switch, load, switch sequence.

sequence in cable, white wires are travelers again. As we found this was permissible with the previous sequence, it is permissible here as well.

Again in conduit both travelers will have to be proper power colors and will run directly from switch to switch passing uncut through the light box.

In the event that two or more lights are to be controlled by a pair of three-way switches (Figure 7-16) any desired number of additional lights may be added by tapping the load of any of the five diagrams in Figures 7-14 and 15. Sequences such as Figure 7-16 a and b cannot be done in three wire cable, but could be done in conduit. Figure 7-16 c and d are no problem in three wire cable.

Wiring Combination Three-Way and Four-Way Switch Circuits

When it is necessary to switch a circuit from more than two locations two three-way switches are used and as many four-way switches (Figure 7-3b) necessary to take care of the additional locations (Figure 7-17). The four-way switch, like the three-way, is an ***either-or*** switch, but it can only be used in a circuit that already contains two three-ways.

As demonstrated in Figure 7-13, three-way switch circuits control loads from either of two locations by allowing either of the switches to connect two alternate electrical paths, one of which will always connect

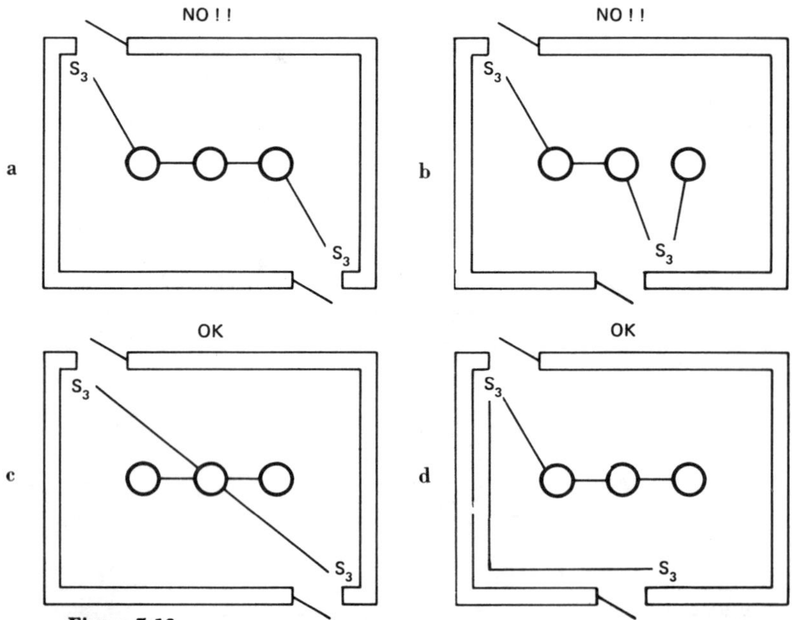

Figure 7-16
3-way switch sequences with multiple loads.

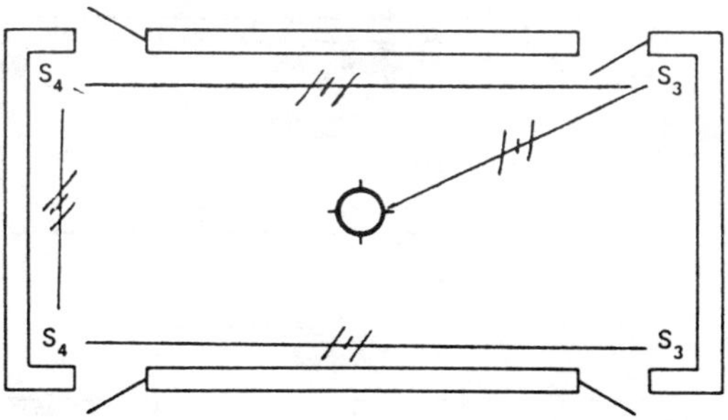

Figure 7-17 Floor plan utilizing 4-way switches. By combining 3-way and 4-way switches, a load can be controlled from any number of locations.

power to the load while the other will not. By inserting four-way switches along the two traveler lines connecting a pair of three-way switches (Figure 7-18), the capability of switching back and forth between the two alternate electrical paths can be extended to as many additional switching locations desired.

When a circuit consists of two three-ways, one four-way, and a load, there are eight alternatives for wiring it in cable, depending on the part sequence, and the point along each sequence at which unswitched power enters the circuit. One possible sequence is to go through all three switches and end at the load. Figure 7-19 illustrates the four alternative ways of wiring this sequence in cable. Figure 7-20 shows the other four alternatives that can be used in cable to wire the circuit when the se-

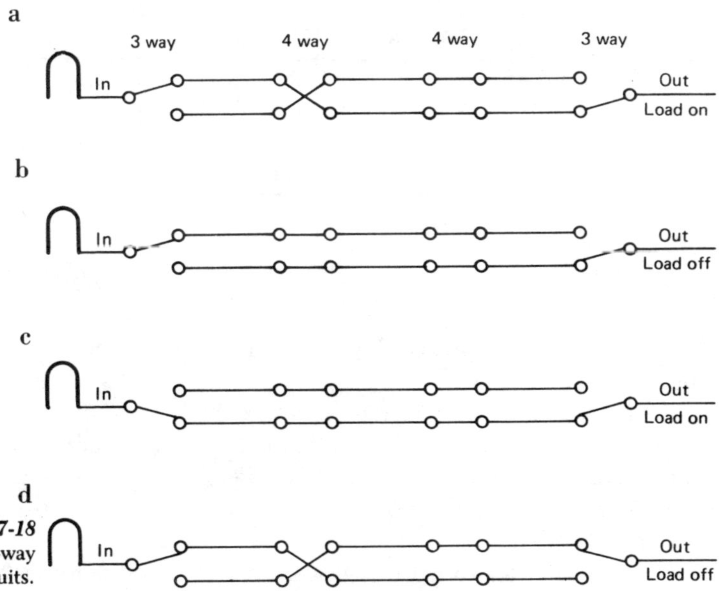

Figure 7-18 Power side of 4-way switch circuits.

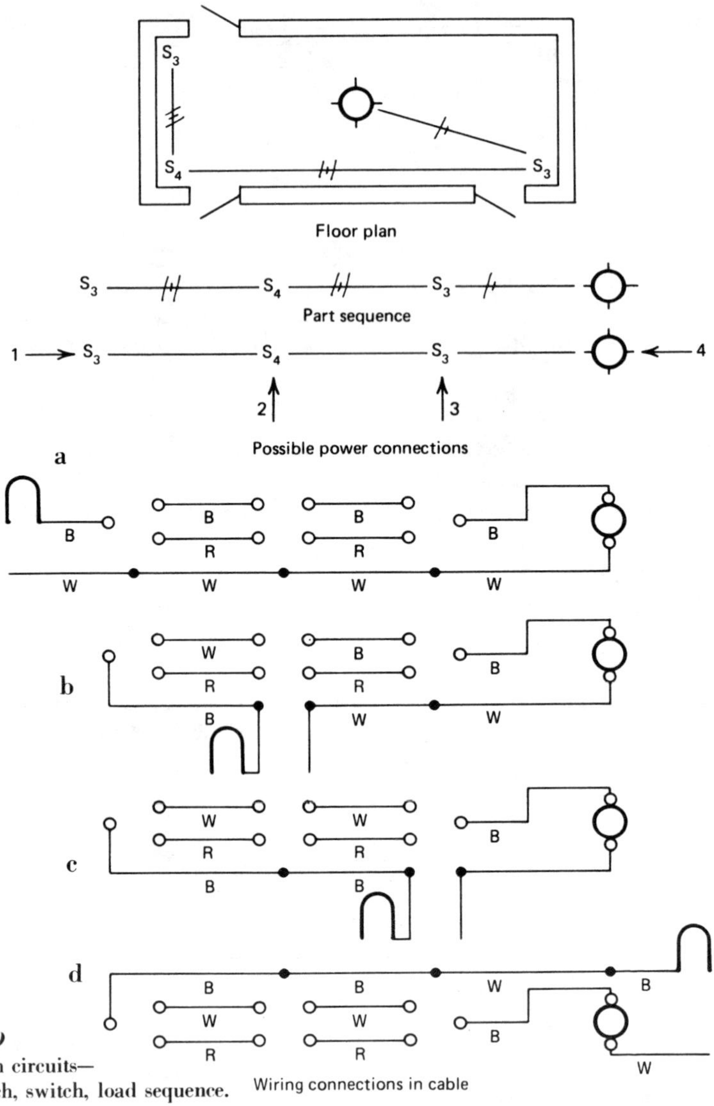

Figure 7-19
4-way switch circuits—switch, switch, switch, load sequence.

quence is changed to place the load part way along the line instead of at one end.

Once again these diagrams of cable show many instances of white wires on the power side of the circuit. All of these are permissible, as well, under the code exception cited above. In conduit, as always, none of these white wires on the power side is allowed—only power colors may be used.

Cable circuits using three-way switches and those using 4 ways have the following similarities:

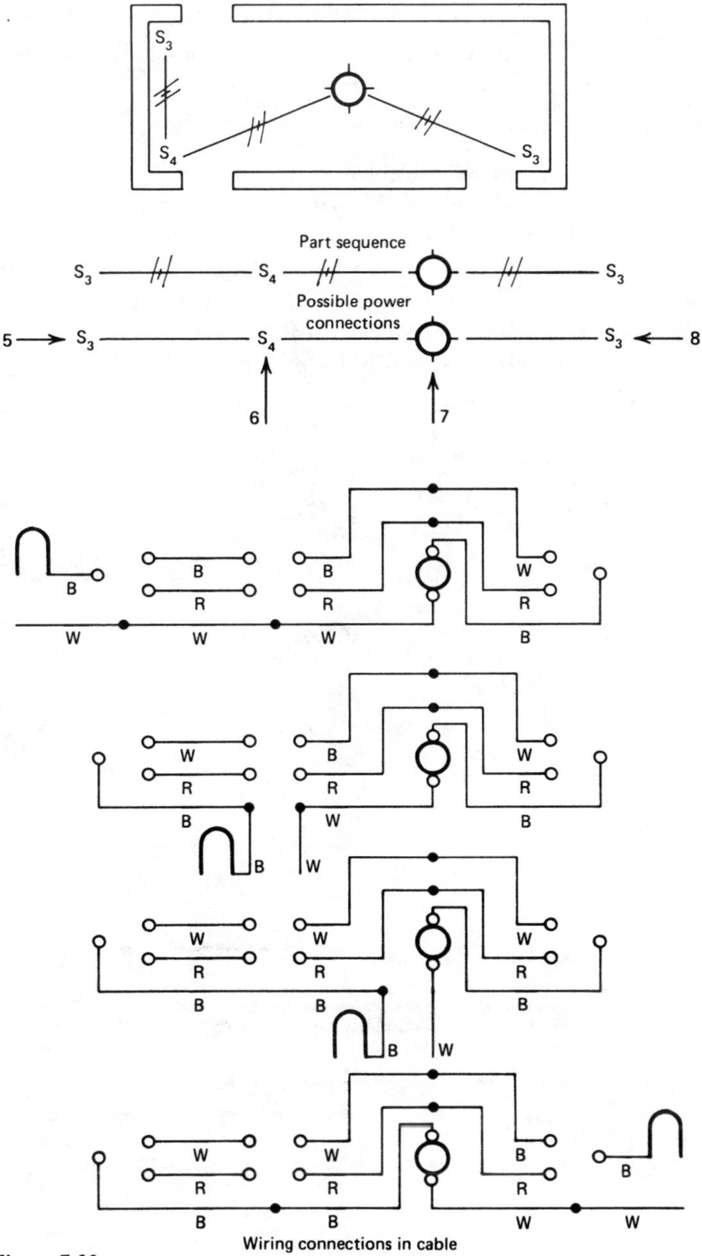

Figure 7-20
4-way switch circuits—
switch, switch, load,
switch sequence.

1. White common goes only to the load.
2. Power into the first three-way and out of the last one are both black.
3. One traveler is always red straight through.

Multiple loads on cable wired circuits can be tapped from the first load using any of the eight alternatives given, as long as it is done in a manner similar to what was done in Figure 7-16 c and d (the electrical sequence allows all loads to be tapped from a single output point). As before, sequences such as figure 7-16 a and b, which require multiple output locations, will not work in cable—in conduit they may, for the necessary extra wires can easily be pulled.

Figure 7-21
Two single pole switches on a single yoke.

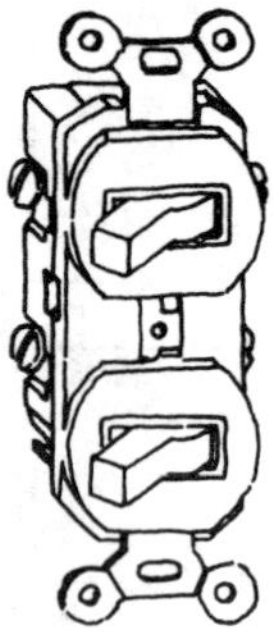

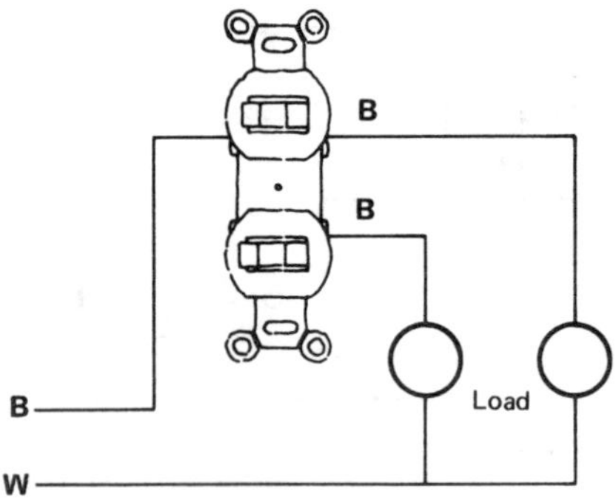

COMMON FEED
Two S.P. switches on same circuit. Each switch controls an independent light.

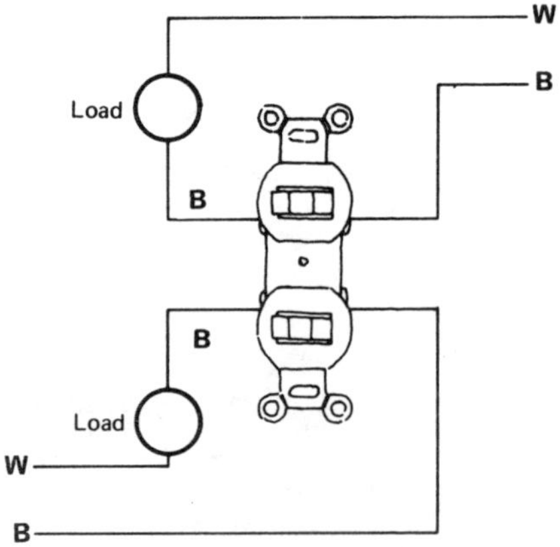

SEPARATED FEED
(Breakoff Tab Removed)
Two S.P. switches on separate circuit. Each switch controls an independent light.

Wiring Dual Single Pole Combinations

Space limitations might require the placement of two single pole switches in an area normally occupied by only one. To meet this need a double single pole switch combination unit (Figure 7-21) is available. While two screw terminals are found on each side of the case, the two sides are not identical. The difference is that while on one side the two terminals are independent, on the other they are connected by a removable link. Because of that link, unswitched power can be supplied to both switches by attaching either of the linked terminals. The two independent terminals on the other side may now be used to feed and control two separate loads.

In the event that the two loads require separate power sources originating at different breakers, the linked terminals can easily be disconnected by breaking the connecting link with screw driver tip or needlenose pliers, depending on shape of the link.

Switch/Outlet Combination

Space limitations can also require a combination switch and convenience outlet (Figure 7-22) that will fit in the space normally occupied by either one or the other. This device also has two screw terminals on each side plus a green hexagonal head ground screw at the bottom. Note carefully the color code on the screw heads; three are brass heads, and one is silver.

White common goes to the silver terminal. The power lead to the load being switched goes on the brass terminal next to the silver common. That leaves two brass terminals on the other side connected to each other by a removable link. When unswitched power is connected to either of the two, the switch will operate the load, and the outlet will be powered at all times regardless of the switch position. If the switch is to control the outlet as well as a remote load, simply reverse the power connections. Put the unswitched power on the brass terminal next to the common, and place the load on the opposite side. If switch and outlet are parts of unrelated circuits, and therefore are to be powered separately, break out the link connecting the two brass terminals and power them individually.

When cable is used, the ground wire goes on the hexagonal green terminal at the bottom. In conduit this connection is unnecessary as there is no ground wire.

Pilot Light Switches

Different combinations of pilot lights with single pole switches are shown in Figure 7-5. The first contains an internal light inside the switch toggle that glows in the *on* position. There are three terminal screws on the case: two brass heads and one silver head. White common goes, as always, to

Figure 7-22
Switch and outlet combined on a single yoke.

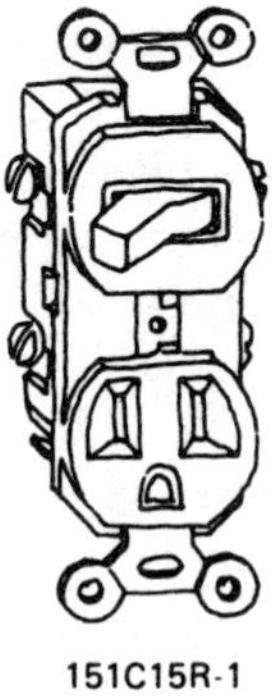

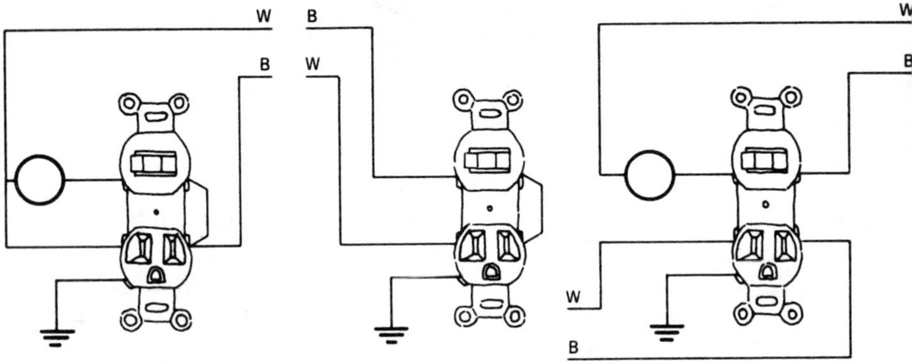

COMMON FEED
S.P. switch controls light only. Power outlet for grounded appliances.

Grounded power outlet controlled by S.P. switch.

SEPARATED FEED
(Breakoff Tab Removed)
S.P. switch and grounded power outlet on separate circuits.

the silver head. Power in and power out go to the two brass heads. Either one can go on either terminal.

Figure 7-5b shows a type of pilot light separate from the switch but internally connected to it. This unit has four terminal screws, a brass head and a silver head on each side. To wire this one, connect power in and power out to the two brass heads. It makes no difference which is which. Now put a white common on either one of the two silver heads. It is unclear why there are two silver terminals, when only one is necessary.

In Figure 7-5c the screw terminals are the same as those found on the single pole switch/convenience outlet combination: one brass and one silver terminal on one side, and two linked brass terminals on the other. As usual, white common goes to the silver terminal.

Normally power is connected to the brass terminal next to the silver and taken out on either of the linked brass terminals on the other side. Wired in this fashion, the pilot lights when the switch is in the *on* position. If the pilot does not shut off regardless of switch position, the power leads have inadvertently been reversed. With this switch it does make a difference which is which.

By breaking out the connection between the linked brass terminals, the pilot light becomes completely independent of the switch and can now be used as the indicator of some completely different activity.

Dimmer Switches

Figure 7-4 illustrates the dimmer switches most commonly encountered. ***A*** with two black leads is a single pole switch and dimmer combination. ***B*** with one black and two red leads is a three-way switch and dimmer. The single pole is connected by splicing unswitched power to either one of the black pigtails and the load to the other. As usual it makes no difference which goes where.

With the three-way switch, the travelers go on the two red leads, and power either ***in*** or ***out*** goes on the black. Remember, only one end of a multiple switch circuit can have a dimmer so care must be taken to locate it in the most convenient place.

Either of these dimmers will fit in the space occupied by the standard switch it replaces. However, the dimmers are definitely bulkier and thus fill more of the interior space in the switch box. If the box is already fairly full of wires, splices, and other bits and pieces, stuffing a dimmer in too can become quite difficult.

As mentioned earlier the dimmers shown are good for loads up to 600 W only. However, most of the time for the loads encountered in residential and small commercial buildings this capacity is adequate. Should greater capacity be required, dimmers fitting standard electrical boxes are available with capacities as high as 2000 W.

Outlets

120 V Duplex

While this simple device is familiar to all, it is necessary to look at several details (Figure 7-23) that might have escaped notice. In the position shown, note that the vertical slots on the left of the device are longer than the ones on the right. The two screw terminals on the left are silver headed while the two on the right are brass. Both pairs are linked. At the bottom is a single green hex headed screw which is internally connected to the two ***U*** shaped holes. These holes are ground connections.

As always, incoming power goes to a brass terminal on the side having the shorter slots. White common goes to a silver terminal on the long slot side. Except when wiring in conduit, put ground on the green

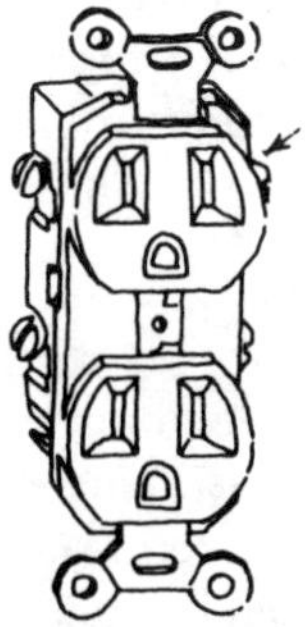

Figure 7-23
Duplex receptacle.

hex headed terminal. In conduit the outlet mounting yoke makes metal to metal contact with the box which in turn is grounded through the conduit.

Quite often half of a duplex outlet is switched while the other half remains always on. To accomplish this the connecting link between the two brass terminals ***only*** is broken out splitting the power side. Now unswitched power is connected to the top brass terminal, switched power to the bottom and common to either one of the silver terminals.

In addition to splitting outlets for the purpose of switching one half while the other remains unswitched, split outlets are often ***split wired***. This means that the power sides of the two halves are connected to two completely separate circuits. The code requires kitchen countertop outlets [NEC 210-52(b)(3)] be wired this way. When this is done, the two circuits feeding such split-wired outlets must be connected to a double-pole circuit breaker or two single breakers with linked handles [NEC 210-4(b)].

When wiring a series of outlets in sequence (Figure 7-24), rather than connecting incoming power and common to one pair of terminals and outgoing power and common on the other pair, thus using all four terminals, it is preferable to splice power to power with a pigtail to one brass terminal, and common to common with a pigtail to one silver terminal. Wired in this manner the continuity of the rest of the circuit is kept independent of the condition of any of its component outlets.

240 V Outlets—Range, Dryer, and Air Conditioning

In residences that have electric ranges, it will be the largest single load in the entire system as it will require a 240V circuit with a breaker capacity of about 50A. While ranges are sometimes wired directly from a junction

Figure 7-24
Duplex receptacle in 2 × 4 box with wiring.

box, more often they are plugged into an outlet. Figure 7-25 shows the outlet normally used for ranges. The two power leads of the 240V circuit connect to the upper slanted slots, while the common goes on the lower vertical slot. In cable, the wire size to a range will normally be #8 for both the power wires and the common. If for some reason it were to be wired in conduit, the common could be reduced to a #10 [NEC 210-19(c), Exception No. 2].

The normal range outlet will fit a standard 2 ×4 outlet box, but it will have to be done with ¾″ knockouts. Assuming the wiring is done in cable, an 8-3 cable with ground will have to be used which will not go in the usual ½″ knockout. Should the job be done in conduit, ¾″ EMT will be needed unless the insulation type is switched to either THHN or THWN. In ½″ conduit using TW, let alone anything thicker, two #8 wires are all that are allowed. The #10 common can not go in as well—see Appendix Tables 3A and 3B.

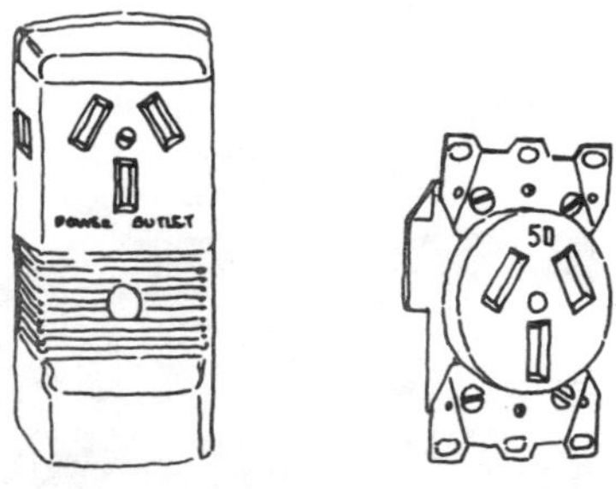

Figure 7-25
Range receptacles.

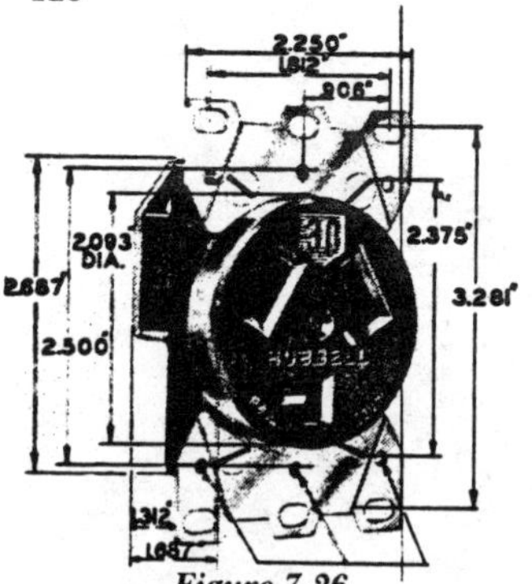

Figure 7-26
Dryer receptacle.
Courtesy of Pass and Seymour, Inc.

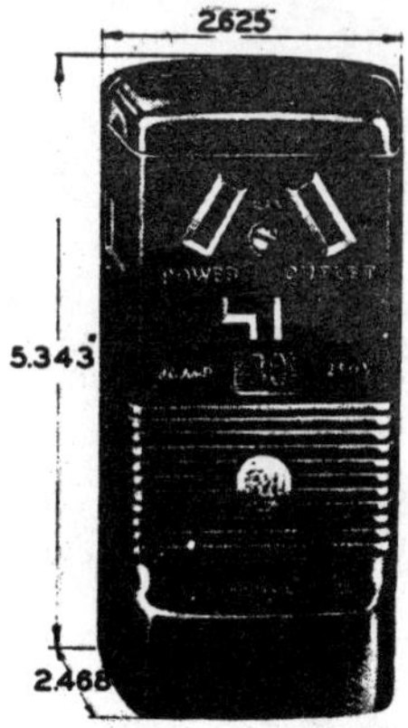

Figure 7-27
Dryer receptacle.

The standard electric clothes dryer outlet (Figure 7-26) looks very similar to the range. In fact the two slanted upper slots are exactly the same. The difference is in the lower slot. Instead of a simple vertical it is now half horizontal and half vertical—rather like an ***L*** upsidedown and backwards.

The dryer will be supplied by a 30A, 240V circuit wired with #10 wire. Again the two power leads go to the upper slanted slots just as for the range. The common goes to the upsidedown ***L***. A dryer outlet also fits comfortably in a standard 2 ×4 box, and the ¾″ knockouts will not be necessary. (Figure 27.)

Note Range and dryer circuits go to those appliances ***only***. Nothing else may be on either of those circuits.

CHAPTER 8

PLANS

When a home or other building is planned, the work of the electrician, like that of most other trades, begins long before ground is broken, and actual construction started. Estimates of required labor and materials must be made first in order to bid on the work to be done. After the job has been secured, labor and materials adequate to accomplish the necessary work must be secured. This requires detailed study of the building plans. Let us then examine an average set of plans for a three-bedroom residence to familiarize us with the kinds of information to be found there. A set of drawings will be found at the back of the book.

Plot Plan

The drawings will start with a plot plan. The plot plan will show the shape and dimensions of the lot, and the size, shape, and location of the building on that lot. In order to insure conformity with local zoning regulations, the distances between property lines and building walls must be given. Somewhere on the plan an arrow indicating geographic north will appear.

On the example given, the heavy building outlines are the outlines of the roof. The dashed lines inside of them are the outlines of the walls. Therefore, the measurements of 30′-0″ on the left and 12′-0″ on the right between the property lines and the dashed lines in the building outline are the required zoning department measurements from property lines to building walls.

Foundation Plan

The foundation plan and associated details show where foundation footings, walls, piers, posts, and slabs are to be located. Also given are footing depths and dimensions, thicknesses of walls and slabs, materials specifications, reenforcing bars required in walls and footings, reenforcing mesh required in slabs, and size and spacing of anchor bolts.

When a building is built over a slab on grade the electrician may wish, or need, to embed some conduits in the slab in order to distribute electrical power as efficiently as possible. Or in order to do his work he may have to go through foundation walls. In either case he will have to know how the foundation is to be built in order to lay out any required electrical work at the foundation level.

Floor Plans

Separate floor plans are made for each floor from the basement (if there is one) up. Floor plans show in scale the locations and sizes of walls, doors, and windows as well as miscellaneous additional structural information. All important features are specifically detailed in the floor plans in addition to being drawn to scale. Residential plan sets seldom include separate electrical plans. Commercial or industrial sets normally do include separate electrical sheets.

On residential plans, the electrical requirements are found on the floor plans designated by means of conventional symbols (Figure 8-1). There are special symbols for various types of lights, outlets, switches, and other electrical devices. These symbols are located on a plan in their proper positions according to the scale of the plan, but are seldom drawn to scale for reasons that will be explained later.

Typical Sections

The typical sections reveal the true structure of the building by stating the sizing and spacing of the frame parts. Sections sometimes, but not always, give interior and exterior finishes as well. The sections in this book do not include finishes.

Studying the typical sections in conjunction with the floor plans, the electrician can decide how he wishes to route his wiring through the building frame.

Exterior Elevations

A plan set will normally include four exterior views. From these the electrician can determine how he will handle the installation of the required exterior lighting, outlets, and other exterior electrical work that may be required.

Other Sheets

Additional sheets may show interior elevations, cabinetry details, or other finish details. Some of this information may also assist the electrician to decide how to handle parts of the electrical installation.

Types of Plans and their Uses

In the course of his work the electrician will have to deal with three different types of plan drawings. These are the floor plan, the cable plan, and the wiring diagram.

Floor Plan

This rendering shows the locations of walls, doors, windows, and other structural features along with dimensions. Various parts of the electrical system are also located and identified on the floor plan (Figure 8-5) by the use of conventional symbols. However, with regard to the electrical system all that appears on a floor plan are symbols to indicate where electrical supply is to be made available, where fixtures, receptacles, and other electrical devices are to be placed, where switches will go, and what each switch will control. What a floor plan does ***not*** show is how any of this is to be accomplished.

Cable Plan

Using the same symbol system as the floor plan, the cable plan shows what parts of the electrical system go where, but it goes one step farther and indicates how the various points are to be connected together into branch circuits as well as how many wires will be required for each connecting run (Figure 8-6) from box to box.

The number of wires entering each box as well as where they originate appears on the cable plan. The connections inside each box appear on the wiring diagram (Figure 8-7). Thus it is cable plans and wiring diagrams together that show how the electrical parts appearing on a floor plan are to be connected and made operational.

Floor Plan Symbols

The standard symbols illustrated in Figure 8-1 are used to show the location, type and function of the various parts of the electrical system. These parts are divided into two general categories: first using equipment and use points, and second controls.

The first category includes current using devices that are directly connected to power wiring such as light fixtures installed on walls or ceilings as well as receptacles into which current using devices can be plugged. The second category includes switches, dimmers, fuses, and circuit breakers, all of which are non-consuming devices.

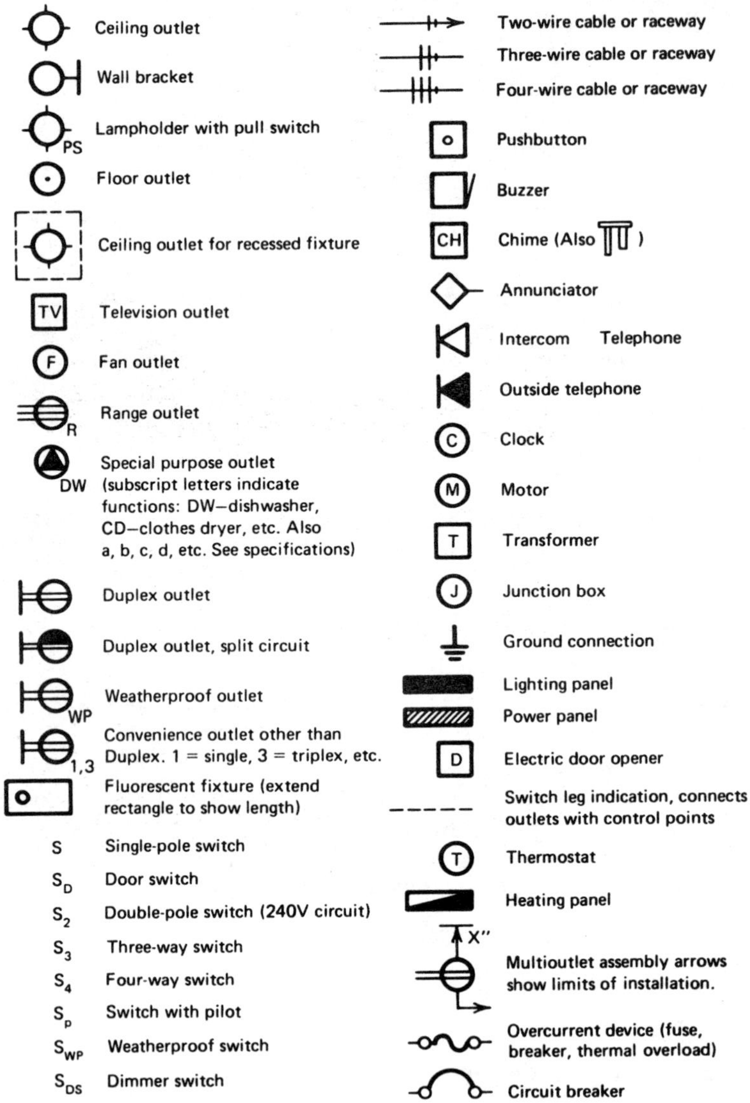

*If there is an arrow on the cable, it indicates a home run to breaker

Figure 8-1 Floor plan electrical symbols.

Light Fixtures

 Indicates a surface mounted ceiling light fixture—incandescent (Figure 8-2)

 Light fixture recessed into the ceiling—incandescent (Figure 8-3)

Figure 8-2
Incandescent surface mounted ceiling light.

Figure 8-3
Recessed ceiling incandescent fixture.

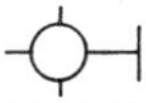 Wall mounted light fixture—incandescent (Figure 8-4)

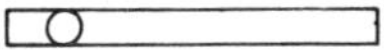 Fluorescent light fixture

Incandescent fixture with pull switch

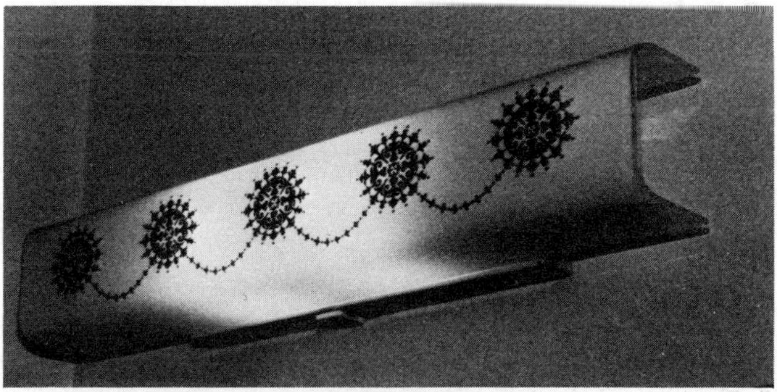

Figure 8-4
Incandescent wall fixture.

Receptacles

 Duplex receptacle (Figure 7-23)

 R Range receptacle (Figure 7-25)

 D Dryer receptacle (Figure 7-26 and 7-27)

 WP Ordinary duplex receptacle, but enclosed in weather-proof box (Figure 6-16)

 A receptacle that is ***not*** a general purpose convenience outlet, but is for some specific purpose as: disposal (Disp.), airconditioner (A/C), or if sub-letters (a) (b), etc. refer to specifications for purpose. Specifications will also give voltage and amperage requirements for listed special purpose outlets.

Switches

S	Single pole switch (Figure 7-2)
S_3	Three-way switch (Figure 7-3)
S_4	Four-way switch (Figure 7-3)
S_{Dim}	Dimmer switch (Figure 7-4)
S_P	Pilot light switch (Figure 7-5)
S_{WP}	Weather proof switch (Figure 7-6)

Other Symbols

The remaining symbols in Figure 8-1 represent parts that are required in comparatively small numbers. For example a home might have 25 or 30 duplex outlets, but only one door chime. While some of the devices represented by these symbols consume power, they constitute minor loads. The others are either safety equipment, such as fuses or circuit breakers, or they are passive parts, such as a panel box or a junction box.

Design of the Electrical System

The first step in the design of a residential electrical system is to locate properly on a floor plan the minimum number of electrical outlets and controls required by code. Then add any electrical equipment beyond code minimums that is desired.

When all the use points have been determined, and the power requirements at those points defined, the points must now be divided into branch circuits. This must also be done in conformity with applicable

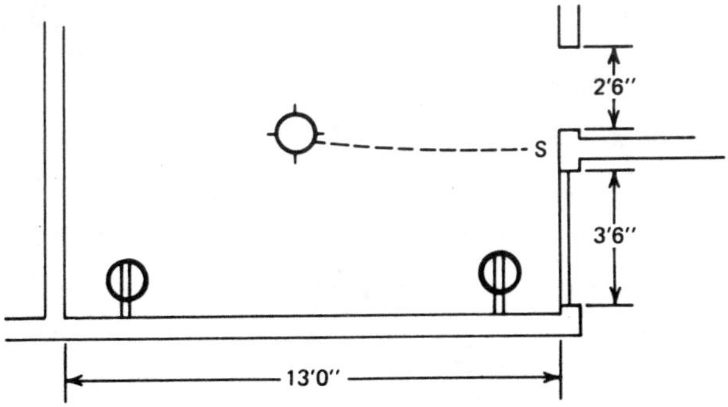

Figure 8-5
Partial floor plan.

code provisions. After the composition of the various branch circuits has been determined, cable plans and wiring diagrams can be prepared to show in detail exactly how the desired results are going to be accomplished.

Since commercial and industrial requirements vary widely, the code requirements relating to them are more general in nature than those governing residential installations. Residential requirements are very detailed and specific so we shall examine them in considerable depth.

Minimum Residential Requirements

1. General purpose, convenience outlets

Every kitchen, dining room, den, recreation room, bedroom–essentially any room other than a hall, stairway, or bathroom–shall have receptacles located so that no point along the floor line is horizontally more than 6 feet from an outlet [NEC 210-52(a)(1)]. The 6 foot count starts at a doorway or other break in the wall. Any separate wall space (as between two doors) that is 2 feet or wider is considered individually. The measurement is permitted to go around corners, thus including parts of more than one wall.

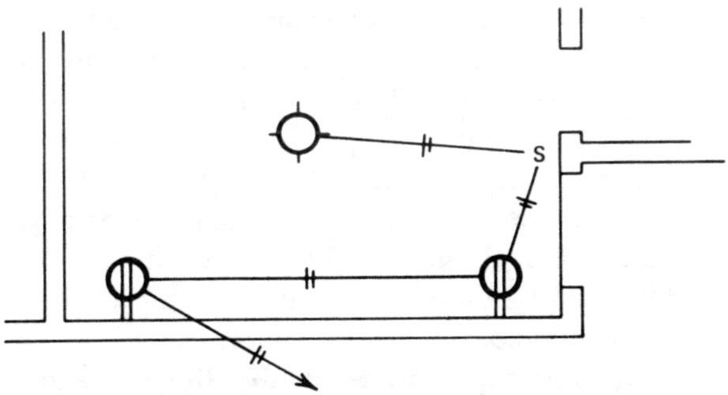

Figure 8-6
Cable plan for Figure 8-5

Looking at Figure 8-8, start measuring at (A) 2 feet into the corner, then up 4 feet to (B). This is a total of 6 feet at which point there must be

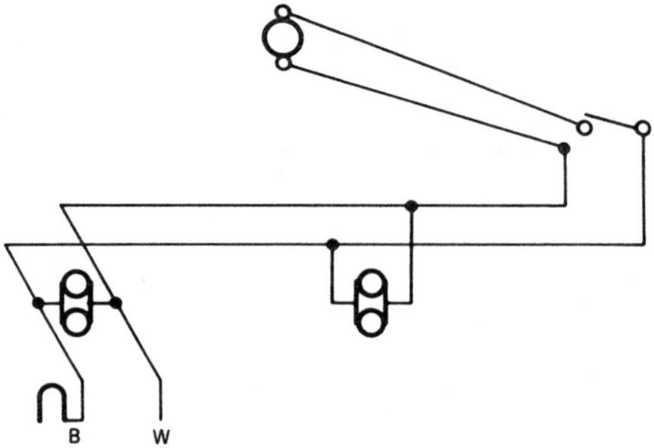

Figure 8-7
Wiring diagram for Figure 8-6.

an outlet. From this point, it is 8 feet into the corner then 4 feet to the right to the next outlet, a total of 12 feet. Along this 12 foot run there is no point more than 6 feet from an outlet; thus it meets code. (C) to (D) is 12 feet, so again no point between is more than 6 feet from an outlet. From (D) to (E) is only 5 feet, no difficulty. The wall section at (F) is only 1½ feet wide. No outlet is needed as this is less than 2 feet. From (G) to (H) around the corner is 7½ feet so somewhere here an outlet is required. (H) back to (A) is another 7 feet so again one outlet is required.

In residential bathrooms at least one wall receptacle is required adjacent to the washbasin [NEC 210-52 (d)]. This outlet, and any others in a bathroom, shall be on a separate 20 A circuit, and must be protected by a ground-fault circuit-interrupter (GFCI) per NEC 210-8(a)(1).

According to NEC 100 (Definitions) (Figure 8-9), a bathroom is defined as a room that has a basin plus either a toilet, a tub, or a shower. A lavatory with basin only is, per code, not a bathroom and therefore exempt from the above requirements. Even if there's a partition between the sink and tub or toilet, a GFCI is still required.

The ground-fault circuit-interrupter (GFCI) required in bathrooms needs some explanation. In Chapter 3 we said that the smallest branch circuit capacity that will be found in any building is 15 amperes. Also mentioned was the fact that 0.1 ampere, or 1/150th of the power in that circuit, can kill a person. If your mind is wandering, even 0.01 ampere will do a good job of getting your attention. This being true, very small defects in electrical equipment can cause serious shocks.

In a bathroom, where a person often stands with bare feet on a wet floor, resistance to ground is very slight indeed; he offers a low resistance path to any stray current leaking from a defective appliance he might be handling. Enough serious shocks have resulted from situations of this kind that the GFCI is now required as a preventive measure.

The operation of a GFCI is very simple and extremely effective. To understand its operation, remember the elementary fact that an alternating current passing through a conductor creates a fluctuating magnetic

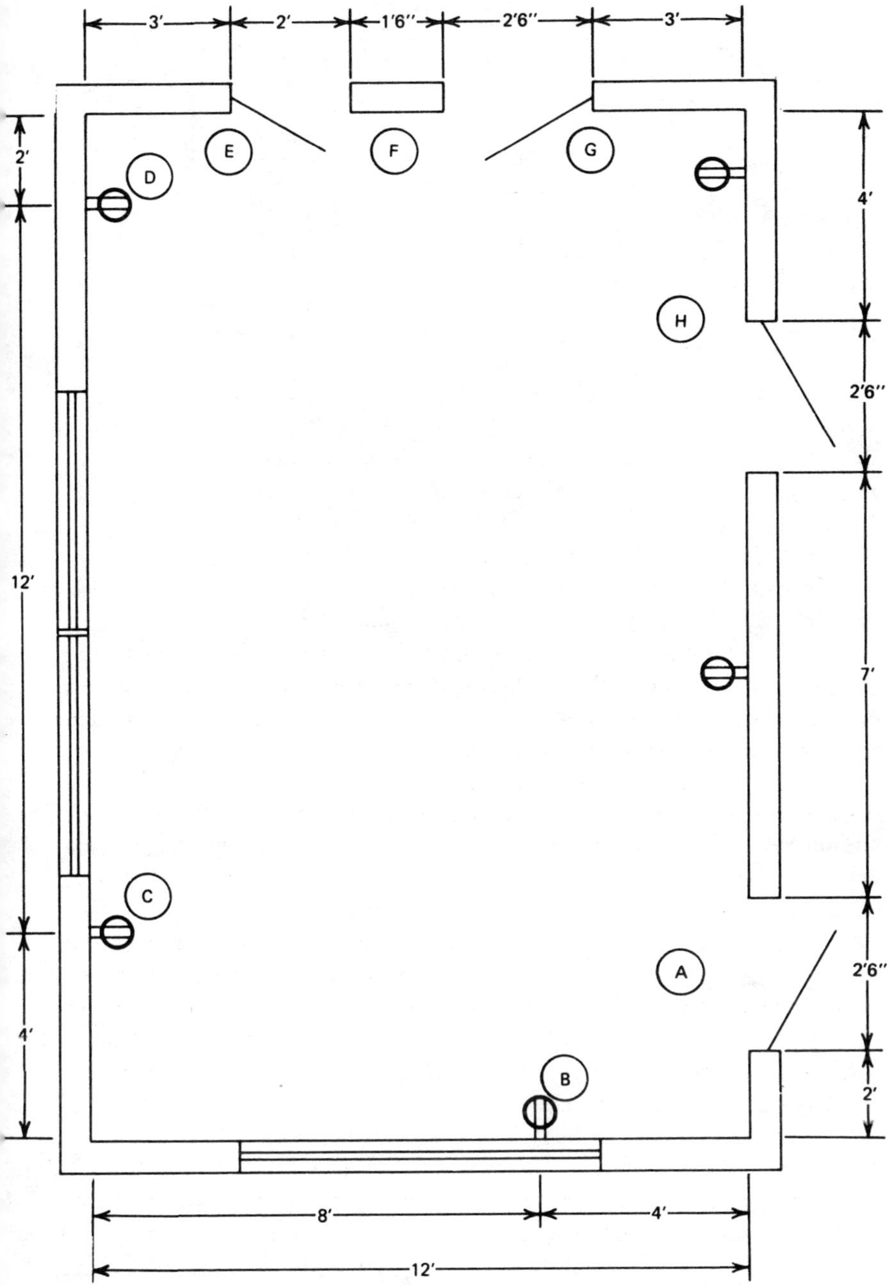

Figure 8-8 Spacing of wall receptacles as required per Code—residential.

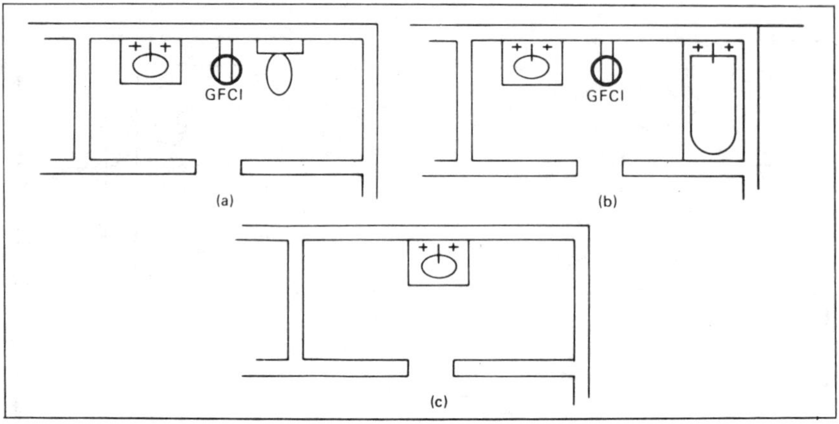

Figure 8-9 What constitutes a bathroom per code: (a) must have outlet and GFCI; (b) must have outlet and GFCI; (c) exempt.

field, which fluctuates with the alternating voltage. That fluctuating field can, in turn, induce a voltage in another conductor.

When a properly operating appliance is connected to a circuit that is protected by a GFCI (Figure 8-10), the currents going in and out are equal and opposite. The magnetic fields they create are thus equal and opposite, and cancel each other out. The coil in tbe GFCI feels no induced voltage and remains inactive.

When a defective appliance containing a small leak to ground is connected to the same circuit, the currents in and out are no longer equal

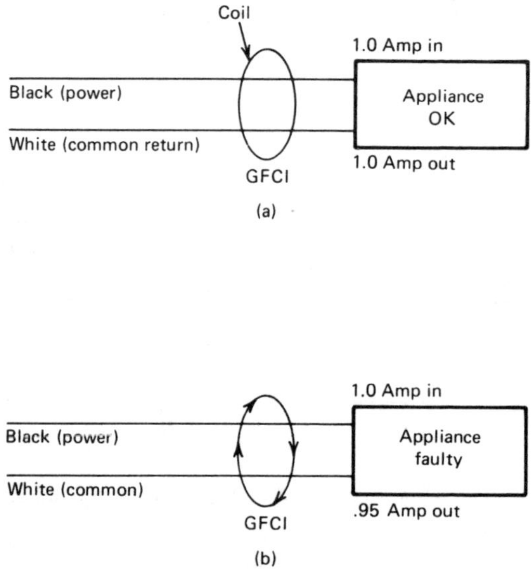

Figure 8-10 How a GFCI operates: (a) no current induced in GFCI coil because equal currents in and out cancel each other out; (b) .05 amp difference between circuit conductors induces small voltage in GFCI which trips it off.

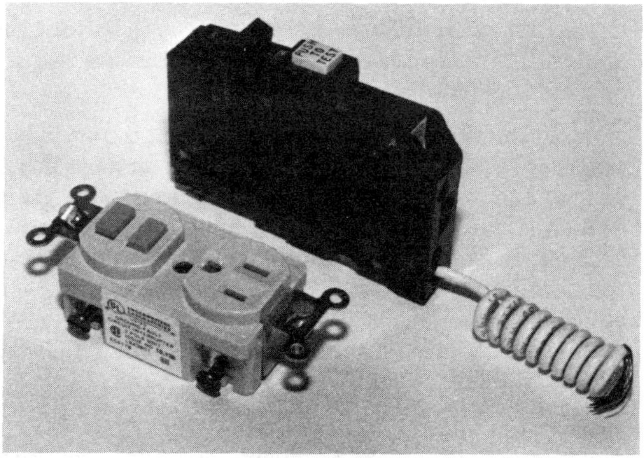

Figure 8-11
Receptacle and breaker types of GFCI.

and opposite; therefore, the magnetic fields they produce no longer exactly cancel each other out. The uncanceled portion of the incoming current leaves a magnetic field to excite the coil in the GFCI which immediately trips it off, and opens the circuit.

The GFCI is available in two basic models (Figure 8-11). One type fits in the same space as a normal duplex 120V receptacle, and includes either one or two receptacles plus a ***Test*** and a ***Reset*** button on its face. Some models are supplied with wire pigtails, others have screw terminals, but all have both incoming and outgoing connections. This (Figure 8-12) makes it possible to use them so as to protect their own outlets plus any number of outlets placed beyond them in a circuit.

The other type of GFCI is combined with a circuit breaker (Figure 8-11). The combined unit fits in the same space as a normal breaker. It looks different from the standard breaker in that it has a yellow or white button on its face marked ***Test,*** and a long pigtail of white wire is attached to it. As we shall see, a normal 120 V, single pole breaker has a single output terminal screw to which the power wire (black, red, blue or what-

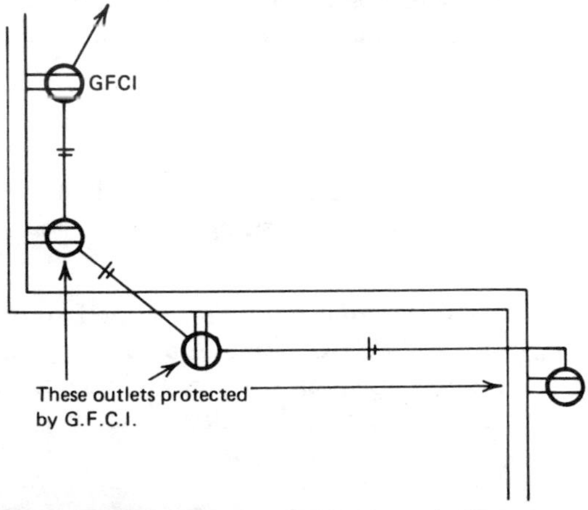

Figure 8-12
GFCI with other receptacles downstream.

ever color other than white, grey or green) from the branch circuit is attached. The GFCI/breaker combination has two screw terminals, one of which is identified by a white dot.

At the breaker box, a branch circuit is normally connected with the power wire going to the output terminal on a breaker, and both the white or grey common and the ground attaching to the ground buss. When the branch circuit is to be protected by a GFCI, the power wire goes to the breaker, but now the common also goes to the breaker and attaches to the terminal identified by the white dot. The white pigtail already attached to the breaker then connects to ground. While a breaker disconnects only the power side of the circuit, a GFCI disconnects **both** sides of the line.

For both one- and two- family dwellings [NEC 210-52(e)], at least one outdoor receptacle, accessible at grade level, is required at the front and back of the dwelling. For each dwelling at grade level, all outdoor receptacles [NEC 210-8(a)(3)] must also be protected by a GFCI. Code defines "grade level access" as being located not more than 6'6" above grade, and usuable without entering the dwelling.

2. Special Outlets

In the basement or attached garage of any one-family dwelling, there must be at least one receptacle outlet in addition to whatever has been provided for laundry use [NEC 210-52 (g)]. These outlets must also be covered by a GFCI [NEC 210-8(a)(2)]. An unattached garage need not be supplied with electricity, but one that is must conform to the same requirements as an attached one.

In addition to all of the above requirements, NEC 210-52(c) also requires that in residential kitchens and dining areas, wherever a countertop exists that is more than 12 inches wide, a receptacle outlet must be provided for small appliances. When a countertop is interrupted by a range, a sink, or a refrigerator, each section shall be considered as a separate space for purposes of this provision. The purpose here is to make it unnecessary to pass small appliance cords across a range or a sink where either heat or water could cause difficulties. These countertop outlets shall be protected by a GFCI.

In a ***dwelling unit***, at least one receptacle outlet must be provided for the laundry [NEC 210-52 (f)]. This means a 120V outlet in addition to any 240V outlet that has been installed for a dryer. It is intended as the power supply for the washing machine, and may be used for an electric iron.

A dwelling unit may be a single-family house, but it could also be a part of a ***multiple dwelling*** or apartment building. In the case of a multiple dwelling in which laundry facilities are provided for the use of the occupants, the special laundry outlet is not required. Also in multiple dwellings where laundry facilities are not permitted the laundry receptacle is not required either.

3. Lighting Requirements

The code requires that in dwelling units, whether single family or multiples, at least one wall switch controlled ***lighting outlet*** be installed in each habitable room plus bathrooms, hallways, stairways, attached garages, detached garages with electric power, and outdoor entrances (NEC 210-70 a). A habitable room, you will recall, is a bedroom, living room, dining room, den, sun room, recreation room, etc.

A switch controlled lighting outlet in this instance means some type of light fixture. It can be fluorescent or incandescent, ceiling or wall mounted, flush or recessed as long as it holds a light. Immediately after requiring all of these light fixtures the code has a general exception. The exception states that in all of the habitable rooms other than the kitchen the light fixture may be omitted, if a switch controlled receptacle is provided.

This means that in kitchens, bathrooms, hallways, stairways, attached garages, and at outdoor entrances a switched ***light fixture*** must be installed, but in the other rooms that fixture is optional as long as there is a switch controlled receptacle into which a lamp can be plugged.

In an attic, crawl space, utility room, or basement, a lighting outlet (some type of light fixture) is required, only if that space is used for storage or contains equipment that may require servicing. Such equipment might be a furnace, water heater, or central air conditioning. When a light is required in one of these spaces, it shall be controlled by a light switch located at the point of entry to that space [NEC 210-70(a)(3)].

The above requirements regarding receptacles and lights are the code minimums. No electrical system in a dwelling unit providing less than these minimums will be approved; however, whatever the owner or the contractor wishes that exceeds these standards is permissible as long as everything that is done conforms to other applicable provisions of the code.

It is appropriate to note also in connection with code minimum standards that while architects and building designers keep up quite well with the current requirements of the Uniform Building Code (UBC), which deals with structural matters, they may at times get somewhat behind on the provisions of electrical, plumbing, and mechanical codes. This means that the electrician must study the plans very carefully to make certain that all code required lights and receptacles are included.

The electrician, not the architect, is ultimately responsible for the correctness of the electrical installation. The electrical contract always includes a clause that states that the electrical system shall be installed in accordance with drawings and specifications, ***and shall be in accordance***

with National Electrical Code and local codes and ordinances. If the architect omits lights, outlets, or switches that are required by code, the electrician must put them in or his job will never pass inspection.

With this in mind let us study Appendix B plans #4 and #5 which are floor plans showing the proposed electrical installation. Sheet #4 is the first floor which includes family room, bath #1, laundry, kitchen, dining room, living room, foyer and garage, with connecting hallways and storage closets.

a. Family Room

Code requires convenience outlets spaced so that no point along the wall at the floor level is more than 6′ from an outlet, also at least one switched lighting outlet which may be a switched receptacle. The lighting requirement is certainly met and exceeded by three recessed and two surface mounted light fixtures all switch controlled. However, the convenience outlet requirement is not quite met. The wall between the family room and bath #1 is given as 6′ to the corner then from the corner along the outside wall it is about another 4′ to the first outlet. A convenience outlet will have to be added on the family room/bath #1 wall.

b. Bath #1

Required: a switched light, and an outlet adjacent to the sink protected by a GFCI. The switched light is there, so is the outlet ajacent to the sink. The GFCI, however, is not specified next to the outlet. These letters must be added to the drawing.

Since this bathroom does not have a window the Uniform Building Code requires a mechanical exhaust vent [UBC 1405 (a)] capable of providing five complete air changes per hour. While an electric fan in such vents is not specifically required, it is what is generally used. In this case the same switch that turns on the light also turns on the exhaust fan. Such combined switching is normal in a windowless bathroom.

c. Laundry

A light—may or may not be switch controlled—and the following are required: at least one receptacle outlet must be installed here. The 220 V outlet for the dryer does not count as the required receptacle. The outlet by the washer will meet the requirement, but not as it is shown. The voltage is wrong. It must be 120 V not 240 V. No problem on the required light.

d. Kitchen

Required: switch controlled light, at least two small appliance circuits, a small appliance outlet on every piece of counter top over 12″ wide, no wall space more than 6′ from an outlet along the floor level. Lighting requirement met. Small appliance circuit and outlet requirements met by properly wiring the outlets shown, except that the two either side of the sink are within 6' of the sink and so must be on GFCI.

e. Dining Room

Required: switch controlled lighting outlet which may be a switched receptacle, and no wall space more than 6′ from an outlet. Both requirements met in this room as shown.

f. Living Room

Required: switch controlled lighting outlet—may be switched receptacle and no wall space more than 6′ from an outlet. Both requirements are met in this room as shown.

g. Foyer

Required: switch controlled light—same requirement as for hallways and stairways. Foyer and hallways correct as shown.

h. Garage

Required: attached garage must have switch controlled light, and at least one receptacle GFCI protected in addition to any laundry receptacles. Plenty of switch controlled lights here plus 2 non-laundry receptacles, but receptacles lack required GFCI protection.

4. Outdoor Requirements

The garage completes our survey of the interior of the first floor, but before leaving this plan sheet, there are a couple outdoor requirements that should appear on it. This is a one family dwelling. Per NEC 210-52 (e) and 210-8 (a) one and two family dwellings must have at least one outdoor receptacle outlet which must be weatherproof, and must be protected by a GFCI. There is one such outlet in the patio, but it lacks GFCI protection which will have to be added.

All outdoor entrances to the house [NEC 210-70 (a)] must have switch controlled lights, and in this case they all do.

Proceeding now to Sheet #5 which is the floor plan for the second floor we find three bedrooms: master, #1 and #2 as well as two bathrooms, hall, stairway, utility closet, furnace room, and outside porch.

a. Master Bedroom, Bedroom #1, and Bedroom #2

Required for each: switch controlled lighting outlet which may be a switched receptacle, and receptacles spaced so that there is no point along the wall more than 6′ from an outlet. In all three rooms both requirements are met. The code does not require lights in closets, and in this case there is one. When closet lights are installed this must be done in accordance with NEC 410-8 which provides that such lights may be installed on the wall above the closet door, or on the ceiling.

An incandescent fixture with *completely enclosed lamp* mounted on the wall or ceiling shall have at least 12" clearance from the **storage area** (Fig. 8-13). Surface mounted fluorescent or recessed incandescent fixtures with completely enclosed lamps shall have at least 6" clearance from the **storage area.**

b. Bathrooms

Required: since both baths have windows the only requirements are the switch controlled lights and the GFCI protected outlets adjacent to the sinks. Both rooms are satisfactory regarding lights, and both have outlets adjacent to the sinks, but both lack the GFCIs.

c. Hall and Stairway

Required: both require switch controlled lights, and both have them.

d. Utility Closet and Furnace Room

Required: both require lights that do not have to be switch controlled. Both of these rooms have switch controlled lights, and this exceeds code requirements.

e. Porch

Required: for second floor porch above grade there are no requirements. Since an outdoor outlet has been provided, it must be weatherproof, and per NEC 210-8(a)3 a GFCI must be added here.

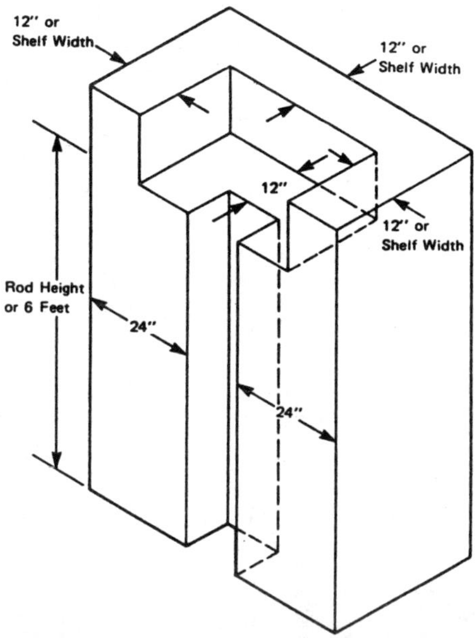

Figure 8-13
Closet light requirements as per Code.

Branch Circuits

Having corrected the original floor plans so as to conform to code requirements, the revised plans are as shown in Figures 8-14, 8-15, 8-16, and 8-17. These plans now not only meet all code requirements, in many respects they exceed them. Nowhere, for example, does the code require that a switch controlled light be subject to control from more than one location. Although such control is not a code requirement it is often a great convenience, and at times a real necessity. It would be ridiculous to turn on a light at the bottom of a stairway in order to be able to find your way upstairs only to have to go back downstairs to turn the light off so you can go back up in the dark. A multiple switching system here is a necessity.

The same sort of condition exists in a long hallway, or a large room with doors at both ends. On the first floor plans the kitchen, dining room, foyer, and garage are all places where multiple switching circuits make good sense.

In many instances code minimums have been exceeded as well regarding the number of light fixtures. The kitchen, family room, and garage are examples.

With the locations and numbers of fixtures, receptacles and switches determined, it is now time to group these elements into branch circuits, and plan how they will be connected to each other and then to the incoming electrical supply. Branch circuit loads are to be computed per NEC 220-3(a) thru (c). NEC 210-20(a) states, "Where a branch circuit supplies continuous loads, or any combination of continuous and noncontinuous loads, the rating of the overcurrent device shall be not less than the noncontinuous load plus 125 percent of the continuous load."

According to Watt's Law: Power (Volt/Amperage) = Voltage X Amperage. The most commonly used branch circuits are 15, 20, 30, 40, and 50 ampere circuits. The following table lists the capacities and permissible loads of those circuits:

Table 8-1

Amperage	*Voltage*	*Wire size*	*Capacity*	*(Volt-amperes) = 125% of permissible load or 80% of capacity*
15	120	#14	1800	1440
20	120	#12	2400	1920
30	120	#10	3600	2880
40	120	# 8	4800	3840
50	120	# 6	6000	4800

Note - Compute residential 240 V circuits as two 120 V circuits added together.

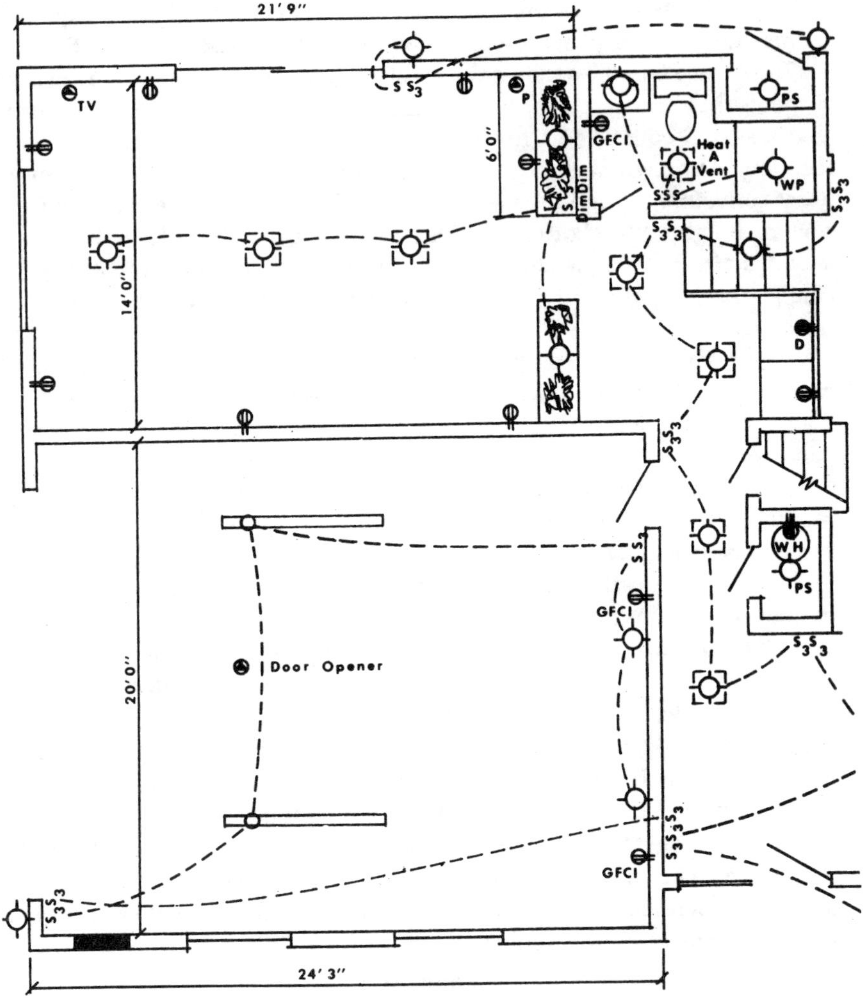

Figure 8-14 Electrical floor plan: family room, bath, garage, utility.

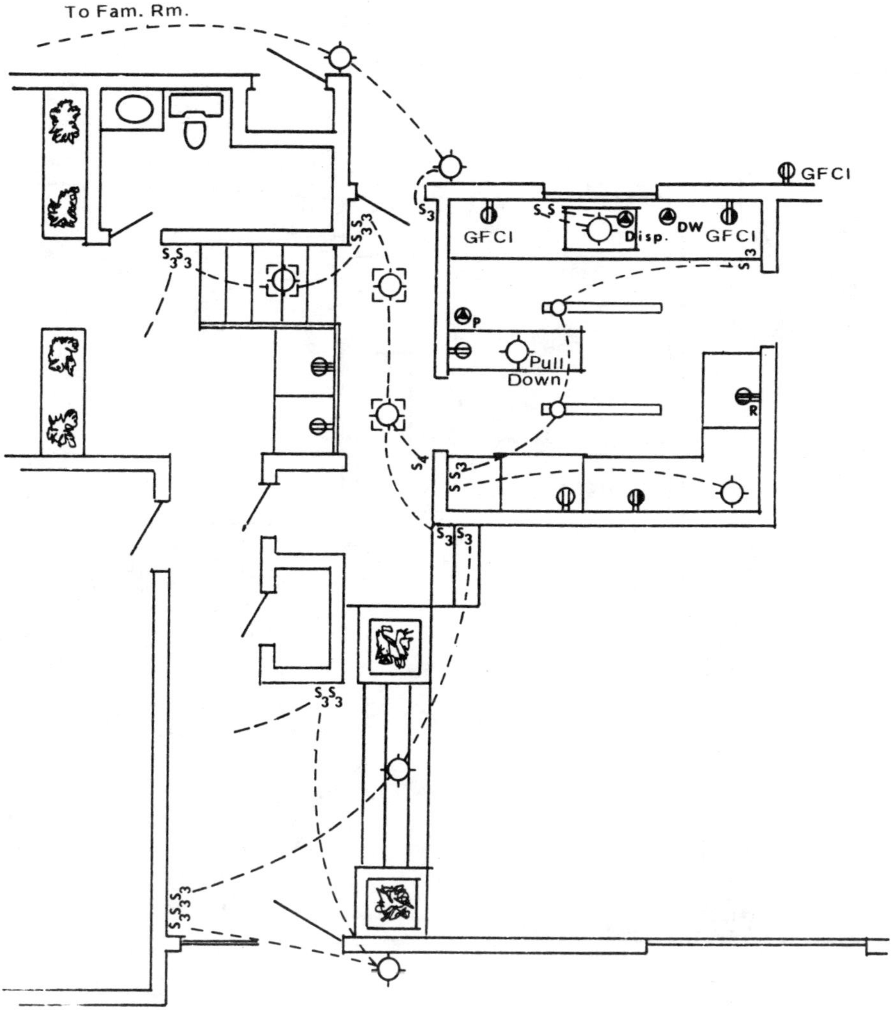

Figure 8-15 Electrical floor plan: kitchen, hall, foyer.

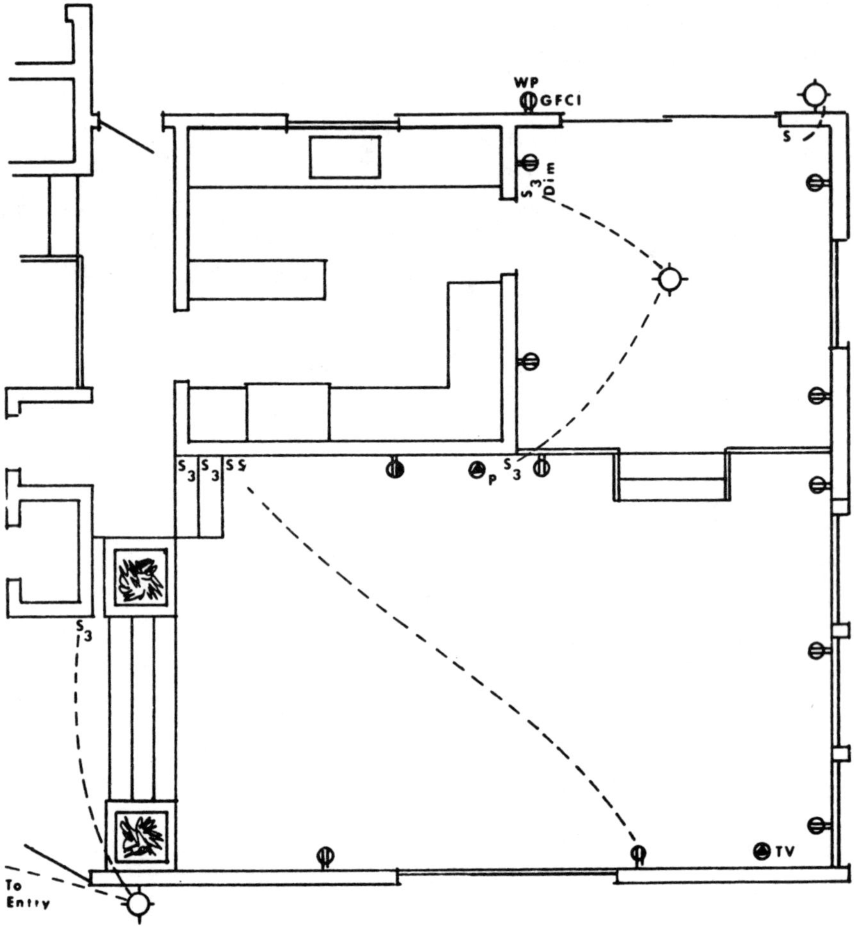

Figure 8-16 Electrical floor plan: dining room, living room.

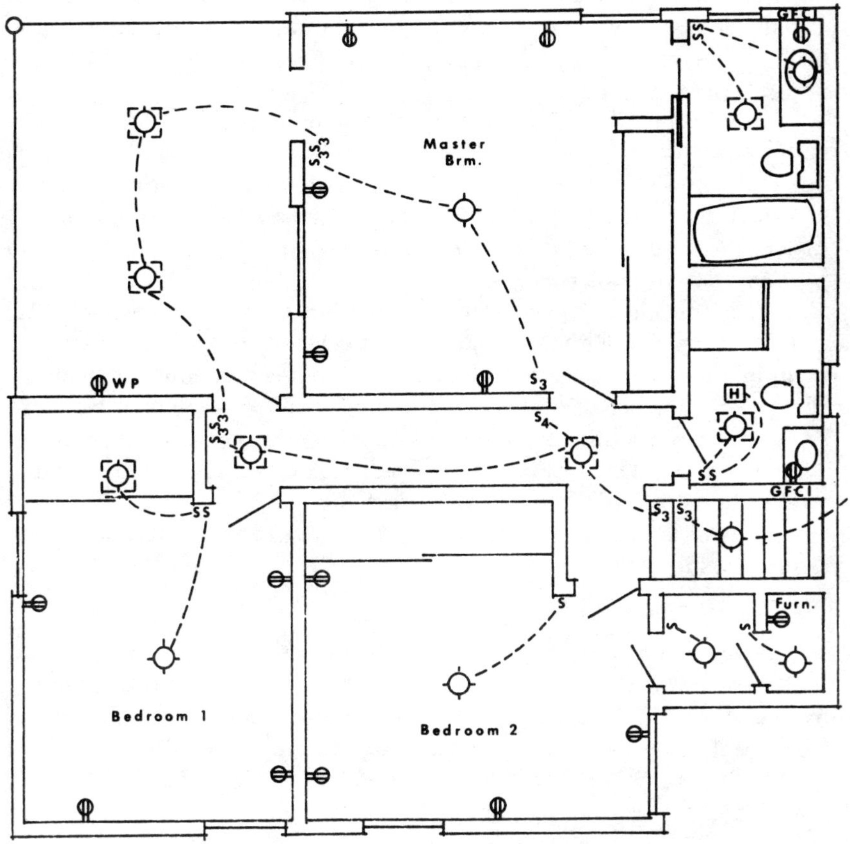

Figure 8-17 Electrical floor plan: second floor.

Most residential lighting and convenience outlet circuits will be rated at 15 amperes. This is because the smallest wire size approved for permanent building wiring, and therefore the least expensive wire to use, is #14 (see Chapter 4). This is the wire size approved in NEC 240-3(d) for a 15 ampere circuit. Since it is the smallest allowable wire size, it is also the easiest to work in terms of cutting, bending, splicing, and running through the building, making it also the most economical material to use in terms of labor time.

The permissible loads in the last column of Table 8-1 are the maximum ***continuous loads*** the code will allow for the circuits listed. It is necessary, then, in order to conform to this requirement, that we know just what the code means by ***continuous load*** as well as how to compute it. Article 100 defines it as *a load where the maximum current is expected to continue for three hours or more.*

Examples: a load consisting of lights turned on at about 6:00 P.M. and off at bedtime around 11:00 P.M. is certainly a continuous load—a vacuum cleaner in use for ½ an hour at a time is certainly not. A water heater, or any other equipment that cycles on and off automatically, has to be considered a continuous load since at any given moment it may be on.

The computation of individual loads for the purpose of determining total circuit loads on general purpose lighting and convenience outlet circuits is based partly on known quantities, and partly on standardized rule of thumb estimates.

a. Fluorescent lights

A label is pasted on the ballast that states the wattage that is drawn by that ballast and the tubes it supplies. If it is inconvenient to open the fixture, figure 10 watts per running foot of tube for each tube. Then add 25 percent of the tube wattage for the ballast load. The total arrived at this way will be a few watts above the load given on the ballast label.

b. Recessed incandescent lights

Such fixtures must be marked permanently inside in letters at least ¼″ high to indicate the maximum lamp wattage for which the fixture is rated [NEC 410-70].

c. Surface mounted incandescents with diffuser—one-or two-bulb

Allow by rule of thumb 200 watts.

d. General purpose convenience receptacle, 180 volt-amperes each. NEC 220-3(b)(9) refers here to volt-amperes, not watts.

The actual loads on general purpose convenience receptacles vary widely. When such receptacles have been located in confirmity with NEC 210-52 so that no point along the floor line in any wall space is more than 6 feet from an outlet, the result is that many never are used at all. Some are blocked by furniture, some are necessarily placed at points where the occupant has no need for electrical power. Consequently the writers of the code have assigned an arbitrary load of 180 volt-amperes to each one on the basis that the average of the loads on all of them will turn out to be less than that.

Appendix 220-3(a) lists code minimum allowances, in volt-amperes per square foot, that must be made in lighting circuits for various types of occupancies. In dwellings the required minimum is 3 volt-amperes per square foot of floor area, excluding porches, garages, or unused or unfinished areas. By the time the various other lighting and outlet requirements discussed earlier have been met, that 3 volt-amperes per square foot requirement has usually been exceeded.

Lighting and Convenience Outlet Branch Circuits

The assignment of the lighting and convenience outlets on our sample plans to 15 ampere branch circuits that conform to code requirements could be done in a number of different ways. One of those ways is shown below. After all lights and receptacles have been divided into branch circuits, cable plans and wiring diagrams will be given for each circuit to show in detail how the system will be assembled.

Starting with the family room on the first floor, Figure 8-14 shows how it will be corrected to conform to NEC-210. The loads are:

7 Convenience outlets (180 W each)	1260 W
3 Recessed ceiling fixtures (100 W each)	300 W
2 Ceiling surface fixtures (200 W each—estimated)	400 W
1 Outdoor wall light at doors (100 W)	100 W
Total wattage	2060 W

The special purpose outlets marked TV, for TV antenna, and P, for telephone, are not loads on the electrical system, and do not connect to branch circuits.

The 2060 watt load of this room is not going to be assigned to a single 15 ampere circuit because as Table 8-1 shows 1440 watts is the load limit for such a circuit. The supply to this room will have to be split over two circuits. The seven convenience outlets plus the outdoor light can go on one circuit which will then total 1360 watts (Figure 8-18). 1360 watts is satisfactory per code, since it is ***less*** than the allowable maximum.

It should be possible to combine the remaining 700 watts of lighting load with some other items since it is so far below the permitted maximum. Fortunately next to the family room is a bathroom which calls for:

1 Heat-A-Vent (300 W)	300 W
1 Medicine cabinet/Vanity light	200 W
1 Shower ceiling waterproof light	100 W
Total wattage	600 W

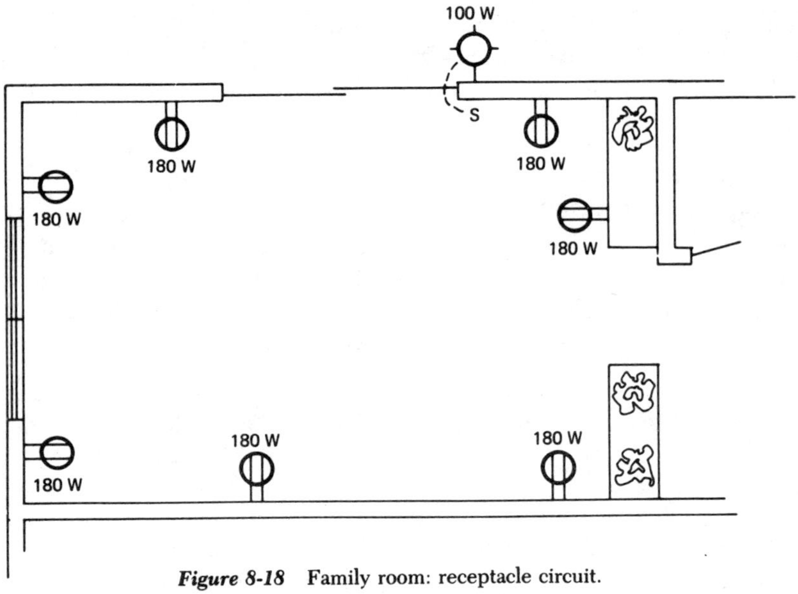

Figure 8-18 Family room: receptacle circuit.

The two now total 1300 W, but next to the bathroom is a utility closet opening to the outside and containing only a 60 watt pull chain light. Adding that to the other two the total then becomes 1360 watts. Figure 8-19 shows this circuit. The GFCI protected outlet in the bathroom has not been accounted for, but will be covered later along with all the other GFCI outlets in the house.

Moving from the bathroom up the short stairs and across the hall to the kitchen we find:

2 Two tube 4′ fluorescents (100 W each)	200 W
1 Recessed ceiling light over sink (100 W)	100 W
1 Surface ceiling light over counter (200 W)	200 W
1 Pull-down over breakfast bar (200 W)	200 W
Kitchen light total	700 W

All other receptacles and outlets in this kitchen are for appliances and per code may ***not*** be included in lighting and convenience outlet circuits. They will be covered later.

This 700 watts of kitchen lighting appears another likely candidate to be combined with other things. The three recessed ceiling lights in the stair and hall outside the kitchen total 300 watts, and the two outside lights total 200 more. Adding these will bring the circuit up to 1200 watts. The ceiling light over the stair from the foyer to the living room certainly looks as though it belongs on a circuit with the living room. However, as we shall soon see the living room has too many receptacles to take that light as well. By adding it to the kitchen-hall circuit the total computed load becomes 1400 watts and the circuit is as given in Figure 8-20.

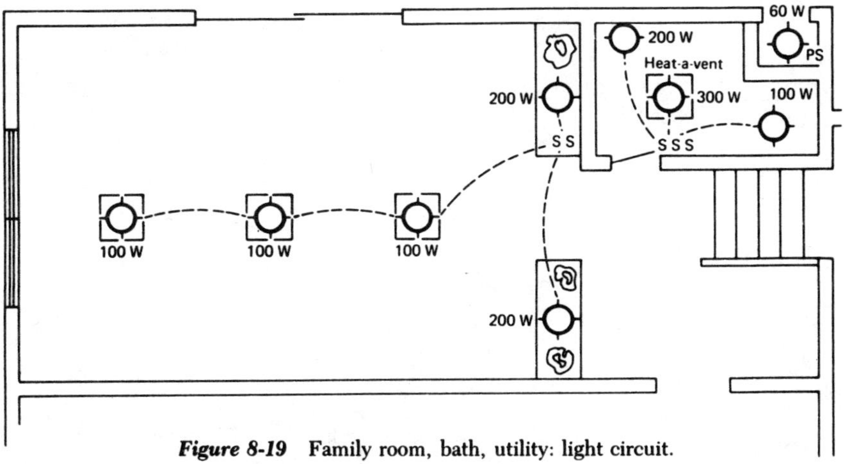

Figure 8-19 Family room, bath, utility: light circuit.

The dining room is an easy one as shown in Figure 8-21.

4 Convenience outlets (180 W each)	720 W
1 Ceiling fixture (200 W)	200 W
1 Outdoor wall light by door (100 W)	100 W
Dining Room total	1020 W

The living room has no light fixtures, but will be illuminated by lamps plugged into the numerous receptacles, two of which are switched. The phone and TV antenna outlets are, again, not parts of the electrical power circuit.

7 Convenience outlets (180 W each)	1260 W
Living Room total	1260 W

Completing the first floor are the garage and the hall leading to it. They contain as noted on Figure 8-23:

2 Two tube 4′ fluorescent fixtures (100 W each)	200 W
2 Surface mounted ceiling lights over workbench (200 W each)	400 W
1 Outdoor light at garage door (100 W)	100 W
4 Recessed ceiling lights in hallway (100 W each)	400 W
1 Outdoor light at entry	100 W
1 Pull switch closet light (60 W)	60 W
Total	1260 W

The two GFCI outlets in the garage are not included in this circuit. They will be dealt with later. Also the laundry outlets for dryer and washer are not included. As we shall see the code specifies how they are to be handled.

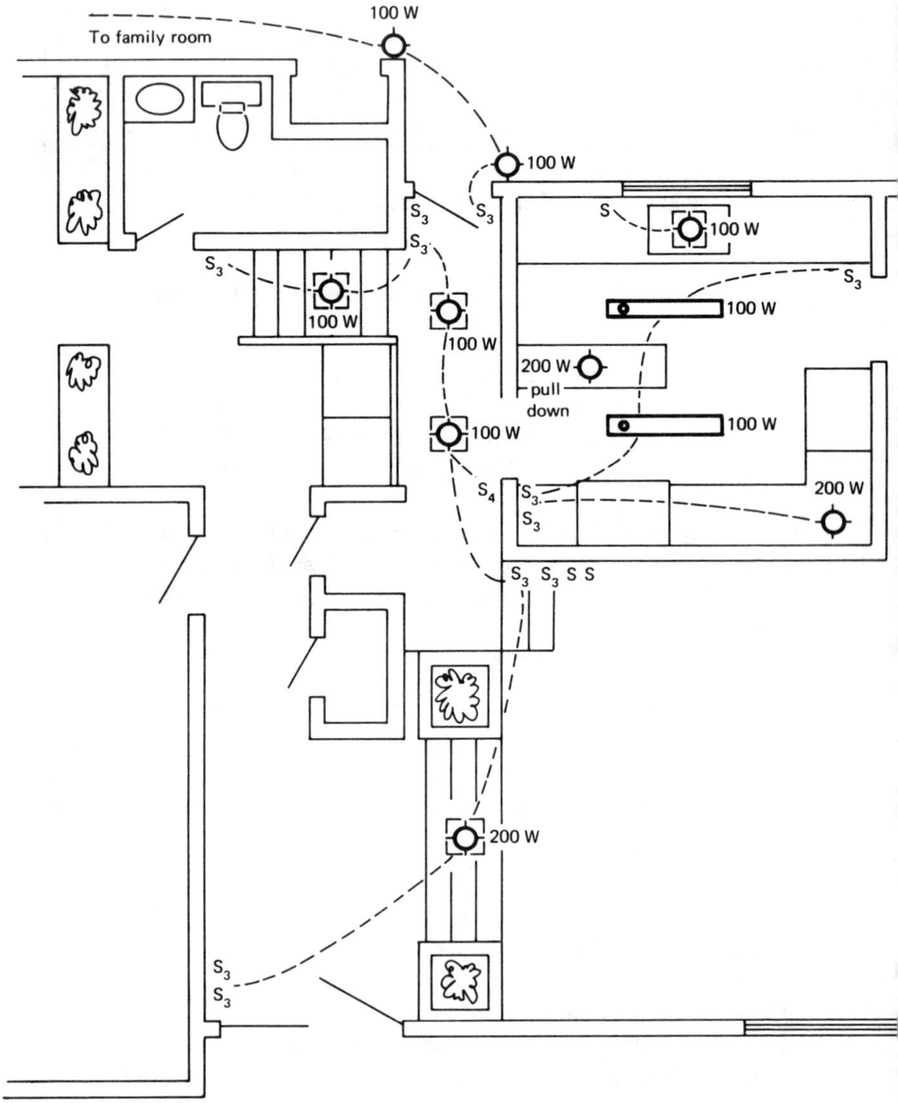

Figure 8-20 Kitchen, hall, foyer: light circuit.

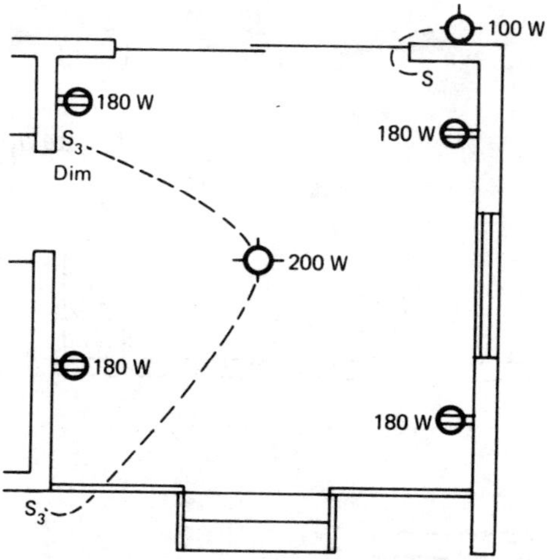

Figure 8-21 Dining room: light and receptacle circuit.

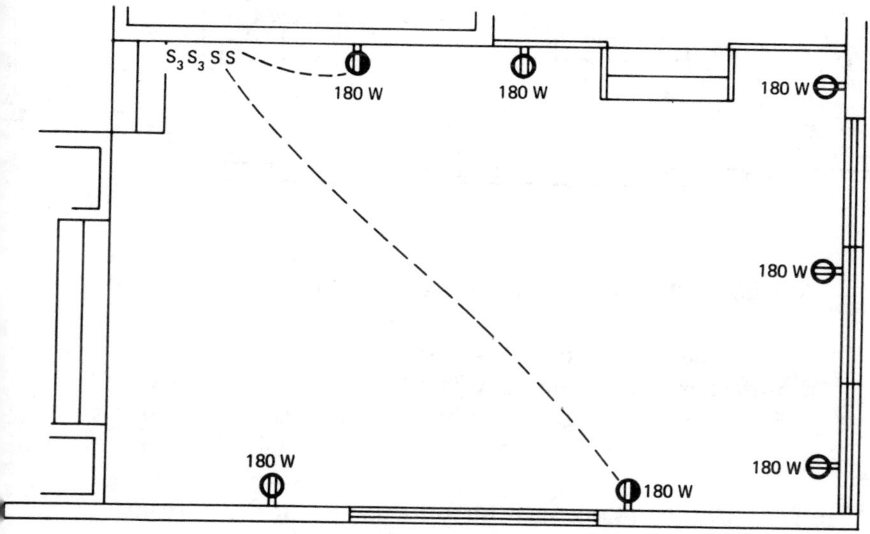

Figure 8-22 Living room: receptacle circuit.

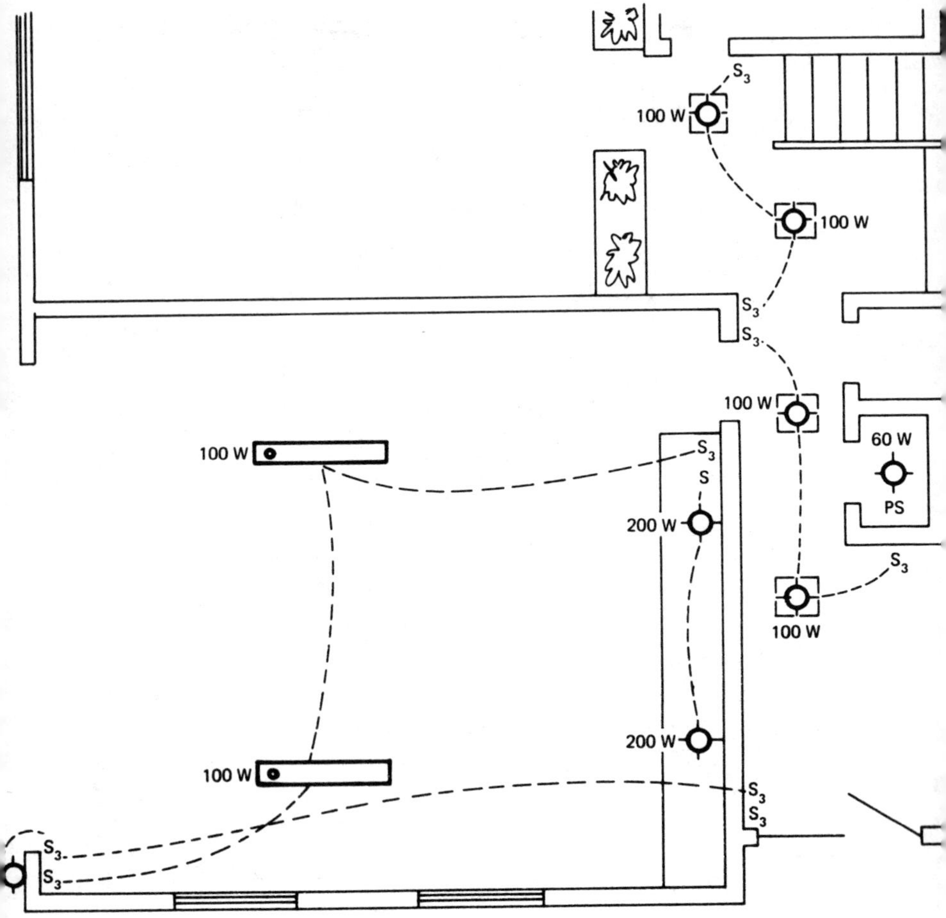

Figure 8-23 Garage and hall: light circuit.

Moving upstairs to the master bedroom (Figure 8-24) we find:

5 Convenience outlets (180 W each)	900 W
1 Surface mount ceiling light (200 W)	200 W
2 Recessed ceiling lights on porch (100 W each)	200 W
Master Bedroom total	1300 W

Since this circuit is already up to 1300 watts the porch weatherproof outlet at 180 watts would put it over the 15 ampere circuit code maximum of 1440. That outlet will go instead on the circuit with bedroom #1.

The master bathroom and bath #2 (Figure 8-25) can be paired on a single circuit. The two required GFCI outlets will be added to the five other GFCI outlets from downstairs, and thus will be omitted from this circuit.

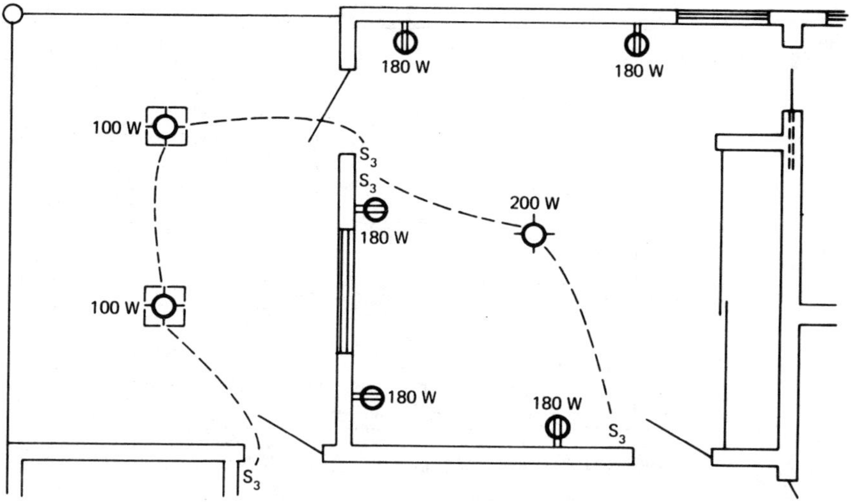

Figure 8-24 Master bedroom: light and receptacle circuit.

2 Recessed ceiling lights (100 W each)	200 W
2 Ceiling heat lamps (300 W each)	600 W
1 Master vanity light (200 W)	200 W
2 Bath #2 vanity 2′ fluorescents (25 W each)	50 W
Total—2 bathrooms	1050 W

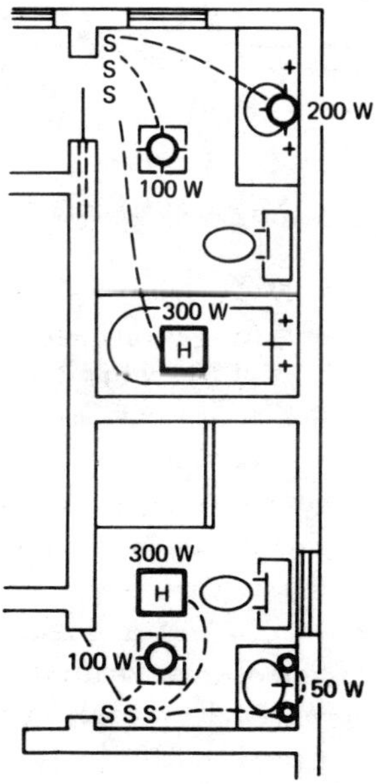

Figure 8-25 Second floor bathrooms circuit.

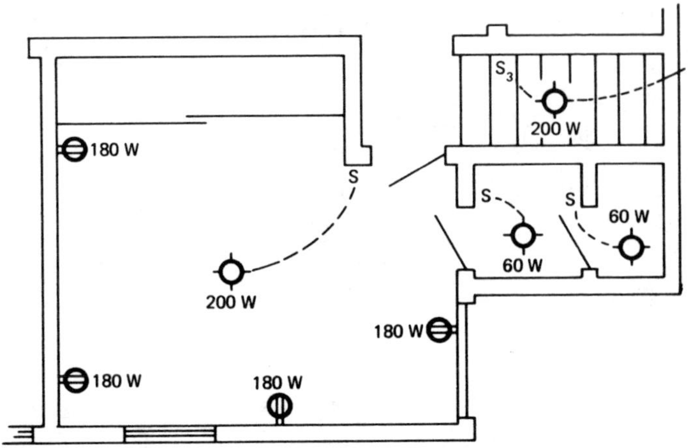

Figure 8-26 Bedroom #2, utility, stairway: light and receptacle circuit.

Bedroom #2 (Figure 8-26) can be combined with the modest requirements of the utility and furnace closets and the stairway light.

4 Convenience outlets (180 W each)	720 W
2 Surface mount ceiling lights (200 W each)	400 W
2 Closet ceiling lights (60 W each)	120 W
Circuit total	1240 W

Bedroom #1 circuit picks up the porch weatherproof outlet (Figure 8-27) and the two recessed hall ceiling lights as well.

5 Convenience outlets (180 W each)	900 W
3 Recessed ceiling lights (100 W each)	300 W
1 Surface mount ceiling light (200 W)	200 W
Circuit total	1400 W

The lighting and convenience outlet circuits for the entire house now total 10, but in the process of setting them up, seven GFCI protected outlets were bypassed. The GFCI is a device that can be obtained, as described above, either in combination with a convenience outlet, or in combination with a circuit breaker. It is also a rather expensive device, so whenever possible, all outlets in a house requiring GFCI protection will be wired so that a single GFCI protects them all in a manner similar to the sequence shown in Figure 8-12. The other way of protecting a number of outlets with a single GFCI is to place them all on one circuit and put the GFCI in the breaker, which is what will be done in this case.

Except for kitchen counter-top appliance outlets, where GFCIs may be required, an average small house will have a minimum of three or

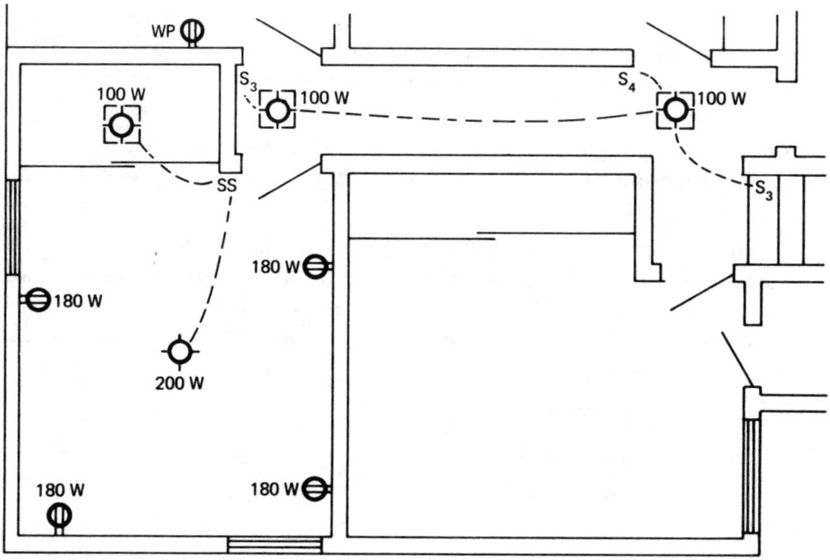

Figure 8-27 Bedroom #2 and second floor hall circuit.

perhaps four outlets that must be covered by GFCI per code: one outdoor outlet, one garage outlet, and one or two bathroom outlets. When there are four, their computed load totals 720 watts. In many instances the load of a single bathroom is 600 watts or less; thus a GFCI can be placed in one bathroom and all the other required GFCI outlets can be wired to it without overloading that bathroom circuit. However, in the house we are working on six outlets require protection. This constitutes a computed load of 1080 watts, which is enough to comprise another circuit.

The lighting and convenience outlet circuits for this sample house along with their computed loads now are:

Family Room Outlets	1360 W
Family Room, Bath #1, & Utility Lights	1360 W
Kitchen, Hall, Foyer Lights	1400 W
Dining Room Outlets & Lights	1020 W
Living Room	1260 W
Garage & Back Hall Lights	1260 W
Master Bedroom Outlets & Lights	1300 W
Bathrooms—2nd floor	1050 W
Bedroom #2, Utility	1240 W
Bedroom #1, Hallway	1400 W
GFCI Protected Outlets	1260 W
Totals	13910 W

NEC 220-3(a) states that a minimum lighting load shall be provided of not less than the volt-amperage listed in Appendix Table 220-3(a) per square foot of each structure being used for any of the purposes listed in that table. For dwelling units the requirement is 3 volt-amperes per square foot. The sample house illustrated here vastly exceeds this minimum requirement; however, the reader should be familiar with the procedure for computing a building's conformity.

The square footage is computed from the outside dimensions of the building, and for dwellings porches, garages, and unfinished spaces may be subtracted. The first floor on our plans is 34′ × 58′ = 1972 sq. ft. less the garage which is 20′ × 24′ = 480 sq. ft. leaving a first floor of 1492 sq. ft. The second floor is 31′ × 31′ - 961 sq. ft. less the porch 14′ × 10.5′ = 147 sq. ft. leaving 814 sq. ft.

Net sq. footage, First Floor	1492
Net sq. footage, Second Floor	814
Total	2306 sq. ft.

2306 sq. ft. × 3 watts = 6918 watts

Computed lighting circuits total 13910 watts which is approximately double what is required under NEC 220-3(a).

Other Branch Circuits

In addition to branch circuits for lighting, several other branch circuits are required in residential structures. The fact has been mentioned that per NEC 210-52 (c) receptacle outlets shall be provided for small appliances on all kitchen counter spaces over 12″ wide. NEC 210-52(b) furthermore specifies that those receptacles shall be connected to ***at least*** two 20 ampere small appliance circuits, and that except for a refrigerator those small appliance circuits shall have ***no other outlets.*** The refrigerator may be on one of these circuits or it may be on one by itself.

The laundry receptacle required in NEC 210-52 (f) is again mentioned in NEC 210-11(c), where it is required that a separate 20 ampere branch circuit supply that receptacle, and that circuit is to have ***no other outlets.***

The above circuit is for a 120 V washer. If an electric dryer is installed, it requires a separate 240 V, 30 ampere circuit which also shall have ***no other outlets.*** The sample plans include an outlet for such a circuit.

An electric range calls for its own 240 V circuit which will usually be a 50 ampere circuit, and it too may have ***no other outlets.***

An electric water heater will require another 240 V circuit which will be 20 amperes unless the heater is exceptionally big. This is another one that can have ***no other outlets.***

While a refrigerator may be connected to one of the two 20 ampere small appliance circuits required by NEC 210-52(b), the dishwasher, garbage disposal, or trash compactor cannot. These appliances often have individual circuits, however, sometimes the dishwasher and disposal are combined on a single 20 ampere circuit.

Electric Heating

Electric heating equipment, other than small portable heaters of the type intended to be plugged into ordinary convenience outlet circuits, will require special branch circuits. Such circuits must be sized according to rated power requirements of the units they supply. Article 424 of the code deals with fixed electric space heating.

Circuits for electric space heating [NEC 424-3 (a)] may be 15, 20, or 30 ampere and either 120 V or 240 V. NEC 424-3 (b) states that branch circuits supplying fixed space heating, with or without a motor, shall be rated at not less than 125 percent of the total load of both motors and heaters. This restates exactly the same requirement mentioned earlier in NEC 210-20(a).

As an example a heater is rated at 1100 watts. If 1100 = 100 percent then NEC 424-3 (b) wants the circuit rated at least 125 percent of that, or at least 1375 watts. Computing in reverse these circuits may be 15, 20, or 30 A at 120 V or 240 V. Looking back at Table 8-1 a 15 A circuit at 120 V can be loaded to 1440 W or 80 percent of its total capacity of 1800 W. This load is only 1375 W, well within Code requirements for a 15A circuit.

General Heating, Ventilating and Airconditioning

Most other types of heating equipment as well as ventilating and airconditioning devices require some sort of electrical supply. The requirements in this area vary widely; consequently the electrician will need to obtain specification information from the people in charge of this area.

This information will be needed at an early stage in order to insure that the necessary circuits are included in the plans, and the necessary power requirements are included in the computations for sizing the incoming service.

In this sample house a gas fired forced air furnace is behind the utility closet off of bedroom #2. To power its blower motor, controls, and intermittant ignition device (IID), a separate 20 ampere appliance circuit will be provided. The intermittant ignition device is a recently developed electrical replacement for the gas pilot light. (It was developed in response to the present emphasis on energy conservation.) It is presently being used on gas ranges, dryers, and water heaters as well as furnaces. You will be seeing a great deal more of them in the future.

Cable Plans

At this point the plans have been checked to insure that all outlets and switches required by code have been included (Figures 8-14, 15, 16, and 17). Then the switches and outlets that are to make up the lighting and convenience outlet circuits were grouped so that the computed loads on those circuits all conform to code loading requirements (Figures 8-18 to 8-27). The next step is to prepare cable plans and wiring diagrams to show exactly how every circuit will be put together.

The switch, fixture, outlet, and other symbols used on cable plans are the same as those used on floor plans. However, the plans themselves are vastly simpler. They are drawn to scale just as floor plans are, but have few if any dimensions or other building construction information. All that is given are walls, doors, windows, etc., and electrical symbols. Their purpose is simply to show the electrician how many wires are to be used in each of the runs from one electrical box to another. Any unnecessary information is omitted.

Figure 8-28 is the cable plan for the lights assigned to the circuit that was laid out in Figure 8-19. Since the load on the circuit has already been found satisfactory, wattages are omitted. The various electrical symbols are connected by solid lines each of which is crossed by two or more hatch lines

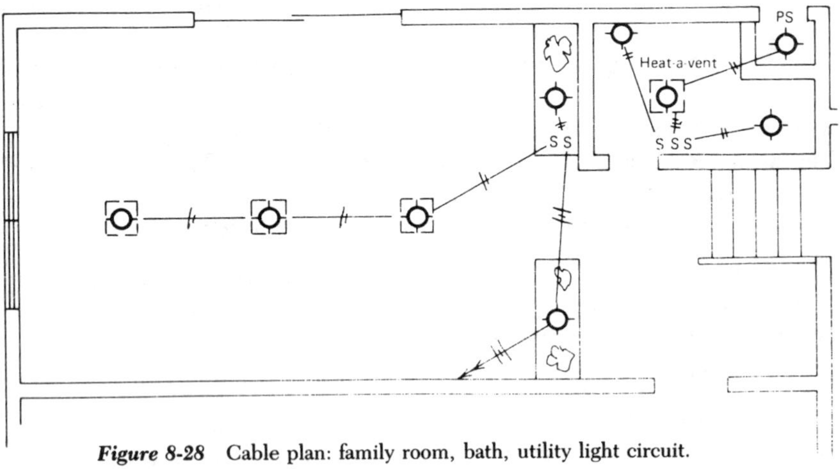

Figure 8-28 Cable plan: family room, bath, utility light circuit.

indicating the number of circuit conductors to be used in that run (see also Figure 8-1). The circuit conductors, remember, are white or grey common, and power (black, red, blue, or other color). The ground wire is not a circuit conductor and does not appear on either cable plans or wiring diagrams because it is normally an inactive part of the system. The house is being wired in NM cable, so the wire runs are cross-hatched by either two or three lines because only two or three wire cables will be used. In any of the conduits—rigid, EMT, or flexible—a good many more wires are commonly used. When cross-hatching a line to show how many wires are in a conduit or cable, the longer hatch indicates a power wire, the shorter one a common. Thus if a conduit line is crossed by four long hatches and two short, it contains four power wires and two commons.

In Figure 8-28 three three-wire runs are shown the remainder being two wire. Note that while the cable plan is intended to indicate how many wires are required in every run from one box to another they are ***not*** to be regarded as indicating the actual route of that wire run through the building.

Wiring Diagrams

Cable plans show how many wires make up each box to box run, and how many cables or conduits enter each box. The wiring diagrams tell us what to do with those wires after they are in the boxes. For these diagrams a different set of symbols is required (Figure 8-29) in order to indicate the terminal points to which connections are made to the devices in each box.

Looking at the family room and using the cable plan in Figure 8-28 as the guide to how many wires enter each box, and the symbols given in Figure 8-29 to show the details of how they are connected, Figure 8-30 is the wiring diagram for that light circuit. Each splice is indicated by a black dot. The number of lines that meet at each splice dot is the number of wires in that splice. The short lines from dots to lights or switches are pigtails.

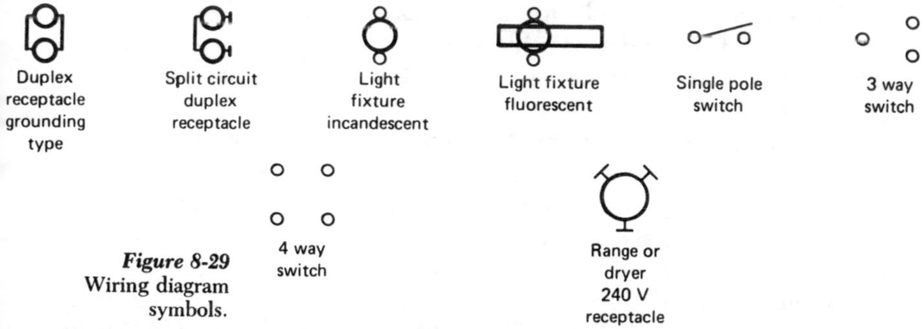

Figure 8-29
Wiring diagram symbols.

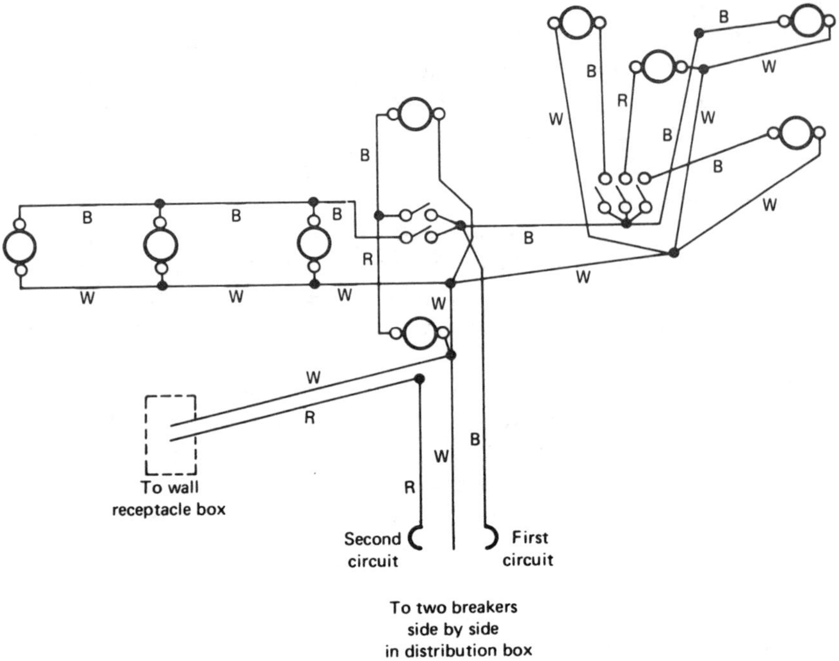

Figure 8-30 Family room, bath, utility light circuit.

For example, the three switches in the bathroom are fed by three pigtails spliced to the incoming unswitched power wire coming from the switch box in the family room. The same splice contains the unswitched black power wire that goes to the Heat-A-Vent and connects there to the cable supplying the pull-switched light in the utility closet.

The home run cable going back to the breaker box contains three wires instead of the two one might expect. Connecting a circuit to the breaker box requires two wires: a power wire protected by the breaker, and a common connected to ground. The safety grounding wire is in the same cable, but does not show on cable plans or wiring diagrams.

The third wire on this cable is the power wire for a second circuit that will supply the wall outlets in the family room. That circuit will share the same common and ground wires in this home run cable. Two circuits can be fed through a three wire cable sharing one common and one ground ***when*** the two power wires are connected to two breakers on the opposite phases of the incoming service.

Looking back to Figures 2-10 and 2-11, the voltage changes illustrated are those that occur in the conductors on ***one side*** of a generator armature rotating in a magnetic field. That set of conductors is always matched by another set directly opposite them on the same armature. The cycle of voltage changes in the second conductor is exactly opposite to what occurs in the first (Figure 8-31). When conductor #1 is at + 120 volts, conductor #2 is at − 120 volts. The difference across the two is 240 volts.

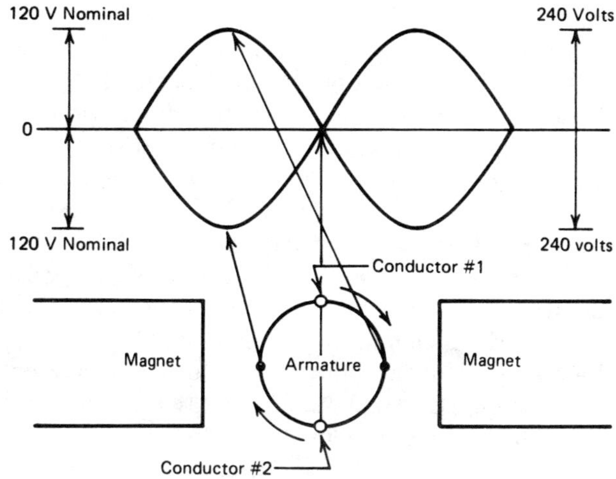

Figure 8-31 240 V single phase current.

This is the source of the 240 volts across the two power wires in the service drop coming into the building, and also the source of the 240 volt power used for ranges, dryers, water heaters, airconditioners, space heaters, and other 240 volt equipment.

When two circuits are properly connected to a three-wire cable (Figure 8-32) the neutral common carries only the current imbalance between the two power lines. If in error the two power lines are connected to the same side of the incoming service, the amperages drawn will add to each other rather than subtract. In most modern breaker boxes any two breakers ***next*** to each other on the same side of the box (Figure 9-11) will be on opposite phases of the supply. Any two breakers ***opposite*** each other will be on the ***same*** phase.

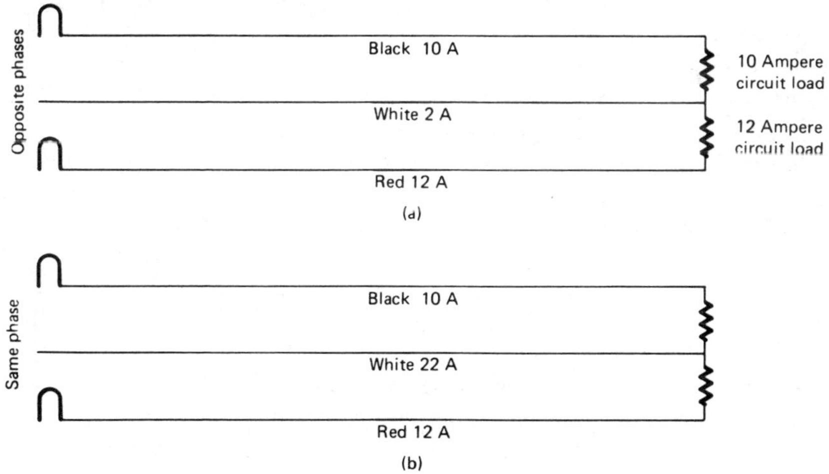

Figure 8-32 3-wire feed to two circuits: (a) correct; (b) incorrect.

Carrying the feed for two circuits via a three-wire cable through the building as far as possible is a common practice and will be done repeatedly in this house. The reason for this is obviously economic. The feed for a single circuit is a cable containing three wires: power, common and ground; thus two circuits will require six wires: two power, two commons, and two grounds. When, as here, two circuits are fed with a three-wire cable the job is being done with only four wires instead of six. There are two separate power leads, but only one common and one ground.

The second circuit in this three-wire supply cable drops down from the light (Figure 8-30) to a wall receptacle box in the family room where it becomes the incoming feed (Figure 8-33) to the receptacle circuit. With the exception of the short branch to the outside light the wiring of this circuit (Figure 8-34) is a simple repetition of splices and pigtails to outlets as the cables circle the room.

Kitchen-Hall-Foyer Light Circuit

The cable plan for this circuit as given in Figure 8-20 appears in Figure 8-35. The feeder from the panel box in the garage comes in at a four gang switch box on the wall between the kitchen and the living room. It again contains two circuits in a three-wire cable as was done in the family room.

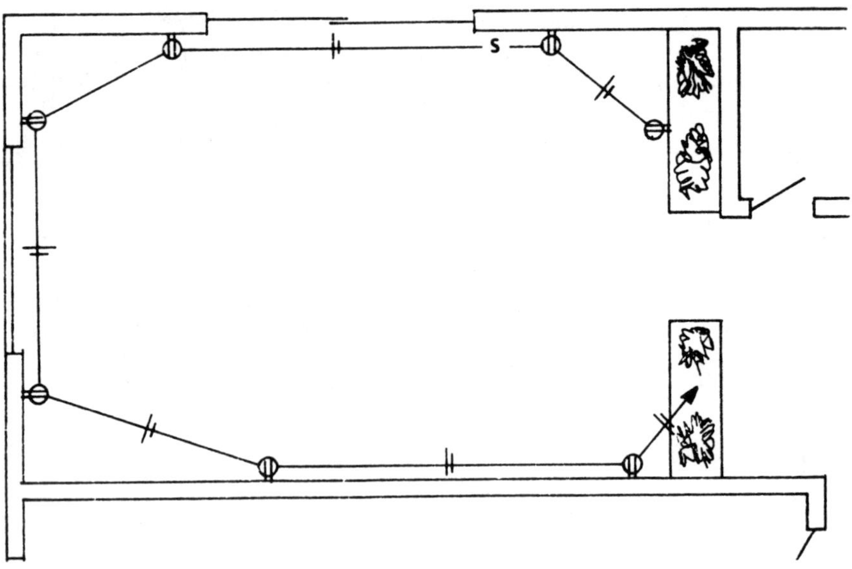

Figure 8-33 Cable plan: family room receptacle circuit.

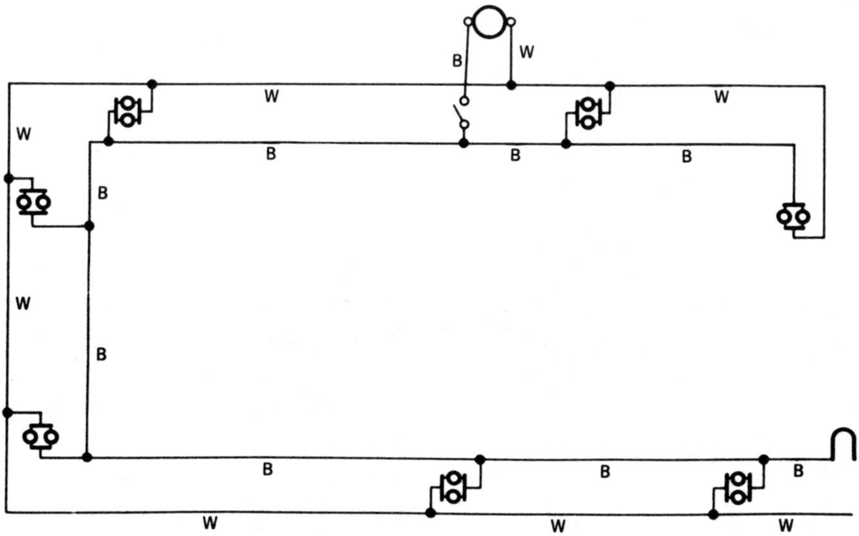

Figure 8-34 Circuit wiring: family room receptacles.

This time the second circuit will take care of the living room which will be discussed later.

The switch circuits indicated in Figure 8-35 are not complicated. They are all examples of circuits shown in Chapter 7. However, there are a great many in this section of the house, making it a trifle difficult to keep track of them. All together there are five multiple switch circuits here. Three of them control two lights each, the other two control single lights. An area with this many switch circuits is more easily understood by analyzing each of them individually. Then deal with the interconnections afterwards.

The wiring diagram for this section is given in Figure 8-36. The upper part of this diagram shows a pair of three-way switches controlling two outside lights. This is the same as circuit **b** in Figure 7-14, except that a second light has been added. Just below this is a light on the stairs controlled by two three-way switches, a circuit the same as **a** on Figure 7-15. The circuit at the bottom of the page controls the light over the steps into the living room in exactly the same way.

The three-way, four-way, three-way circuit leading to the two recessed hall lights is similar to Figure 7-19c, with a second load added, and the kitchen fluorescents are fed by a circuit based on Figure 7-14 **a**.

Dining Room Lights and Outlet Circuit

By comparison with the last one, this circuit is extremely simple. The one three-way switch circuit is a repetition of Figure 7-15 except that one of the

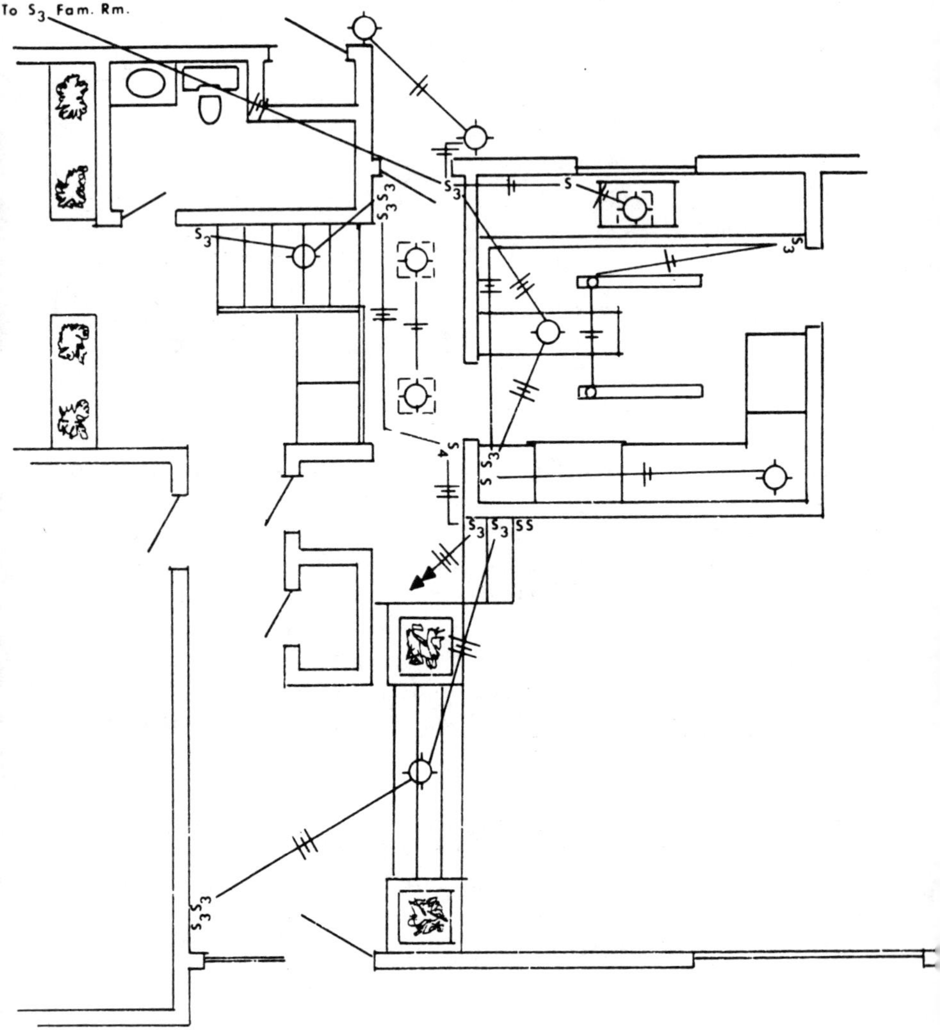

Figure 8-35 Cable plan: kitchen, hall, foyer light circuit.

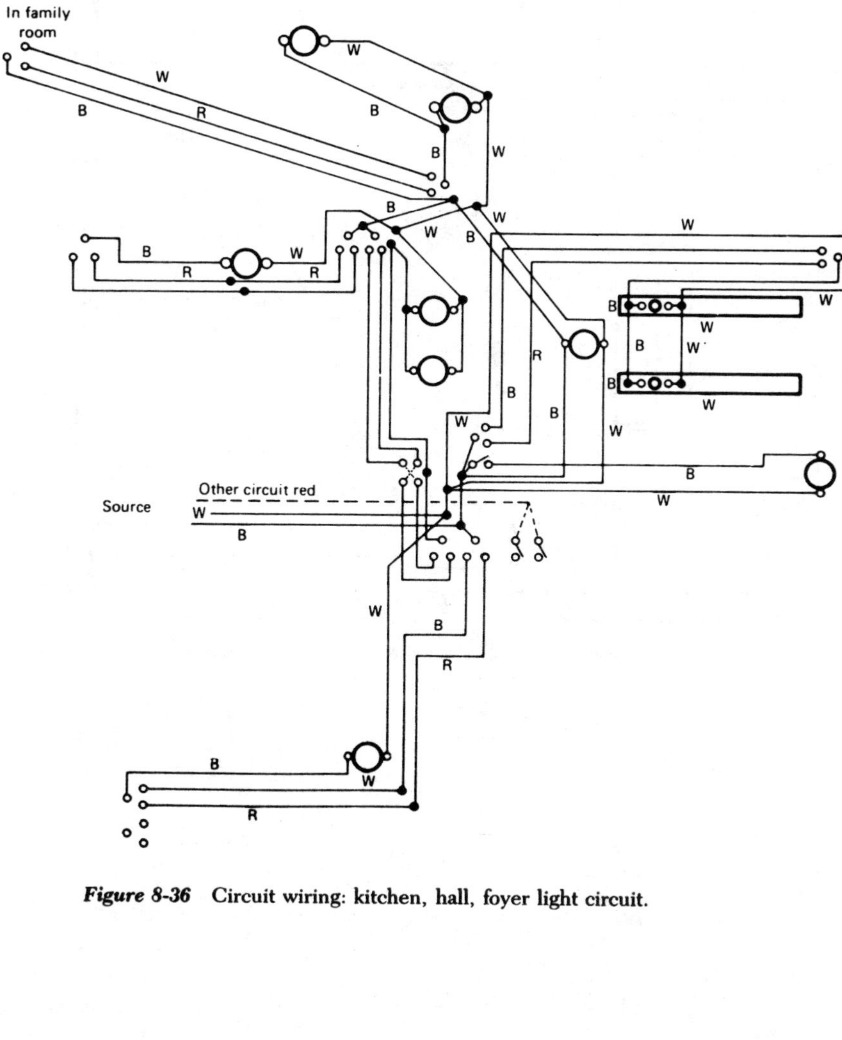

Figure 8-36 Circuit wiring: kitchen, hall, foyer light circuit.

three-ways is a dimmer switch (Figure 7-4 b). Figure 8-37 gives the cable plan with the wiring details shown on Figure 8-38. Remember that the three-way dimmer switch has no screw or push-in terminals, but has three pigtails of wire attached. Incoming unswitched power goes to the black one while the two travelers—black and red—go to the two red pigtails. The rest of this circuit consists merely of routine connections of duplex outlets with a single pole switch to an outdoor light at the end of the line.

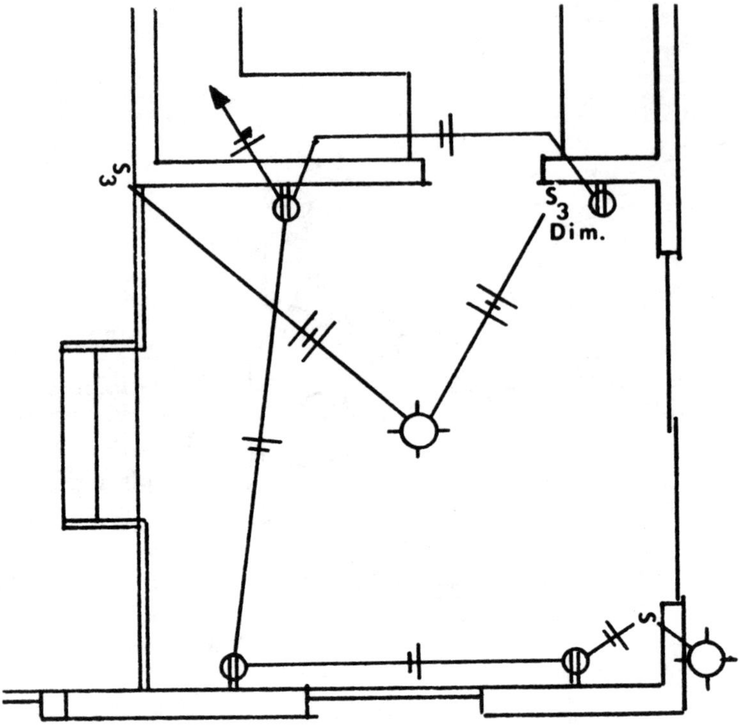

Figure 8-37 Cable plan: dining room receptacle and light circuit.

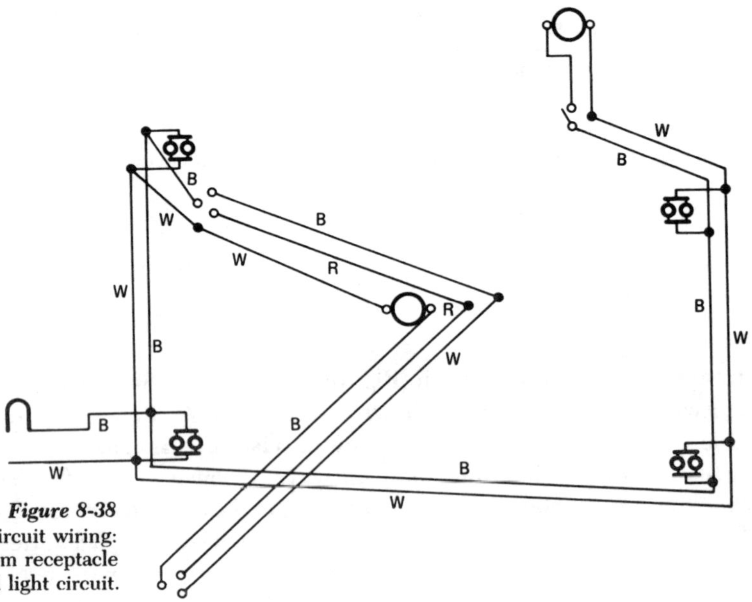

Figure 8-38 Circuit wiring: dining room receptacle and light circuit.

Living Room Wall Outlet Circuit

The power for this circuit comes to the living room switch box in a three-wire cable along with the power for the kitchen, hall, foyer light circuit detailed in Figures 8-39 and 8-40. This is a very simple circuit to wire consisting only of seven duplex receptacles. Two of these are split with their bottom halves switched.

Garage and Hall Light Circuit

This is another circuit containing a great deal of multiple switching. Figure 8-41 shows five pairs of three-ways. Power enters this circuit (Figure 8-42) at a box containing two three-way switches. Those two three-way circuits are powered at that point; however, to power the remainder of the lines an unswitched power wire is carried along with the switched power up to the ceiling fluorescents. This unswitched line along with the common comes down on the other side of the garage to go through and power the hall lights. It also returns to the garage from the hall to power the workbench lights.

The use of three-wire cable to carry both switched and unswitched power is common practice as shown with the split receptacles in the living room, and in the family room light circuit as well (Figure 8-31). The hall circuit uses three-wire cable as well.

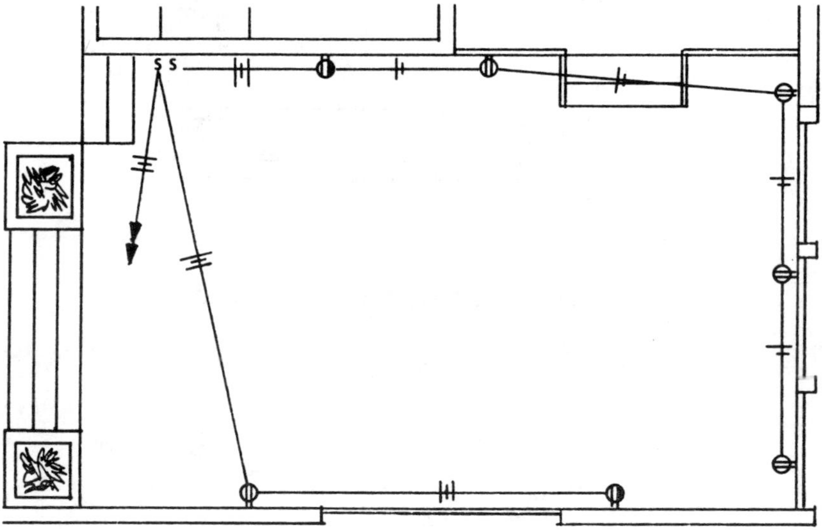

Figure 8-39 Cable plan: living room receptacle circuit.

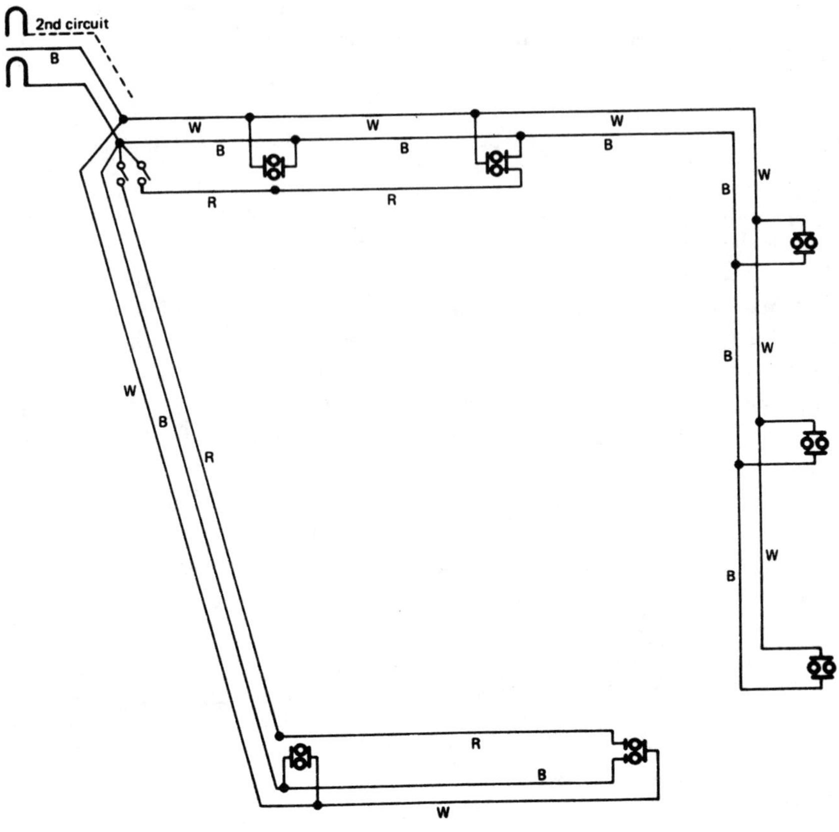

Figure 8-40 Wiring: living room receptacle circuit.

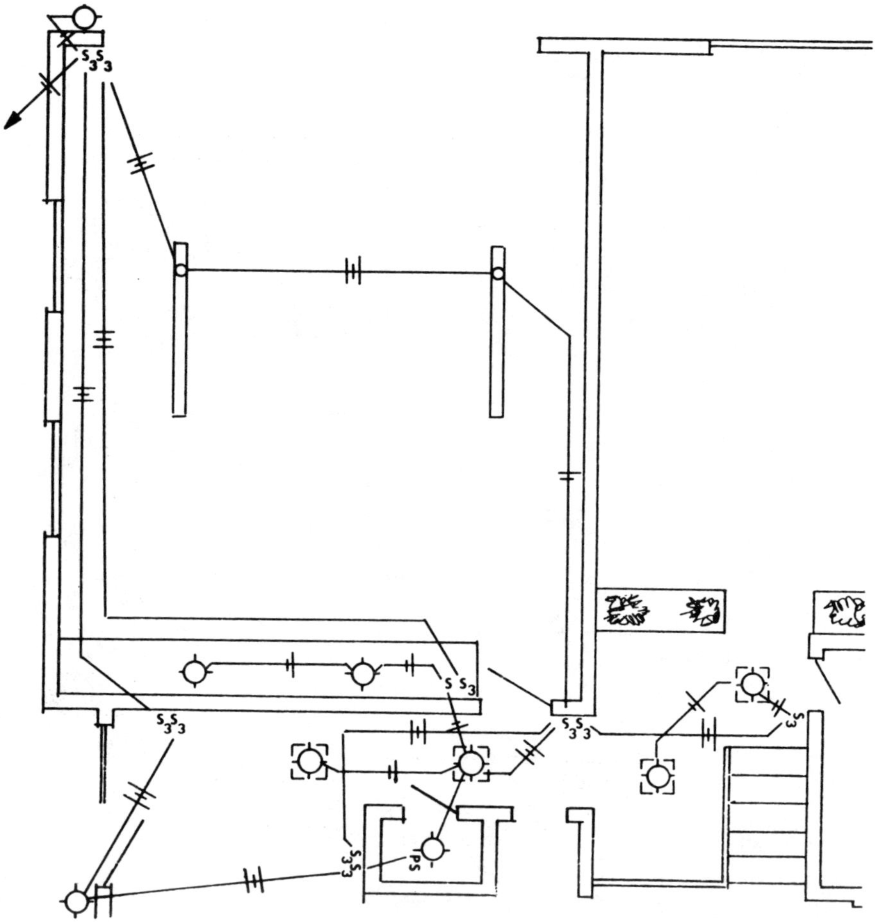

Figure 8-41 Cable plan: garage and hall light circuit.

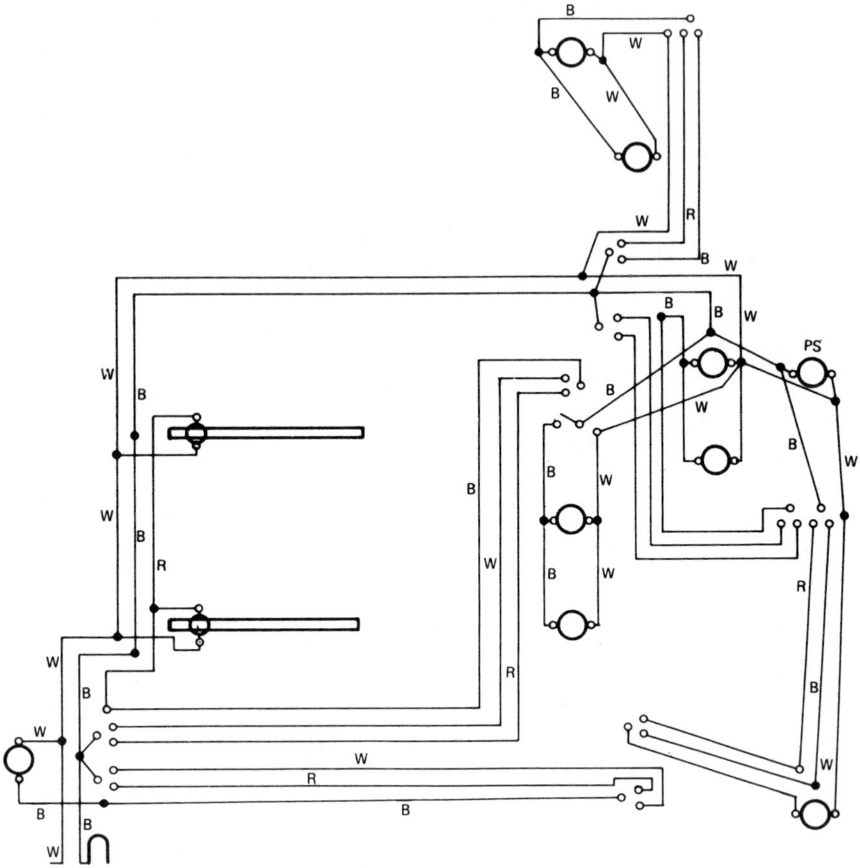

Figure 8-42 Wiring: garage and hall light circuit.

Master Bedroom and Porch Lights and Receptacles

The cable plan, Figure 8-43, shows two three-way switch circuits plus five duplex receptacles with the power entering at one of the wall receptacles. Both three-way switch circuits by now are familiar arrangements as seen in Figure 8-44. Power to the light switches comes from the receptacle below them. To avoid the labor of cabling around the porch door, and a framed corner as well, the power feed to the two receptacles on the far wall runs under the floor from the same box that feeds the switches.

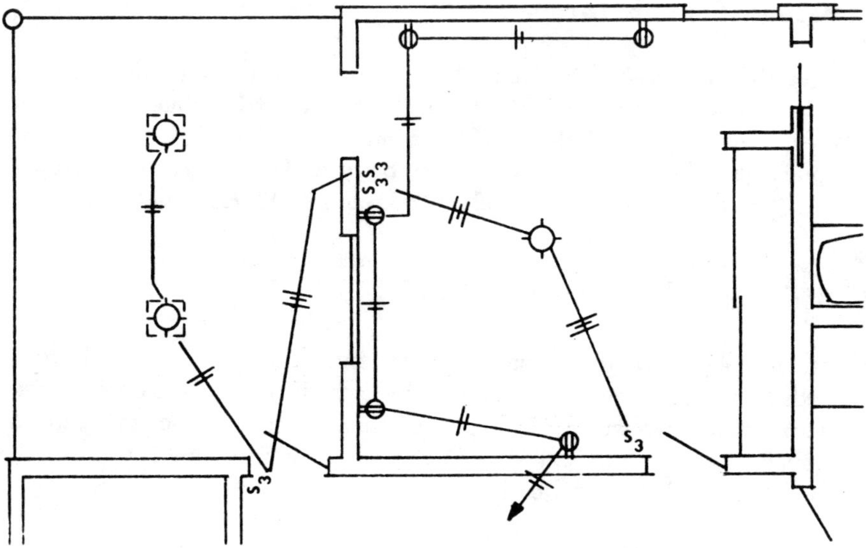

Figure 8-43 Cable plan: master bedroom and porch circuit.

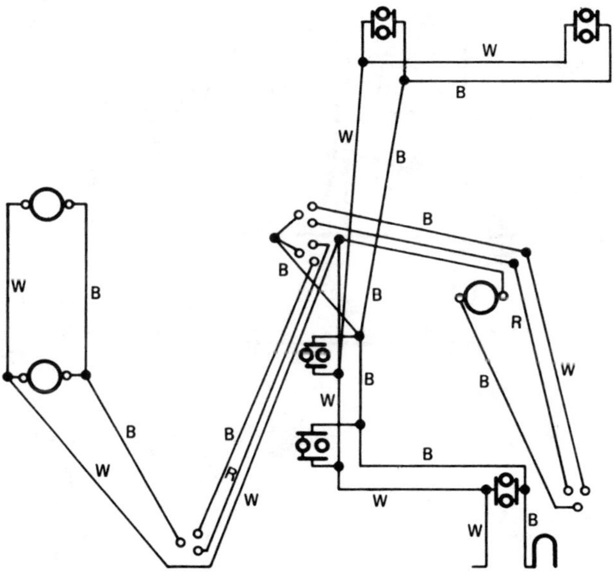

Figure 8-44 Wiring: master bedroom and porch circuit.

Second Floor Bathrooms

Note that the cable plan, Figure 8-45, does not show the required GFCI outlets that appeared very clearly on the final electrical floor plan of this area, Figure 8-17. While they will assuredly be installed, they do not appear in either Figures 8-45 or 8-46, because they are part of the separate GFCI circuit that carries all of the required GFCI outlets together.

Bedroom #2, Utility and Stairs

Figure 8-47 shows the incoming power entering through the wall from bedroom #1. In the two lighted utility closets, note in Figure 8-48 the wiring of the two wire cables from switches to lights. These are standard ***switch loops,*** per NEC 200-7(c). White is ***not*** common, but is power into the switches. Black is properly power back from switches to loads.

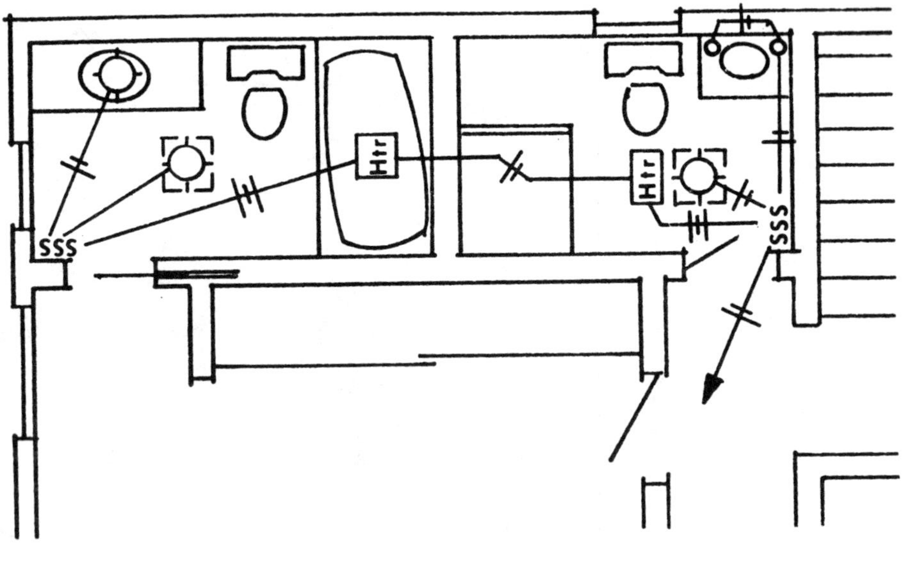

Figure 8-45 Cable plan: second floor bathrooms circuit.

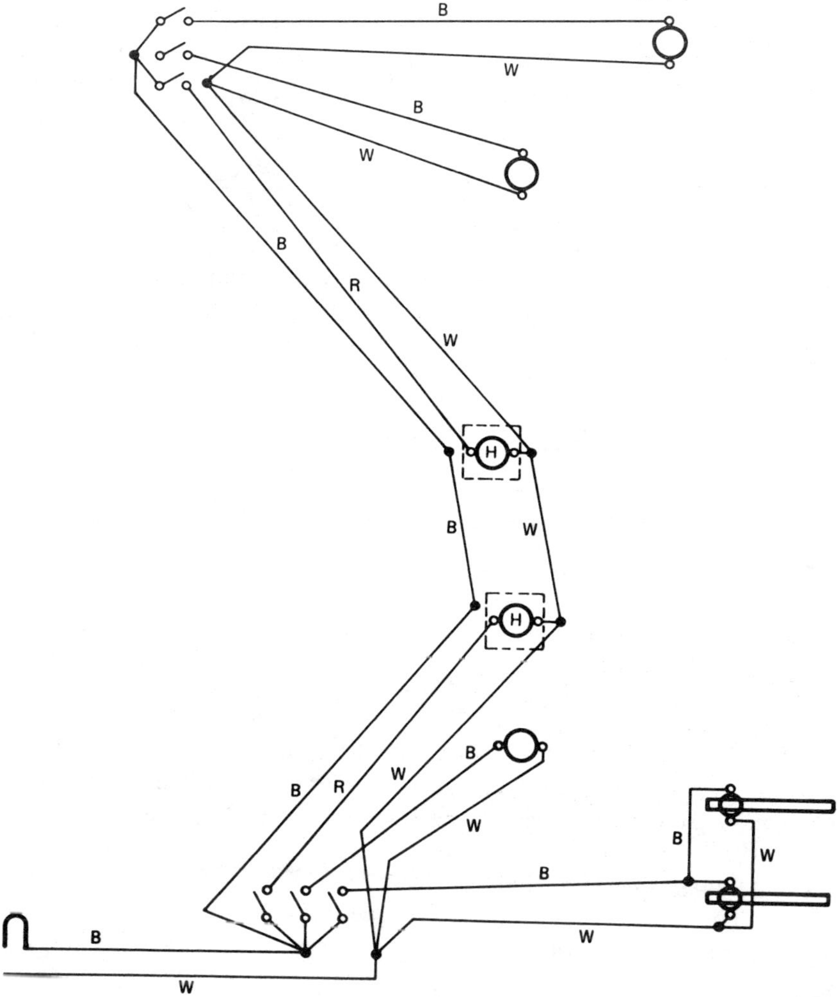

Figure 8-46 Wiring: second floor bathrooms circuit.

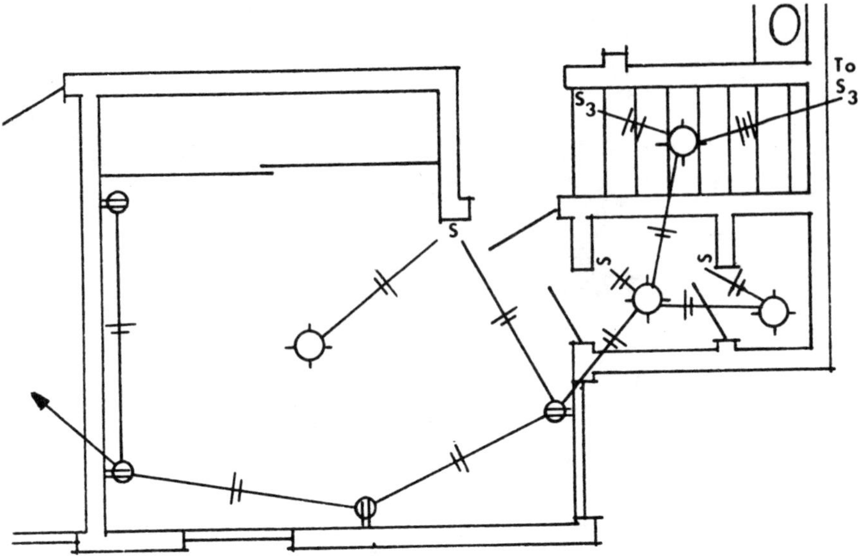

Figure 8-47 Cable plan: bedroom #2, utility, stairway circuit.

Bedroom #1 and Hall

The incoming power supply to bedroom #1 is another case of two power circuits carried by a three-wire cable. The three-wire power cable enters via the lower left hand wall receptacle (Figure 8-49). One power line feeds, this room and the hall lights, the second goes to the lower right receptacle box, and from there through the wall to feed the bedroom #2 circuit. There is nothing unusual or exotic in the wiring of this circuit as will be seen by examining Figure 8-50.

GFCI Circuit

As mentioned earlier, GFCIs can be installed in any receptacle by merely replacing a normal duplex receptacle with a receptacle type GFCI. Generally this is not done because as of this writing the GFCI receptacle is far too expensive. Normal procedure, then, is to use one GFCI receptacle, and wire the other receptacles requiring GFCI protection to it. However, that first GFCI may be part of a circuit that already has a partial load on

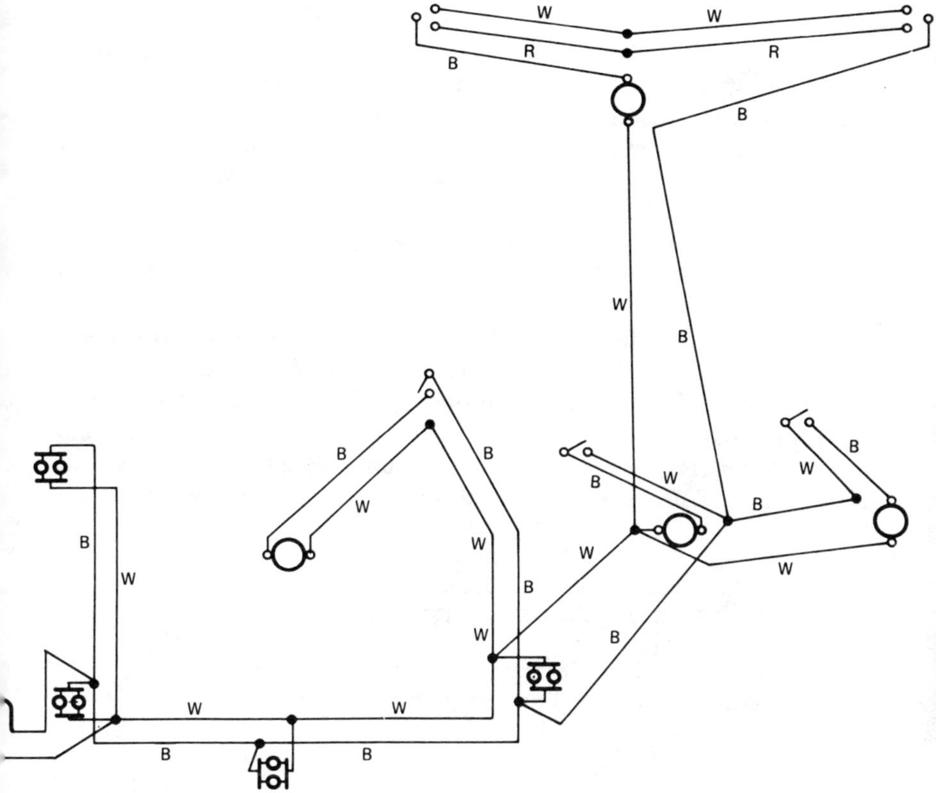

Figure 8-48 Wiring: bedroom #2, utility, stairway circuit.

it. If that load is small enough that all the remaining GFCIs can be added without overloading the circuit, the receptacles may be wired together.

In this case two receptacles in the garage, one in the first floor bathroom, one outside the dining room, and two in the two upstairs bathrooms all require GFCI protection. This makes a total of six and constitutes a computed load of 1080 watts, an amount appropriate for a separate circuit. Cabling for this is given in Figure 8-51, and wiring in Figure 8-52.

Appliances

The circuits for fixed appliances as well as the code-required small appliance circuits have been placed together on Figures 8-53 and 8-54. With the exception of the forced air furnace on the second floor, all are physically close to each other.

The range is the largest single load in the building. It is provided with a 50 A, 240 V circuit. Such a circuit per Appendix Table 310-16 should be wired with #6 copper. However, in accordance with Table 220-19, it is permissible to compute the branch circuit load of a single range at 80 percent of its rating. If the nameplate rating of a range were the full 50 A then 80 percent of that would be 40 A—a load that can be handled by #8 copper. Hence it is permissible to wire for a range from a 50 A, 240 V breaker using #8 copper. ***Do not wire any other 50 A circuits anywhere else using less than #6 copper.*** The range is a special case. Only the range outlet may be on this circuit.

The dryer requires a 30 A, 240 V circuit. It is wired in #10 copper from a 30 A, 240 V breaker. No other outlets may be on this circuit.

Ranges or dryers may plug into receptacles as described in Chapter 7, or they may be directly wired. Water heaters are normally directly wired. Depending on heater size and requirements the circuit will usually be a 20 A, 240 V wired with #12 copper. A larger than average water

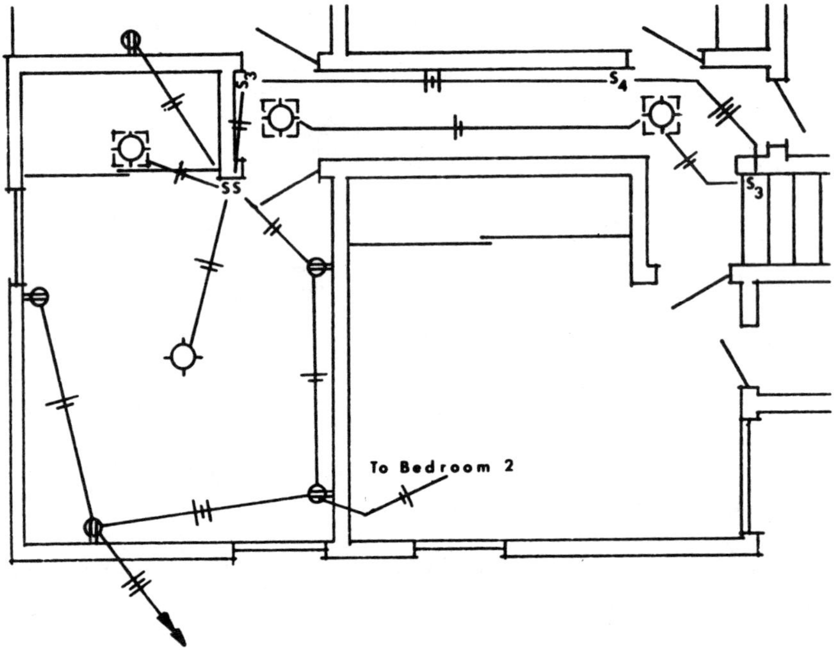

Figure 8-49 Cable plan: bedroom #1, hall circuit.

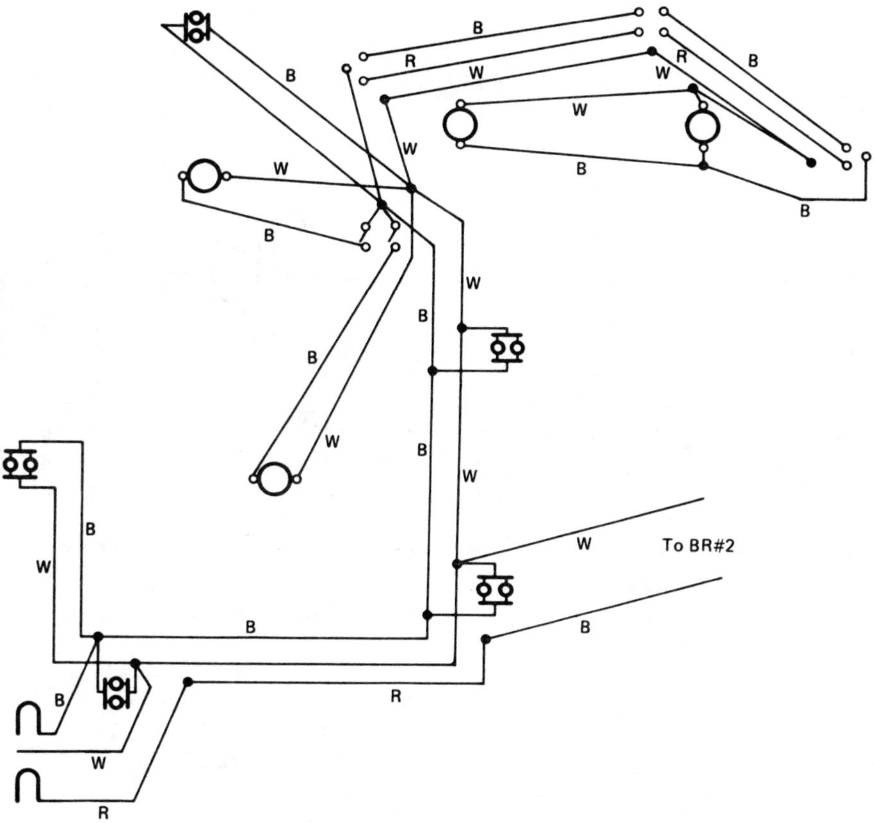

Figure 8-50 Wiring: bedroom #1, hall circuit.

heater might require a 30 A, 240 V circuit, in which case it is wired with #10 copper. Again, no other outlets may be on the same circuit with the water heater.

The two code-required small appliance circuits [NEC 210-52(b)] feed four split receptacles on the kitchen countertops. This is another case of taking two circuits from the breaker box via a three-wire cable. When the two halves of a duplex receptacle are powered by two separate circuits fed through two different breakers then, per NEC 210-4(b), those two breakers must be linked so that any defect that trips either one of them trips them both simultaneously. What this means in simple terms, is that those two circuits must originate at the breaker box from a linked double breaker. These circuits will be 20 A and will be wired with #12

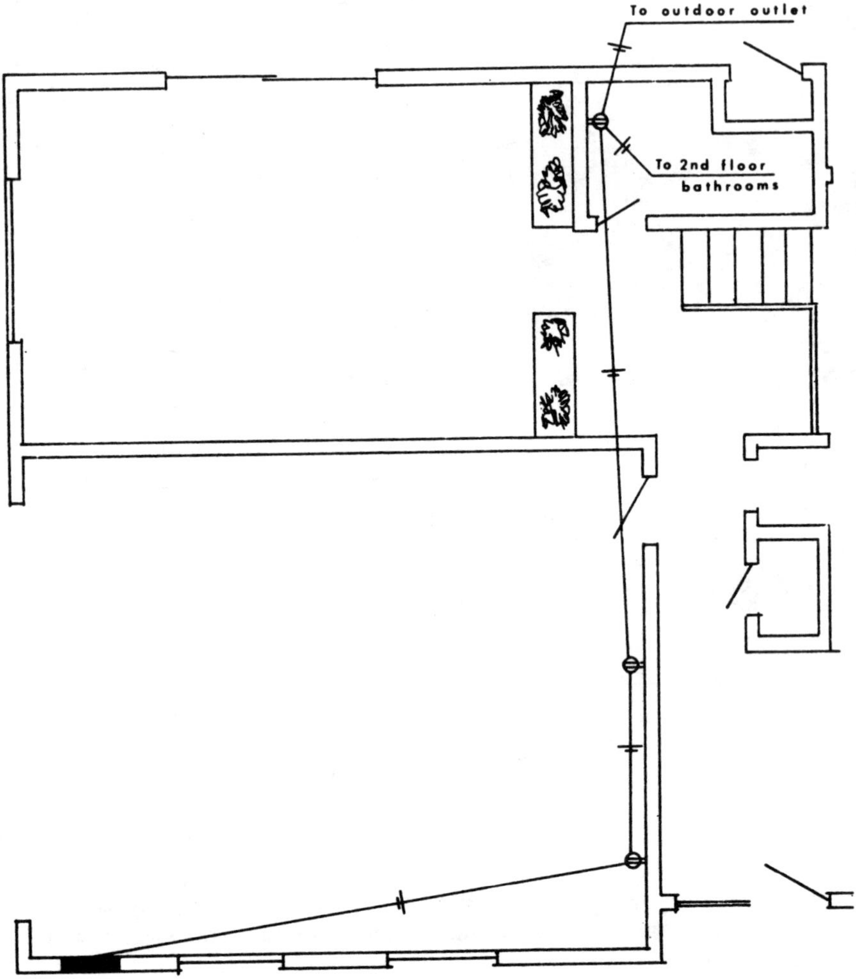

Figure 8-51 Cable plan: GFCI circuit.

copper. Since two of those outlets are within 6' of the sink, they must be GFCI protected [NEC 210-8(a)(6)]. Use of two linked GFCI breakers here will protect the two required outlets plus two more not requiring protection.

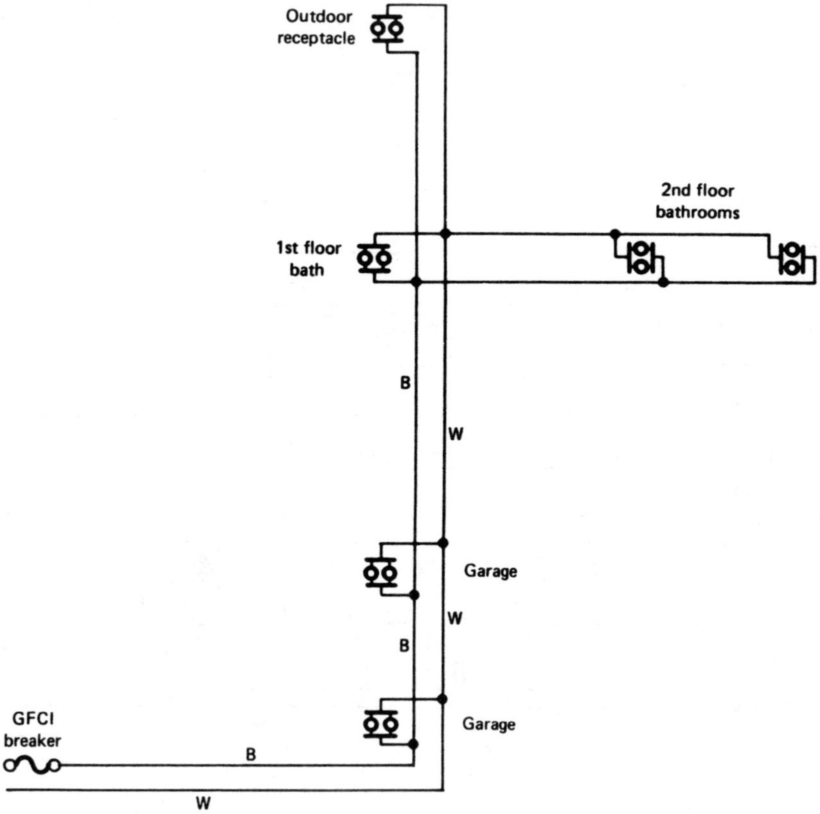

Figure 8-52 Wiring: GFCI circuit.

The refrigerator here has its own circuit at 20 A, to be wired in #12 copper. While according to NEC 210-52(b) the refrigerator may be attached to one of the required small appliance circuits just discussed, quite often, as in this case, it is separated from everything else.

While code specifically permits refrigeration to be attached to a small appliance circuit, the dishwasher, disposal, or trash compactor are not so mentioned. Dishwasher and disposal are sometimes placed together on a single circuit, and sometimes separated as in this case. This home has no trash compactor, but if there were one, it would be on another circuit terminating under the kitchen counter.

In the laundry, separate from the dryer, is the 20 A circuit required by NEC 210-52(b), which supplies the receptacle required by NEC 210-52(f). Again, as we have seen so many times before, here is a single item referred to in two different places in the code, each reference supplying only partial information.

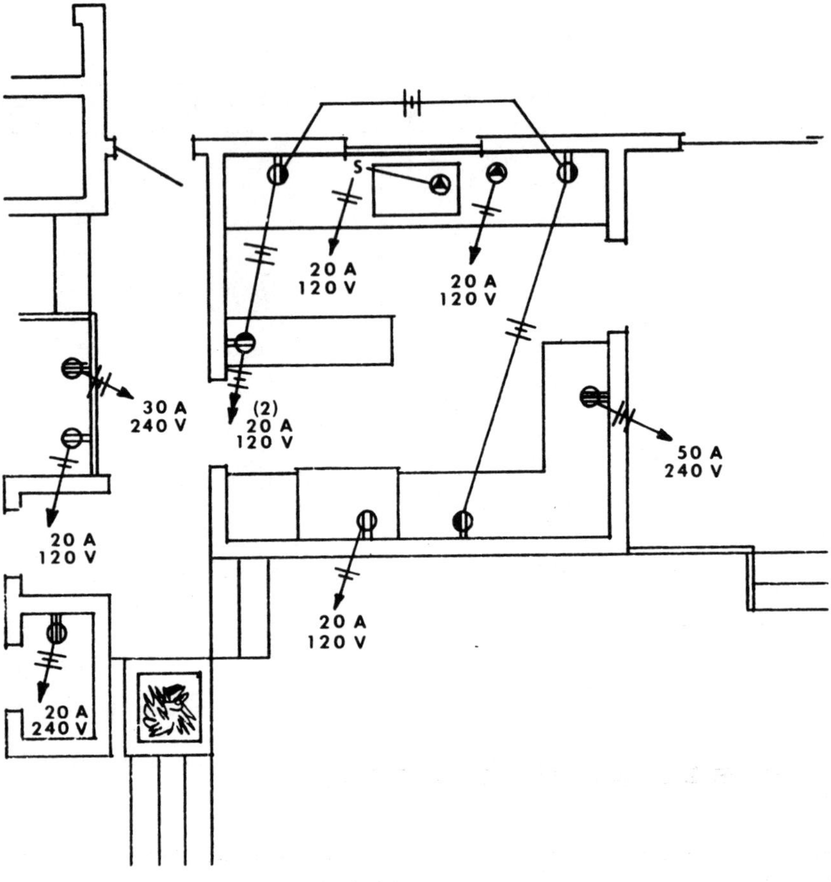

Figure 8-53 Cable plan: appliance circuits.

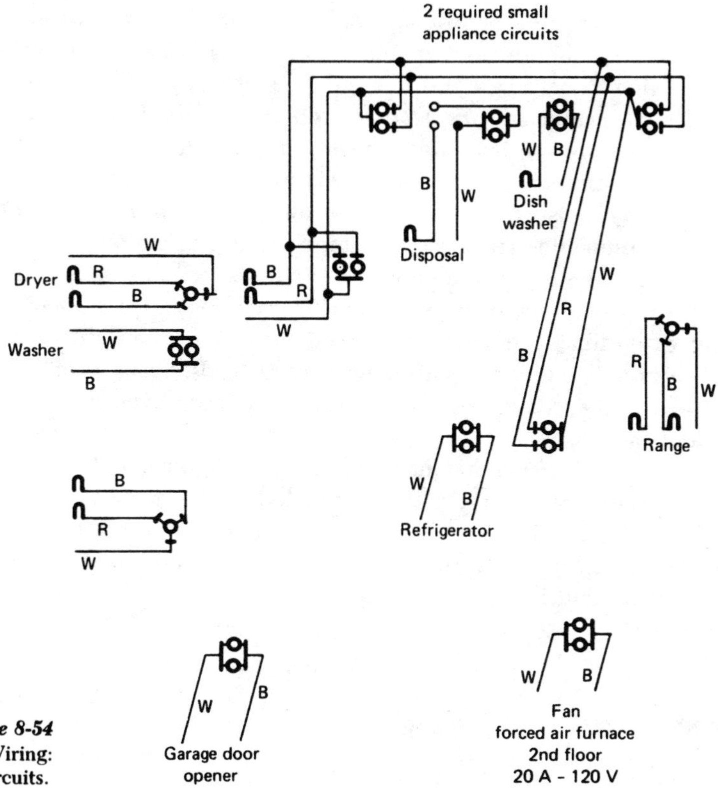

Figure 8-54
Wiring:
appliance circuits.

Service Load Computation

The design of this electrical system is very nearly completed. The architect's floor plans were studied and revised as necessary to conform to code requirements regarding lights, switches, and receptacles. Then circuits, also in keeping with code requirements, were designed to supply these various use points. Cable plans and wiring diagrams, detailing exactly how power is to reach each of those locations, were next prepared. All that remains is to determine how much power must be brought into the building to supply all of the various uses adequately.

This total is ***not*** calculated by adding the capacities of the various circuits. Such a total would be much too large. For example, the computed load on a branch circuit cannot exceed 80 percent of the circuit rating, so the total of the circuit capacities is already going to be 20 percent higher than the computed loads, which themselves are purposely higher than the actual loads.

The writers of the code are also aware that in a dwelling all the ***actual*** loads are never on at once. Try to picture such an absurd situation: every light is on; so are the TVs and the stereo; every burner on the range and oven are at starting load as well as the dryer; all the counter top appliances are on; both water heater and refrigerator just cycled on; the disposal is running; the dishwasher is on drying cycle; someone is ironing; someone else is running an electric lawnmower on the outdoor outlet; all heating units plus any airconditioning equipment are on, and so is the hair dryer and the electric toothbrush. Obviously a total of actual connected loads is also going to be greatly in excess of the real power requirements at any given time. Consequently the code provides several different ***demand factors*** to be applied to different types of loads. The size of the service required is then determined by totalling all loads after applying these demand factors.

As will be seen in Chapter 9, present requirements are such that no new residential service may be installed with a capacity under 100 A at 240 V (NEC 230-79 c). The reason for computing the load on the system in accordance with code instructions is to insure that this capacity is adequate, and if not what capacity is required.

Lighting Load Demand Factors

For computing feeder demand, the lighting load consists of all of the various 15 A lighting and convenience outlet circuits. In this house they are 10. To this total the small appliance circuits and the laundry circuit may be added (NEC 220-16 a & b).

In our present example the lighting load is computed as follows:

Lighting Circuits

Family Room Receptacles		1360 W
Family Room & Other Lights		1360 W
Kitchen, Hall, Foyer Lights		1400 W
Dining Room		1020 W
Living Room		1260 W
Garage & Back Hall		1260 W
Master Bedroom		1300 W
Bathrooms—2nd Floor		1050 W
Bedroom #2		1240 W
Bedroom #1		1400 W
	Subtotal	12650 W

Other

Two Small Appliance Circuits (@ 1500 W each)		3000 W
Laundry Circuit		1500 W
GFCI Circuit (see Note)		1500 W
	Subtotal	6000 W
	Total	18650 W

NOTE In this instance the GFCI protected receptacles are not part of another circuit, but are completely separate. Since the reason for the GFCI protection of these receptacles is that they are intended to power appliances that may be defective, some jurisdictions would classify this as a small appliance circuit. There are six receptacles. If rated as general purpose receptacles at 180 W each, they would total 1080 W. Considered together as a small appliance circuit they are rated at 1500 W. The difference is not great.

Appendix Table 220-11 shows that for dwelling units the first 3000 W of lighting load is to be taken at 100 percent. Then the load from there on up to 120,000 W is figured at 35 percent. Here the total load is 18650 W.

3000 @ 100%		3000 W
15650 @ 35%		5478 W
	Lighting Load Demand	8478 W

Dryer Demand

The load for a dryer (NEC 220-18) shall be 5000 watts (volt-amperes), or the nameplate rating, whichever is larger. Demand factors for dryers are listed in Appendix Table 220-18. A single dryer should be computed at 100 percent, which in this case will be 5000 watts.

Range, Oven, Cooktops and Cooking Units Rated Over 1750 Watts Demands

Appendix Table 220-19 dealing with these units can be a bit confusing. However, an average residential oven-range combination will be rated at approximately 12 kW (12,000 watts), making the demand for that unit 8 kW (8,000 watts) as given in Column A. If the unit has a nameplate rating over 12 kW, then the demand given in Column A must be increased by 5 percent for each kW or major fraction thereof over 12 kW.

Suppose a large oven-range combination carries a nameplate rating of 14,750 watts, or 14.75 kW. The demand for this unit would be figured as 8,000 watts for the first 12,000 watts (12 kw) of nameplate rating, plus 15 percent of that 8,000 for the additional 2,750 watts of nameplate rating above 12,000.

Column A demand for 12 kW	8 kW or 8000 W
Plus 15% of 8 kW	1.2 kW or 1200 W
Total demand	9.2 kW or 9200 W

When, as often happens, a counter mounted cook-top and one or two wall mounted ovens are located in the same room, and supplied by the same branch circuit, the demand shall be computed by adding the nameplate ratings of the individual appliances together, and then treating the total as though it were a single unit.

In this house we shall assume an average oven-range combination rated at just under 12 kW producing a demand, per Table 220-19 Column A, of 8 kW or 8000 watts.

Fixed Appliances

In addition to range and dryer, appliances served by small appliance circuits, the house contains other electrically powered appliances. These might be a water heater, disposal, dishwasher, trash compactor, attic fan, garage door opener, or exhaust fan. For all of these items the demand is determined by adding the nameplate ratings together (NEC 220-17) and taking 75 percent of the total. This applies to four or more fixed appliances. In this house there are:

Water heater		3200 W
Disposal		764 W
Dishwasher		1000 W
Garage door opener		720 W
	Total	5684 W
	75% of 5685 W	4263 W

Fixed Electric Space Heating, Airconditioning

Fixed space heating means permanently installed electric baseboards, radiant ceilings, or other non-portable heating elements. These are all to be computed at 100 percent of their rated connected loads per NEC 220-15.

NEC 220-21 states that when two dissimilar loads exist that will not be in use at the same time the smaller of the two may be omitted from load calculations. Thus when a building contains both electric heating and airconditioning, the smaller of the two may be omitted. If there is airconditioning, and oil, gas, or coal heating, then the airconditioning is figured at 100 percent and the smaller electrical requirement of the heating system is omitted.

In this house neither electrical heating nor airconditioning are included. The only electrical requirement for the heating system is the blower for the forced air gas furnace. This will require 1100 watts and will be figured at 100 percent.

Sizing Service Entrance

The various individual demands on the service entrance have now been calculated. They may now be added in order to determine the amperage that this service will be required to deliver.

Lighting Load	8478 Watts
Dryer	5000 Watts
Range	8000 Watts
Fixed Appliances	4263 Watts
Furnace	1100 Watts
Total load on system	26,841 Watts

$$\text{Amperage} = \frac{\text{Watts}}{\text{Volts}} = \frac{26{,}841}{240} = 111.84 \text{ Amperes}$$

The required minimum 100 ampere service will not suffice after all. This service will have to be increased to 125 amperes. The size for the service entrance wires will be #2 copper in accordance with Table 310-15(b)(6).

CHAPTER 9

THE SERVICE ENTRANCE

The service entrance is the section of the electrical system through which the electrical supply to the building is connected to the distribution network inside. Normally only one service entrance is allowed per building (NEC 230-2). Specific exceptions are allowed to this one service rule, but they are seldom encountered in the types of small buildings discussed in this work.

In addition to the service entrance limitation, there are also restrictions dealing with situations where one service is used to supply two or more buildings (NEC 230-3). It is permissible to run wiring from the service entrance on building #1 through the interior of that building, then out and across to building #2 (Figure 9-1), when both buildings are occupied or managed by the same people. If, for example, building #1 is a single family residence, and #2 is its garage, both are ***occupied*** by the same people; thus the supply for #2 can go through #1.

Let us suppose that building #3 is either a business or a residence, and #4 is also one or the other, but under different ownership. In this case each one should be supplied by a different set of incoming wires. If for some reason that is not convenient or feasible, then building #4 can be fed from building #3, only if the wires to #4 remain completely ***outside*** of #3. They can be run along a wall as shown, but must remain on the outside of that wall all the way to the point where they jump to #4.

Overhead Service Entrance

When the incoming wires bringing the electrical supply to a building pass above ground from a power pole to the building (Figure 9-2), this is called an ***overhead service.*** The wires coming to the building are called the ***service drop.*** These wires will terminate in connections to another set of

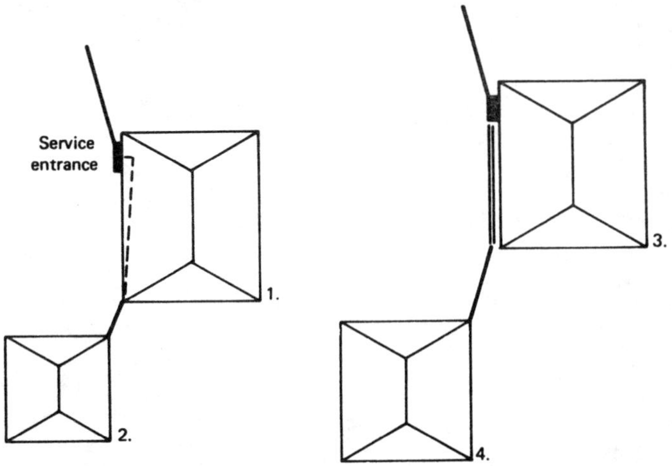

Figure 9–1
As per NEC 230-3: (a) building #2 may be fed from building #1 when under the same management by routing the feed lines *inside* building #1. (b) When under different management, building #4 may be fed from building #3 only when the feeders run *outside* building #3.

wires that hang down from the service head. This piece is alternately termed the weather head. The wires hanging from the service head are termed ***service entrance wires,*** and must be marked as approved for this purpose [NEC 310-11(b)(1)]. The service head itself must be a raintight device (NEC 230-54), and is to stand above the point at which the service drop wires attach to the service entrance wires (Figure 9-2). The service entrance wires ***must***, therefore, extend down from the head.

The point of attachment of the service drop can never be less than 10′ above grade (NEC 230-26). It may be more if necessary to conform to the additional clearance requirements dictated by NEC 230-24 and illustrated in Figure 9-3.

When the service drop terminates at a pipe mast that extends above the roof of the building the NEC requirements can become just a bit confusing. NEC 230-24a states at the outset that a service drop must have a vertical clearance of at least 8′ from all points of a roof over which it passes. That sounds very simple and definite (Figure 9-4), but it is quickly followed by ***Exception No. 2,*** which allows a reduction in clearance to 3′, if the voltage does not exceed 300V, and the roof has a slope of at least 4″ in 12″. That, too sounds simple and definite, but read on. Exception No. 3 follows Exception No. 2. This exception permits a further reduction in clearance to 18″ when the voltage is not over 300V and the mast is located so that no more than 4′ of service drop conductors pass over the roof. This time nothing is mentioned about the roof slope.

If making sense of these seemingly contradictory provisions appears somewhere between difficult and impossible, be of good cheer. Most overhead service drops that terminate on a roof do so within the 4′ limit covered by Exception No. 3. For this reason most roof top masts need to provide only an 18″ clearance for the service drop wires. Note that the 18″ mentioned is the required clearance for the ***wires*** making up the service

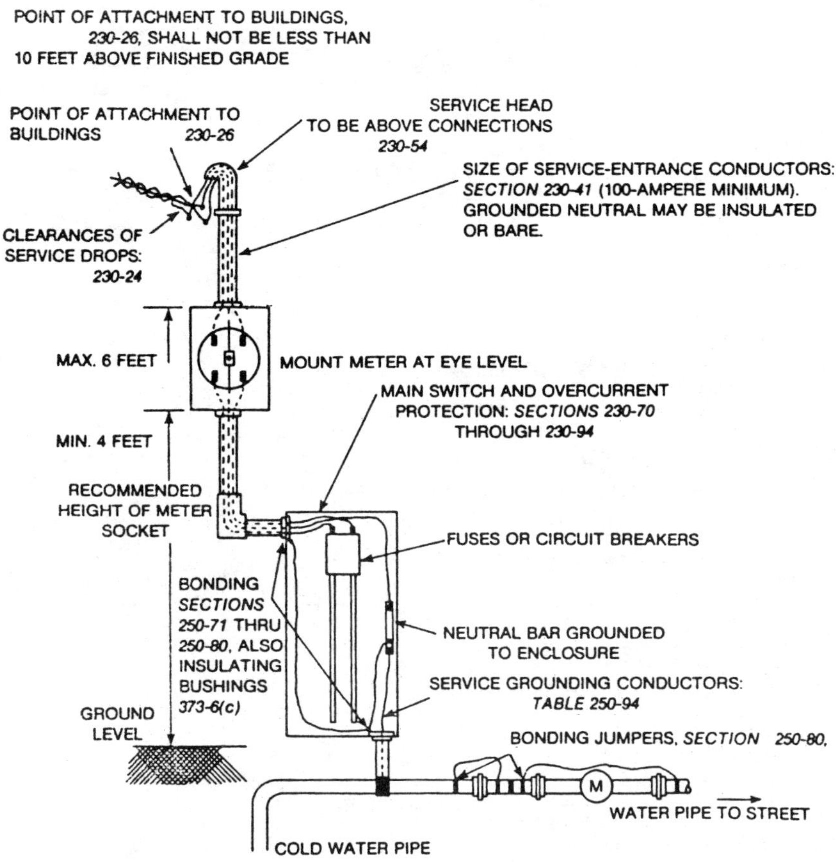

Figure 9–2 Typical overhead service installation with references to Code provisions relating to such installations. *Reprinted from* Electrical Wiring Residential, *Delmar Publishing Company.*

drop. The mast and weatherhead are going to need to be enough higher than that to allow for the service entrance wires to come down from the weatherhead to make connection to the drop wires while they maintain that 18″.

When an overhead service drop terminates on the side of a building instead of on the roof (Figure 9-5), the attachment point still has to be at least 10′ above grade, but it must also be at least 3′ horizontally from any windows, porches, or fire escapes. If it comes in above the top of a

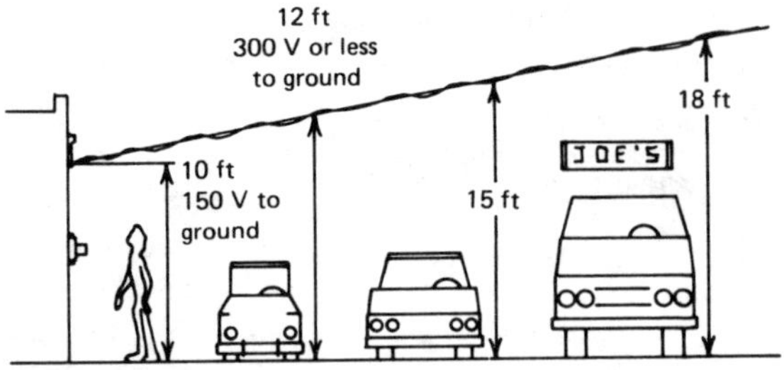

Figure 9–3
Minimum clearances under service drop required by NEC 230-24.

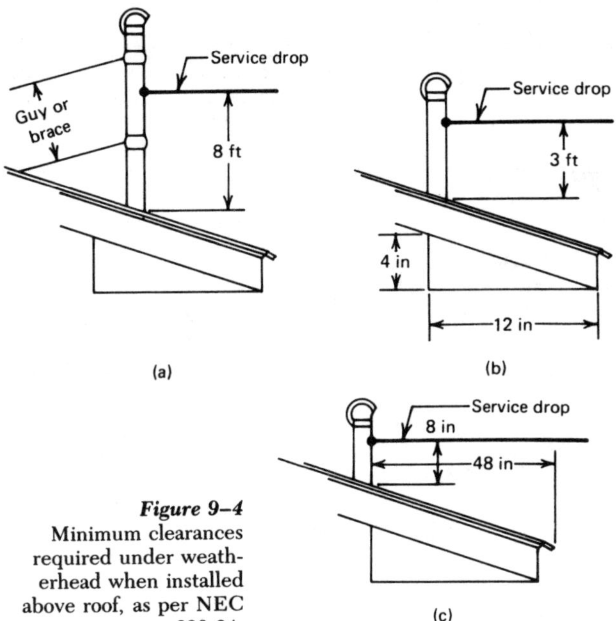

Figure 9–4
Minimum clearances required under weatherhead when installed above roof, as per NEC 230-24.

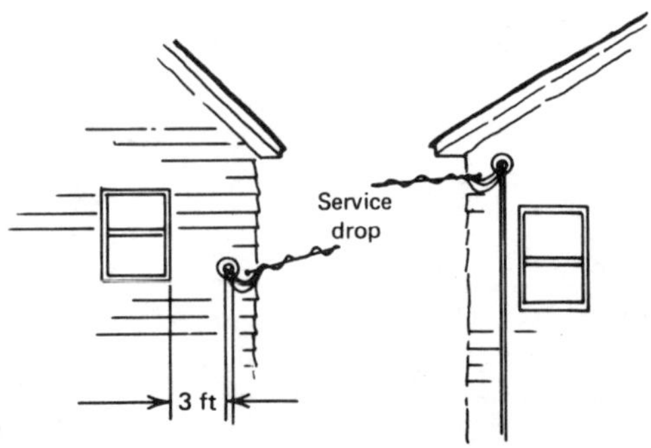

Figure 9–5
When the weatherhead is on the side of a building, it must be three feet away from any windows horizontally, or it must be above the top of a window, as per NEC 230-24.

window, the 3′ horizontal clearance is not necessary. The reason for these requirements is to insure that service drop and service entrance wires are safely out of reach of any people.

From the point where the service drop is attached, power is transmitted through the service entrance wires to the electric meter. The utility company owns this meter. It is placed in the meter box and sealed by it; however, the meter box itself is supplied and owned by the building owner. It is normally mounted on an outside wall of the building in order to be visible to the meter reader as he might not have access to the interior. Also for his convenience, it should be mounted at a height above grade (Figure 9-2), somewhere between 4′ and 6′.

The service entrance conductors leading from the service head down to the meter must be correctly sized in accordance with the power requirements of the service being supplied. The size of the service is determined by load computations as discussed in the last chapter. In the case of a one family dwelling a minimum supply of 100A is now required (NEC 230-79c).

In wiring the service entrance for a dwelling, the entrance conductors are permitted to be one size smaller than the rated amperage of the service according to the ampacity ratings shown in Table 310-15(b)(6). Sizes permitted for residential occupancies only are as follows:

Copper	*Al & Cu clad Al*	*Service Amperage*
4	2	100
3	1	110
2	1/0	125
1	2/0	150
1/0	3/0	175
2/0	4/0	200

From the meter the power goes to the required ***service disconnecting means*** (NEC 230-70). Here again we encounter one of those vague parts of the code. Section 230-70 clearly requires that

> Means shall be provided to disconnect all conductors in a building or other structure from the service entrance conductors.

As that section continues one tends to accept it as calling for A means which would translate to a single main switch or main breaker.

However, immediately following, Section 230-71 very clearly and specifically permits the ***service disconnecting means*** to consist of up to six switches or breakers. Although six are permitted, in normal field practice, the majority of small building electrical services contain a single main disconnect, which most commonly is a breaker. When it is not a breaker, it will be a fused knife switch, or perhaps a pull-out twin fuse holder.

The service disconnecting means may or may not be incorporated in the same box in which the electric meter is mounted. In the warmer and dryer areas of the country, a single combination outdoor box housing the meter, the main breaker/disconnect, and many if not all of the branch circuit breakers is commonly used. In the colder and damper areas a more successful arrangement is an outdoor meter box. From there the supply goes to a separate indoor main breaker and distribution box. In any case the separate indoor distribution box is preferable in terms of protection from the elements as well safety from meddlesome or mischievous tampering. Remember, locking an outdoor box to prevent tampering is not a good idea. A breaker box contains a group of safety devices which should be accessible at all times. Should it be necessary to get in there in a hurry the time spent finding the key and unlocking it could be important.

Underground Services

In recent years the installation of underground services has become more common. In fact, in some areas all new services must be underground. Typical underground service entrances are diagrammed in Figure 9-6. This method involves a somewhat different distribution system on the part of the utility company. When the overhead method is used, the utility's final step-down transformer is mounted on a power pole from which it reaches the building via a service drop. The underground method permits the power company to mount the final step-down transformer on or under the ground. Power is then delivered to the building through an underground ***service lateral*** that runs from the transformer to the building, then up and into the meter box.

When the transformer is mounted at ground level, it stands on a concrete slab (Figure 9-6a) and is protected by a heavy steel casing. In other cases it may be placed in concrete underground vault (Figure 9-6b). Either way in the event of stormy weather conditions that endanger power poles, the transformer on or under the ground is in a far safer place.

The service laterals from the transformer may terminate either inside or outside the building wall (Figure 9-7). In either case the building must supply a meter box in which the utility can mount its meter and it should be the same 4′ to 6′ above grade that it was for the overhead service, and for the same reason.

From the meter box through the remainder of the building electrical system there is no difference between overhead and underground services.

The advantages of the underground service entrance are first that it is less conspicuous and looks better. The wires dangling in air, the service head, and the mast all together do not exactly constitute an aesthetic triumph. Not only are the transformers safer from bad weather, so are the

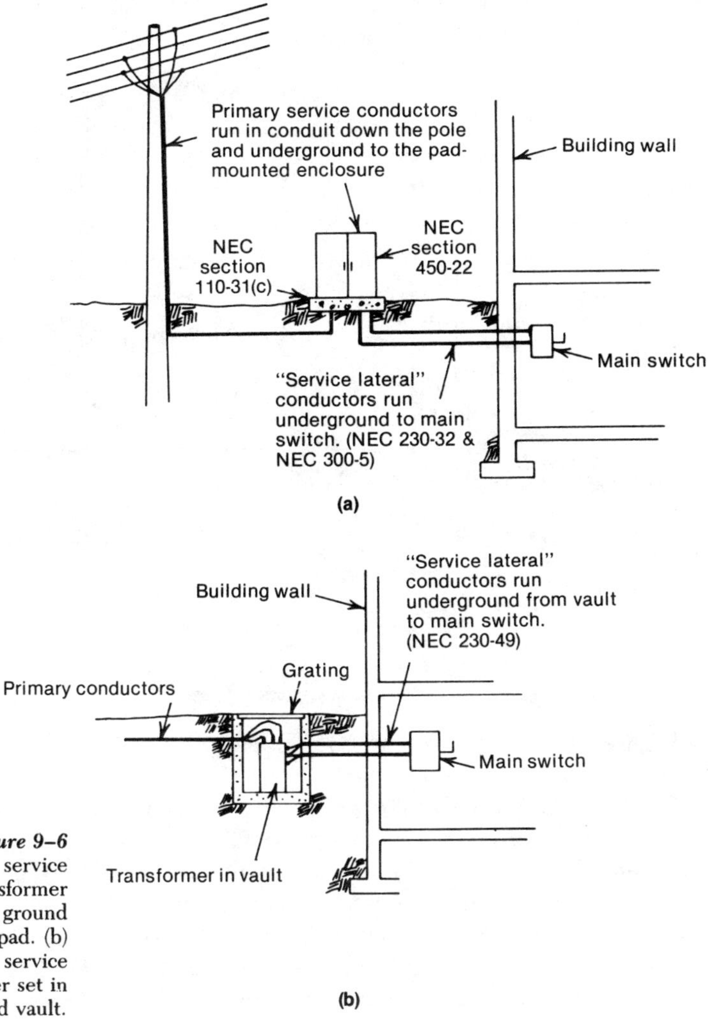

Figure 9–6
(a) Underground service with transformer mounted on the ground on a concrete pad. (b) Underground service with transformer set in underground vault.

wires. The major disadvantage is that when it is necessary to increase the capacity of the building's electrical system, it is much simpler to replace an overhead service drop than to dig up and replace a buried service lateral.

Meter Box

Regardless of which way the electrical supply reaches a building—overhead or underground—it will be metered when it gets there. Consequently the service entrance wires will terminate at a meter socket

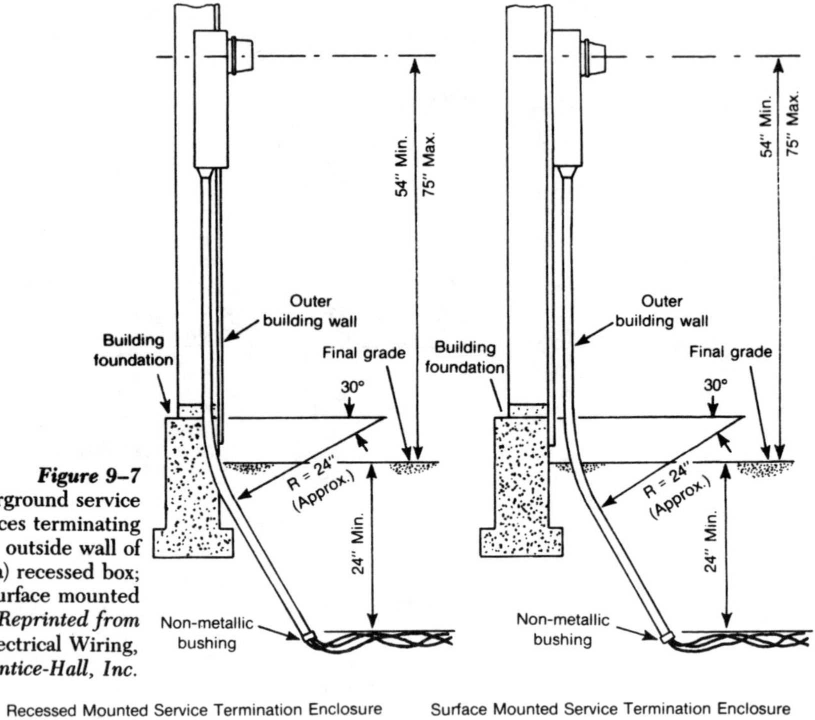

Figure 9–7 **Underground service entrances terminating on outside wall of house: (a) recessed box; (b) surface mounted box.** *Reprinted from* Electrical Wiring, *Prentice-Hall, Inc.*

(Figure 9-8). This socket will be mounted in a weatherproof metal box suitable for outdoor mounting.

For a 120V-240V, single phase service the service drop or service lateral will consist of three wires. Two are technically termed ***un-grounded***—in lay terms ***hot.*** These will each read 120V above ground, and 240V apart from each other. The third wire is the grounded common return. When these are attached to the service entrance wires the one used as the common will have white insulation, or more often it will be identified with white tape at both ends.

When the service entrance wires reach the meter box the two hot leads go to the top, or line, terminals on the meter block. The common marked by white insulation or tape goes to the grounded terminal. From the bottom, or load, terminals on the meter block two more leads connect the now metered supply to the required service disconnecting means discussed previously. Most of the time this is going to be a single main breaker (occasionally a fused knife switch) that cuts power from everything on the load side of the meter.

When the meter box (Figure 9-9) is a part of a combination unit containing the meter and the main breaker and a number of the branch circuit breakers, the meter is internally wired to the main breaker which, in turn, connects directly to two phase buss bars from which the individual branch circuit breakers draw their power. When the meter box is

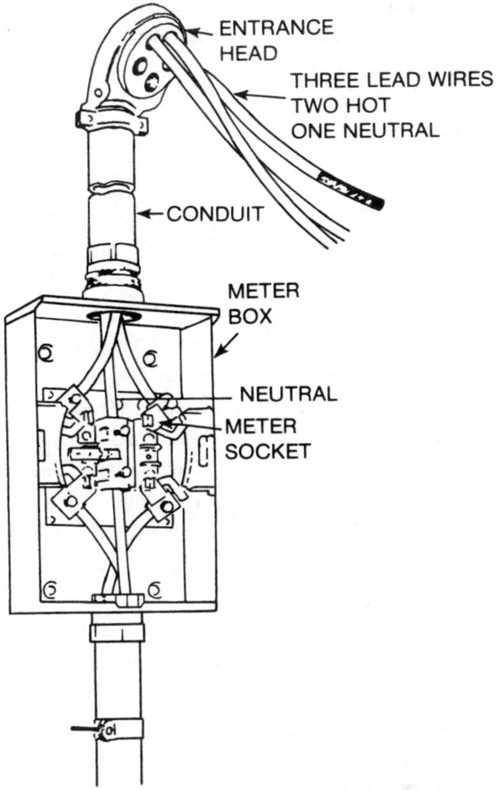

Figure 9–8
Typical installation of outdoor meter box with feeders going on to indoor distribution box.

Figure 9–9
Typical combination outdoor meter box and distribution box.

separate from main and branch circuit breakers (Figure 9-10), it will be connected to the distribution panel usually by means of a metal conduit containing the two hot leads and the common.

Breaker Boxes

The size of the box required to enclose the branch circuit breakers must be determined on the basis of the number of breakers required. This, in turn, was determined during the planning of the system discussed in Chapter 9. However, there are a couple general code limitations that should be remembered in this connection.

One family residences having six or more two-wire branch circuits must be supplied by at least a 100A service (NEC 230-79c). The standard 100A service panel contains 12 breaker spaces in addition to the main breaker. Thus whether that many breakers are needed or not, 12 spaces will be the minimum number in a new residential panel. At the other end of the spectrum the maximum number of breakers that may be mounted in a single distribution panel, whether for residential or other use, is specifically limited to 42 (NEC 384-15). This limitation is not a cause of everyday difficulties for the average electrician.

All wiring installed in buildings shall be protected against overcurrent by breakers or fuses appropriate to the amperage for which such wires and their insulation are rated (NEC 240-3). This was discussed in Chapter 4, and NEC Table 310-16 is included in the Appendix for reference. However, nowhere is there any requirement stating that all of those required overcurrent devices in a single building be centralized in a single panelbox. Nor is there any requirement that all the panelboxes containing overcurrent devices be located together. This permits the use of as many sub-panels as may be useful, convenient, or economical to be placed about the building in such a way as to make them handy to the areas they protect.

Grounding

All AC building wiring circuits and systems must be grounded [NEC 250-20(b)]. Each branch circuit, as we shall see in Chapter 11, must contain a continuous ground return path throughout. That branch circuit path terminates at the neutral bar (Figure 9-11) in the main panel box. From there the building electrical system is tied to true ground by means of an appropriately sized grounding conductor.

This conductor must be made of either copper, aluminum, or copper clad aluminum (NEC 250-62). It has to be resistant to any corrosive conditions around it, or be suitably protected from any such conditions.

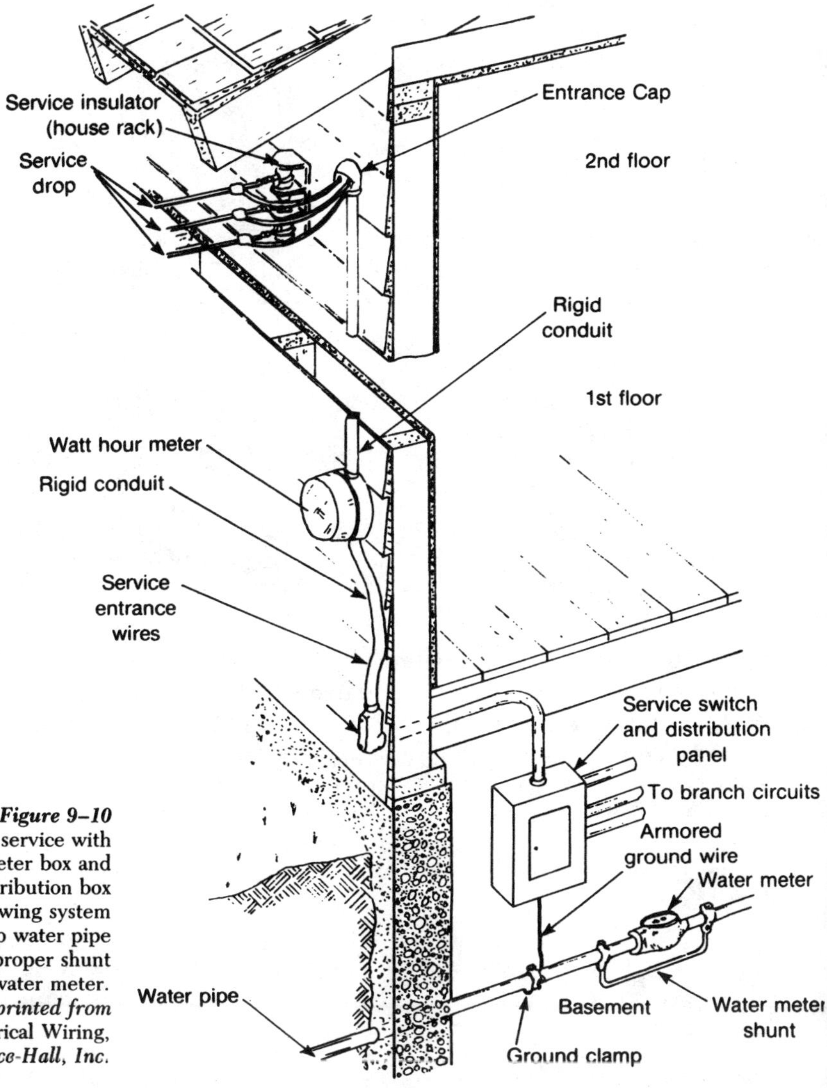

Figure 9–10
Overhead service with outdoor meter box and indoor distribution box showing system grounded to water pipe and proper shunt around water meter. *Reprinted from* Electrical Wiring, *Prentice-Hall, Inc.*

It makes no difference whether it is solid or stranded wire, and it may be either insulated or bare; however, it ***must*** be in ***one continuous length***. There may be no splices anywhere between the connection to the ground buss in the panel box and the point where it is attached to the grounding electrode system. At that point, the attachment must be made using a pressure connection or clamp (NEC 250-70). It may not be soldered.

Grounding electrode conductors smaller than #6 shall be protected against physical damage by rigid metal conduit, intermediate metal conduit, rigid non-metallic conduit, EMT, or cable armor [NEC 250-64(a)]. The flexible cable armor is by far the most commonly used material for

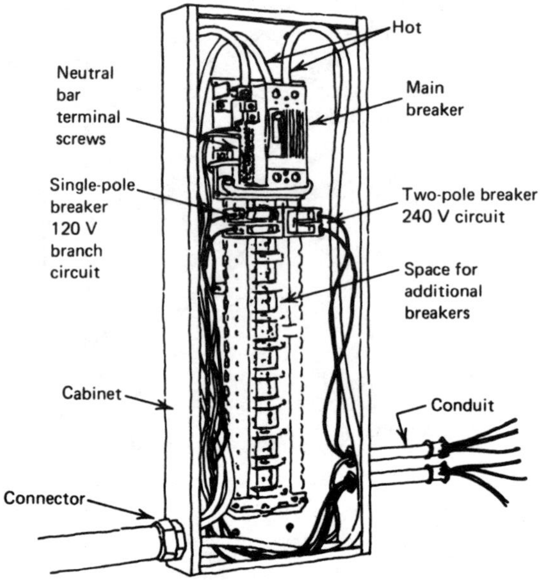

Figure 9–11
Typical breaker box—both common (white circuit wire), and grounds all terminate at neutral bar at upper left of box.

this purpose. When the grounding electrode conductor is larger than #6, it must be similarly protected only if exposed to physical damage. In any case, whatever is used must be secured to the building construction in accordance with the standard requirements for the material used. These requirements are discussed in Chapter 5.

Sizing

Grounding electrode conductors should be sized in accordance with Table 250-66 in the Appendix. Regardless of size of the service, the grounding conductor may never be smaller than #8 AWG, if copper is used, or #6 if aluminum is used.

Grounding Electrode System

The code requirement for a grounding electrode ***system*** is comparatively new(NEC 250-50). It first appeared in the 1978 code. Prior to that time a single system ground sufficed. That ground was commonly the underground metal water pipe coming into the building.

A single grounding electrode is no longer considered adequate. Present requirements call for the following items, if available, to be bonded together to form the grounding electrode system:

a. Metal underground water pipe in direct contact with earth for a distance of at least 10'.

b. Metal frame of the building if it has such a frame and that frame is grounded.

c. A concrete encased electrode which can be either a minimum of 20' of at least 1/2" steel foundation reinforcing bar, or at least 20' of bare copper wire no smaller that #4 AWG embedded in the foundation concrete.

d. A ground ring consisting of at least 20' of bare wire no smaller than #2 AWG buried in the ground no less that 2 1/2' deep.

If b and c above are not available at the job, dig a trench to install d. However, circumstances could be such that digging that trench is not possible. If there is nothing but the underground metal water pipe to use for grounding the system, supplement the ground with the following other electrodes (NEC 250-52):

a. Other local metal underground systems or structures such as piping or tanks. Be sure that the underground metal you use is not protected by paint or a non-metallic material that isolates it from the earth. Isolation makes it useless as a grounding electrode.

b. Rod or pipe – either one must be a minimum of 8' long. If pipe is used it has to be at least 3/4" trade size. Steel pipe must be galvanized or coated with other metallic protection. If rods are used, ferrous metals must be at least 5/8 " diameter while non-ferrous rods need to be only 1/2". Either pipe or rod may be driven (Figure 9-12a) anywhere in the range from straight down to 45° off vertical. If this cannot be done, the alternative is to bury the entire 8' length a minimum of 2 1/2" deep (Figure 9-12b).

c. A plate electrode may also be used as an additional alternative. It will have to be at least 2 square feet in area. If ferrous metal is used it needs to be 1/4" thick, or if non-ferrous metal, 0.06" thick will suffice.

The use of a metal underground gas piping system is now specifically forbidden. Under the Plumbing Code such systems require a plastic wrapping that is precisely the kind of non-conductive coating that makes a gas pipe useless as a grounding electrode.

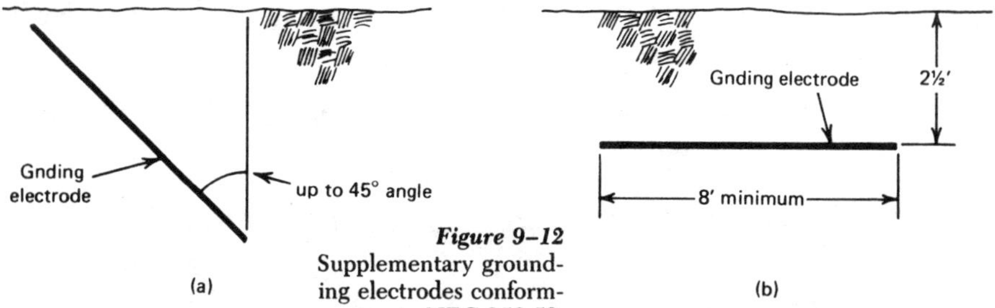

Figure 9–12
Supplementary grounding electrodes conforming to NEC 250-52.

In the case of the plate electrode the code says nothing about galvanizing or other protection of ferrous metal nor does it specify any burial depth. Since ferrous rod or pipe must be galvanized or otherwise protected it seems likely that an inspector will insist on a plate electrode being similarly protected. It also would seem safe to bury it at least 2½′ the same as pipe, rod, or a #2 AWG copper wire ground ring.

The basic emphasis of the 1978 requirements is that the traditional method of grounding the building electrical svstem via a single grounding electrode is now deemed insufficient. At least two, and preferably more, grounding electrodes bonded together are required to form a more effective grounding electrode system. Such additional electrodes shall be located no less than 6' apart.

Overcurrent Protective Devices

Overcurrent protection is a required part of every branch circuit in every electrical installation (NEC 210-20). This protection is provided either by fuses or by circuit breakers most of which are mounted in the distribution boxes that form a major part of the service entrance equipment described above. The purpose of the required overcurrent protection is to prevent damage to the conductors or other parts of the electrical system due to the overheating that would result from an overcurrent. Overcurrent refers to the condition resulting from excessive current drawn through a circuit designed to carry only a certain amperage. This situation occurs for one of two reasons: overloading or an unintentional and unexpected direct connection of power to ground through no or little resistance. The latter condition is called a ***short.***

As an illustration of overloading, in Chapter 1 during the discussion of Watt's Law, an example was given of a case in which three appliances were connected to a 20 ampere appliance circuit. A 20 ampere circuit at 120 volts can deliver 2400 watts of power (I × E = P). The three appliances in the example required 2800 watts. To deliver 2800 watts at 120 volts the current will have to be 23.33 amperes. This is 3.33 amperes more than the wire insulation, the switches, or the receptacles were intended to carry. Any or all of them could be damaged by the excessive current. In this case the required overcurrent protective device would shut off the circuit before damage could occur.

Cases of overloading generally involve relatively small excesses of amperage over and above the rated capacity of the circuit. By contrast shorts involve gigantic excesses of amperage above the rated capacity of a circuit. Current flow can jump up to hundreds or even thousands of amperes at the first peak of the first half cycle after the short occurs (Figure 9-13). Consequently, overcurrent protective devices must react with incredible speed to avert serious damage to system components since that first peak of the first half cycle occurs in 0.00416 seconds.

Overcurrent protective devices are of two basic types: fuses and circuit breakers. Fuses function by the melting of a metal link in response to overcurrent. Since it has no moving parts a fuse can open the circuit in less than one-half cycle. Breakers, dependent on mechanical movement, require about one and a half cycles. The fuse has a considerable advantage in terms of speed of response, but has the disadvantage that once blown, it must be replaced. In the case of some cartridge fuses the fusible link can be replaced and the cartridge reused. The breaker reacts more slowly, but when tripped needs only to be reset by moving a switch handle.

Fuses

In older electrical installations around 40 to 50 years old and older, the standard overcurrent protective device was the Edison base plug fuse (Figure 9-14). This fuse has a threaded base like an incandescent light bulb. The top is transparent, leaving the fusible metal strip visible for easy inspection. When this type fuse is blown the transparent top will be considerably discolored, instantly identifying the bad one.

Plug fuses are available in 15, 20, and 30 ampere sizes, all of which fit into the same socket. This has proven to be a very bad feature, because it allows the uninformed to overload a 15A circuit and blow a few fuses.

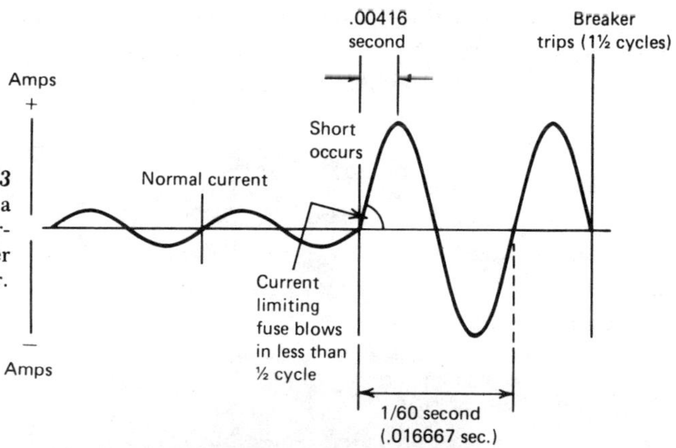

Figure 9–13
In case a short occurs, a fuse will open the circuit considerably faster than a breaker.

Figure 9–14
Edison base plug fuse.

Then without removing the overload, he stops the fuse trouble by simply using a larger fuse. The larger fuse will pass more current than the circuit components can handle safely. Many disastrous fires have resulted from this kind of substitution.

Consequently the ***S type*** fuse was developed (Figure 9-15). S type fuses consist of two parts: the fuse and the threaded adapter into which the fuse is screwed and which in turn screws into the standard fuse socket. Each fuse size has an adapter with a different inside thread, and all adapters are designed to be non-removable once placed in a fuse socket. Thus once a 15A, or a 20A, or a 30A adapter has been put into a socket only that size fuse will be accepted. In many jurisdictions inspection departments require that all old type plug fuses be replaced with S types in order to stop people from mis-sizing their fuses.

As has been noted earlier, many electrical devices draw a considerably higher current when starting than when operating. This starting current as a momentary surge presents no danger to the electrical system, but due to the very rapid response characteristic of fuses, it will often blow a fuse. To allow a brief time for starting surges to occur without blowing fuses, a ***time lag*** or ***slow-blow*** fuse can be used (Figure 9-16). The time lag fuse is a dual element device that will withstand the moderate overcurrents of starting surges, but will blow if the large overcurrent caused by a short occurs.

To fuse circuits over 30A one of the types of cartridge fuses must be used. There are two types: the ferrule contact and the knife blade contact

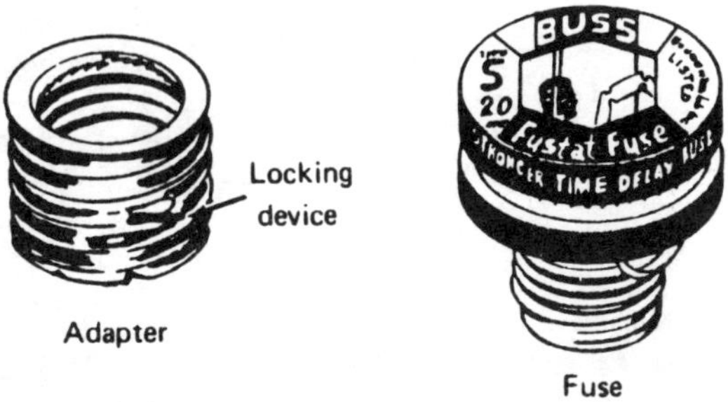

Figure 9–15
S type fuse. Standard Edison base plug fuses should all be replaced with S type fuses to prevent accidental over-fusing (replacement of a blown fuse with one of too high a rating).

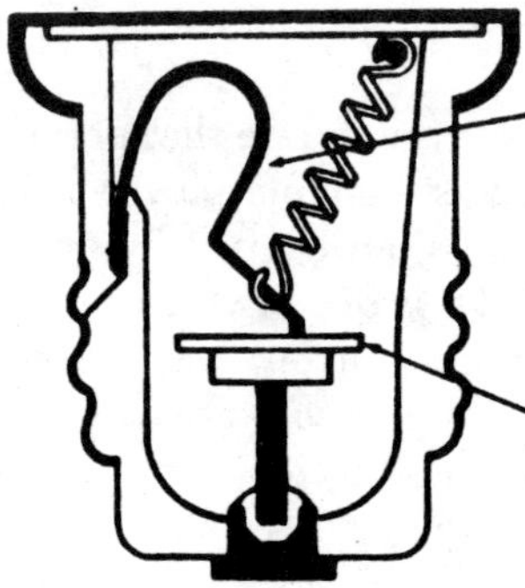

Figure 9–16
Time-lag plug fuse. This type will stand temporary overloads such as the momentarily heavy load of a motor when it starts.

(Figure 9-17). Either fits into the appropriate fuse clip. These are inserted and removed with a fuse puller (Figure 3-14). To put your fingers or metal pliers into a cartridge fuse holder is unwise.

Blown cartridge fuses cannot be visually detected the way bad plug fuses can. They must be tested. A suspected fuse can be pulled and tested with an ohmmeter. If there is continuity through the fuse, it is good. It can also be tested in the circuit with the voltmeter. With the circuit on, put one probe at each end of the fuse. If the line voltage shows on the meter, the fuse is blown. If the meter reads O across the two ends of the fuse, but reads 120V (or whatever the line voltage is) from either end to ground, the fuse is good.

Most cartridge fuses when blown must be replaced. However, one type (Figure 9-18) has removable ends allowing access so that the fusible link inside can be replaced. While cartridge length and diameter increase with amperage there may be more than one amperage at any particular cylinder size. This, again as with plug fuses, opens the possibility of replacing a blown fuse with one the wrong size. Be particularly careful to replace a blown fuse with another of the proper size.

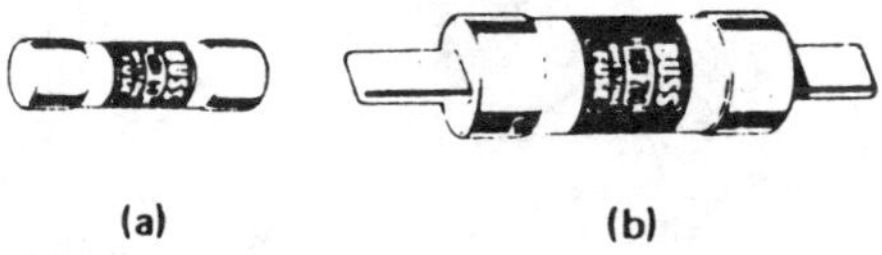

Figure 9–17
Cartridge fuses: (a) Ferrule type; (b) knife blade type. These may be found in main system disconnects and also in the fused disconnects for large equipment such as a central air conditioning system.

Figure 9–18
Replaceable link cartridge fuse.

Circuit Breakers

The circuit breaker is a device that provides overcurrent protection, and also may be used as a simple power switch. It has the great advantage over fuses in that, when tripped by overcurrent, it can be simply reset almost immediately, ***providing*** the cause of the overcurrent has first been corrected. If the overcurrent was caused by a short that has not been corrected, the breaker absolutely will not hold in the ***ON*** position. If the overcurrent was caused by an overload that has not been corrected, it will usually go back to the ***ON*** position after a few minutes' cooling, but a few minutes later it will trip again.

The heart of a circuit breaker (Figure 9-19) is the bimetallic strip, which is made of two different metals fused together. When heated these two metals will not expand at the same rate which will cause the strip to bend. This strip acts as a latch holding two spring loaded contacts together. When a current greater than the breaker was intended to carry flows, the bimetal strip heats, bends, and releases the latch. This separates the contacts and interrupts the current flow.

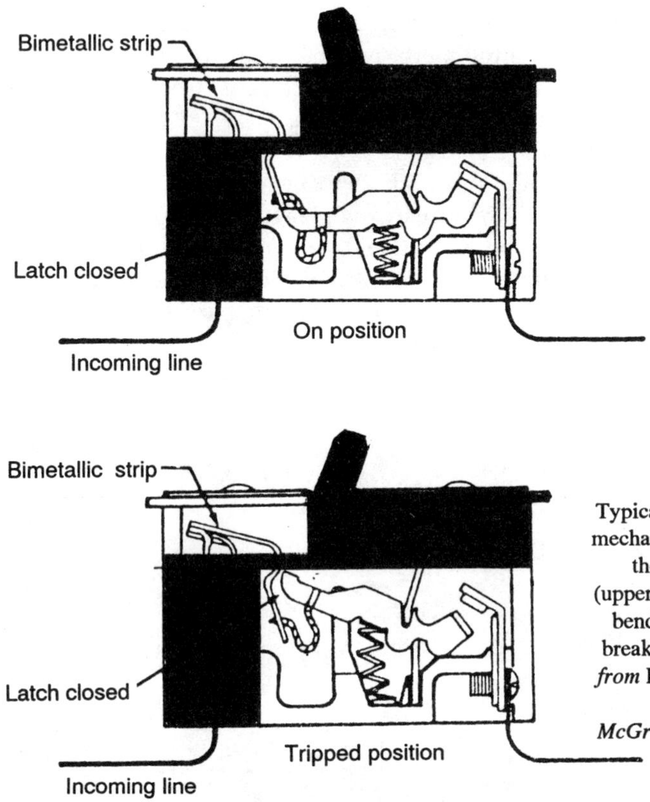

Figure 9-19 Typical circuit breaker mechanism. Heating of the bimetallic strip (upper left) causes it to bend, which trips the breaker off. *Reprinted from* Electrical Wiring Fundamentals, *McGraw-Hill Publishing Company.*

The time required for the bimetal strip to heat allows it to act in much the same way as a time lag fuse; it will carry momentary starting overloads without tripping. The handle on the breaker also allows it to be used when desired as a switch providing an easy way to shut off a single circuit for repairs without disturbing the others.

CHAPTER 10

ROUGH WIRING

With plans and calculations completed, and the building framework in place, rough wiring can be started based on the information given on cable plans. Sequence of the work varies depending on the materials being used; therefore, cable, flexible metal conduit, EMT (thinwall conduit), and surface raceways (Wiremold) will each be discussed separately.

Concealed rough wiring as described below in new work is normally done with at least one side of each wall open. This gives the electrician access to frame members and the internal voids between them. Boxes can then be conveniently mounted, and box to box wire runs placed and secured. When the work is completed, the open wall also allows the electrical inspector easily to determine that cabling, greenfield, or conduit work has been correctly installed and fastened in accordance with all NEC and local code requirements.

Nonmetallic Cable

NM cable was mentioned in Chapter 3 relative to the cable stripper, used to cut outer sheathing to access conductors inside. In Chapter 5 the uses of NM cable as authorized by code were noted as well as various code imposed limitations that must be observed when using this material in building wiring.

Mounting of Boxes

A wiring project to be done in NM cable starts with the installation of the boxes where receptacles, switches and fixtures will be placed, and where all splices and connections must be made (NEC 300-15). For NM wiring either plastic or metal boxes may be used. Since plastic is less expensive, it is most widely used.

Taking one circuit at a time, first determine from the cable plan how many boxes, and of what types and sizes, will be required. As an example look at the bedroom #1 and hall circuit as cabled in Figure 8-49. There are five duplex receptacles. Each needs a 2 x 4 device box. The four-way and two three-way switches call for three more, making a total of seven. The ceiling light in the center of bedroom #1 mounts on a 4″ octal box, and the two single pole switches by the door go in a two gang 4 × 4. The three recessed lights need no boxes; they are self contained. Boxes required for this circuit then are:

2 × 4 device	7
4 × 4 (2 gang)	1
4″ octal	1

Before settling for the above, first check two details. There are five two-wire cables entering that two gang switch box. Cleverly, we remember from Chapter 6 that code limits the number of conductors in boxes based on their cubic volume. By consulting Appendix Table 370-16(a), a 4″ square × 1¼″ deep is approved for only nine #14 conductors, 4″ square × 1½″ deep is good for 10 #14 conductors, and 4″ square × 2⅛″ is good for 15. There are five two-conductor cables coming in giving a count of 10 plus five grounds which together count for one more making it 11. The two switches count as two more making the total for the box 13. Thus the 4″ square box is fine as long as the 2⅛″ deep version is used. The shallower ones will be too small for the number of conductors.

The 2 × 4 device box for the four-way switch should also be checked. Two three-wire cables count to six conductors and one more for the two grounds making it seven. By adding the switch the total becomes eight. None of the 2 × 4 device boxes listed in Table 370-16(a) can be used. Look now at Table 370-16(b) which requires 2 cubic inches per conductor when the wire size is #14. This means a box with internal space of 16 cubic inches is necessary. Single gang boxes in plastic similar to those shown in Figure 6-18 and having internal volumes of 18 and 20 cubic inches are readily available. To avoid any possible confusion or doubt the volume of such boxes are clearly marked inside on the bottoms (Figure 10-1).

With the correct quantities and sizes of boxes in hand, they are now attached to the building frame according to the locations given on the floor plan. As was mentioned earlier dimensions for box positions are not given on the plans. However, the plans are drawn to scale; thus a very close approximation of the desired dimension can be obtained with a scale ruler. The most commonly used scale for residential and other small buildings is ¼″ = 1′. A triangular ruler called an ***architect's scale*** is available at any drafting supply house. This ruler has, in addition to the

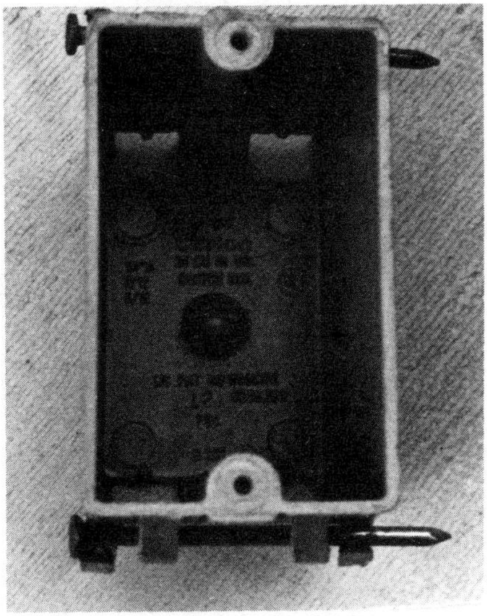

Figure 10–1
Plastic box with internal cubic volume marked inside.

¼″ scale, all other scales an electrician is likely to encounter. It is not an expensive item, and it is very durable. (The author uses one that is at least 30 years old.)

The reason that exact dimensions for the locations of electrical boxes are not given—except in special cases—is that while great accuracy in box location is not necessary for proper operation of the system it is a great convenience for the electrician to have the freedom to move a few inches one way or another to mount on an existing stud or joist. This freedom of movement saves a great deal of unnecessary labor. For example the code requires that the first convenience outlet must be no more than 6′ from a doorway (NEC 210-52). Thus the last stud that is less than 6′ from the door will meet code. Whatever its exact dimension may be is unimportant. Similarly if the plan calls for a pair of switches beside a door on the handle side, it is unimportant whether they are 3″ or 5″ or 6″ away from the door as long as they are close, and on the correct side.

Code does not dictate the proper heights for mounting boxes above the floor; however, customary usage does. Figure 10-2 gives the heights on center normally used for most boxes in a residence. Special purpose boxes for TV antenna or telephone outlets will usually be set, like convenience outlets, 12″ on center off the floor. Special purpose outlet for a window air conditioner will usually be 12″ high as well. Range, refrigerator, disposal, dishwasher, or compactor is variable. Height of outlets for these units is a matter of convenience as none are visible. Dryer and laundry outlets are also variable. Sometimes they are deliberately placed low so the machines may hide them. Other times they are raised above the machines for accessibility. Remember, that while code says nothing about box height from the floor, **NEC 370-20 discussed in Chapter 6 is very definite about box depth relative to finished wall surface, no more than ¼" setback.**

Figure 10–2
Customary heights of switches and outlets above floor.

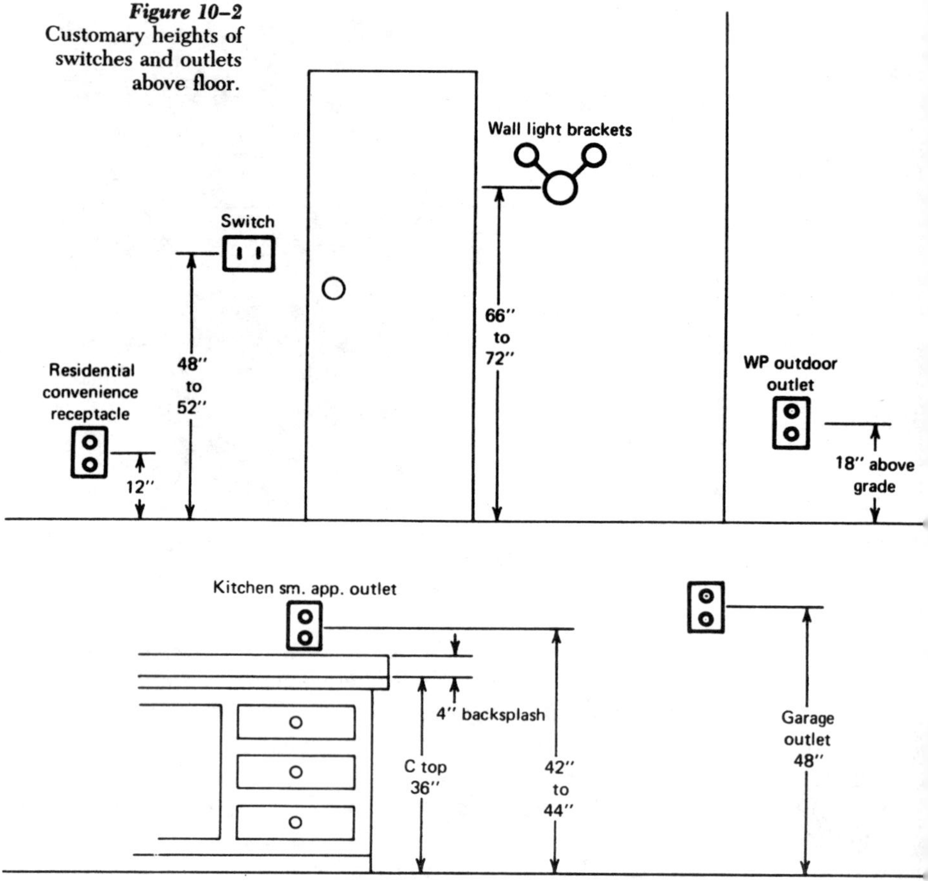

In commercial or industrial structures outlets are placed as needed along the walls, and at whatever height is convenient. Code makes no requirements regarding general purpose power receptacles. However, the minimum lighting volt-amperage stated in Appendix Table 220-3(a) must still be provided in accordance with the type of occupancy.

The boxes are physically attached to studs and joists with nails. Most plastic boxes are supplied with nails. Metal boxes with attached brackets for stud mounting may be secured with 1″ roofing nails. Huge nails are unnecessary as there is no strain on a box once it is in place. Knock-outs that are to be used should be knocked out before mounting each box. It is much easier to do it then.

When a precisely dimensioned mounting location is given on plans (the exact position for a ceiling chandelier in a dining room might be so noted), then either a metal or a plastic box on a bar hanger must be used (Figure 6-11). This will permit precise centering of the box at any desired point.

Running and Fastening Cables

With all boxes properly placed, the wire runs as shown on the cable plan are now installed. When metal boxes are used, the NM cable must enter through a box connector that also serves as a clamp, firmly securing the cable to the box at that point [NEC 370-17(b)]. Some boxes, both plastic and metal (Figures 6-3 and 6-20), are equipped with internal cable clamps. In all cases where the cable is secured to the box by means of a clamp, that cable must next be secured to the building frame using a staple (Figure 5-2) or strap within 12" of the box (NEC 336-18).

Inside the clamp in metal or plastic boxes, or the knockout in a plastic box without clamps, the cable sheathing shall be stripped so as to leave at least 6" (NEC 300-14) of the conductors free for splicing or for connection to devices. Inside each cable clamp, or plastic box knockout, the cable sheathing shall extend into the box [NEC 370-17(c)] no less than ¼". For maximum ease when making splices and working in boxes the sheathing will extend into the boxes no *more* than the required ¼".

> **IMPORTANT** When counting the number of conductors in a box note that an ***internal*** cable clamp must be counted as one conductor. The combination box connector and clamp [Figure 5-3(a) and (b)] that fits in the round knockout of a metal box does ***not*** count as one conductor.

The first staple or strap outside the box, whether at 12" or at 8", will probably be to the same stud or joist as the one to which the box is attached. From there (Figure 10-3) until within required stapling distance from the next box, the cable is secured every 4½' when running vertically through walls, or horizontally across ceilings or under floors while along the side of a joist. When the cable runs horizontally through holes in studs or joists, stapling is not necessary.

When a cable is to run across a crawl space at an angle to the floor joists, rather than parallel to one of them, it is not necessary to drill the joists. The cable may be stapled to the bottom of the joists. However, when a cable is to run across an attic at right angles or diagonal to the joists, the requirements depend on the intended use of the attic, and how access to it is obtainted. The distinction between ***accessible*** and ***inaccessible*** attics was discussed in Chapter 5 along with the specific code requirements for running cables across each.

While the code does not require drilling when making a diagonal cable run across an accessible attic in such cases, it is recommended. Putting in the otherwise required guard strips and staples will call for as much or more labor than drilling plus the material cost of the guard strips. In addition, if at some later date the owner wants to add an attic floor, the

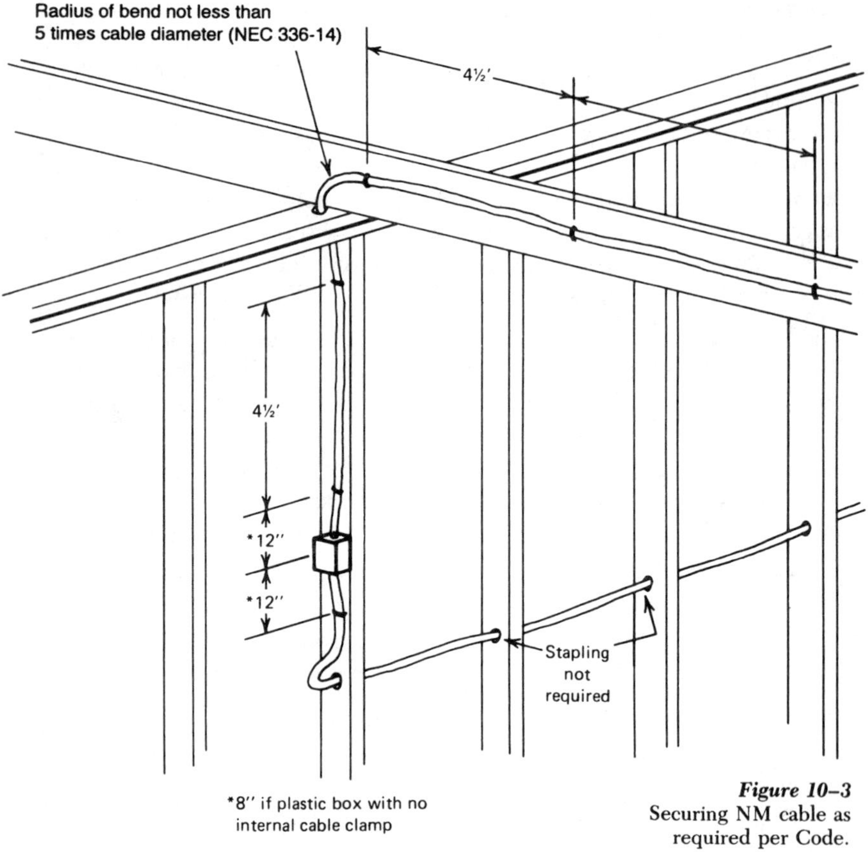

Figure 10–3
Securing NM cable as required per Code.

tops of the joists are clear. Granted, diagonal cables between the joists will cause some inconvenience for the fellows who install attic insulation. The advantages of drilling over using guard strips outweigh the disadvantages.

Rough wiring is complete when properly sized boxes are installed and all cables connecting those boxes have been run and fastened in place correctly. Since at least 6" of every conductor in every box must be clear of sheathing (NEC 300-14), strip it before fastening the cable in place rather than after—it is vastly easier this way.

Bx Cable—Type AC

In conjunction with BX cable, only metal boxes may be used. In other respects the box mounting procedure for BX is similar to that followed with NM. Horizontal distances are found by scale measurements from the floor plans, and heights are conventional unless noted otherwise. The same regulation regarding the relationship between box front and finished wall surface also applies.

Box Mounting

The basic work sequence is again to mount the boxes properly first, then install the required box to box cable runs. Each cable is connected to boxes at both ends using box connectors in which the cable is secured by means of a screw clamp (Figure 5-7). Because of its metal armor, BX cable is much less flexible than NM cable. Consequently both straight and right angled box connectors are available to allow the electrician necessary freedom to bring cables in and out of boxes in limited spaces.

Remove desired knockouts before mounting boxes, and remember any knockouts removed by mistake must be plugged (NEC 370-18) (Figure 6-25).

Running AC Cable

The metal armor of BX is easily cut with an ordinary hacksaw (Figure 10-4). Cut through a loop of the armor as shown at approximately a 45° angle, then twist the armor to break it apart. Since BX already contains its insulated conductors the primary consideration when cutting it is to avoid damaging that insulation. It will take a bit of practice to develop the knack of successfully cutting the armor without damaging the insulation. Be surprised if you *do* get it right the first time, not if you don't.

After cutting BX and before clamping on the cable connector, do not forget to install the required fiber bushing that slips inside the cut end of the armor. This bushing is required because when cutting the armor the sharp edges left might damage the insulation on a conductor in such a way that the damage is not readily visible. The bushing protects the insulation on the conductors from direct contact with the armor, which may damage it. The inside opening of the BX box connector is specially enlarged so that the electrical inspector can easily see whether that bushing is actually in place.

The requirements for fastening BX cable are similar to those for NM cable. It must be fastened within 12″ of each box, and every 4½′ thereafter until within 12″ of the next box. It may be passed horizontally through holes in the studs of a wall without fastening every 4½′.

Since the requirements for protection of BX cables in attics are the same as for NM cables, the drilling recommendations made above for NM cable apply as well.

While code stipulations regarding NM and BX cables are generally very similar, a significant difference is that BX may be embedded in

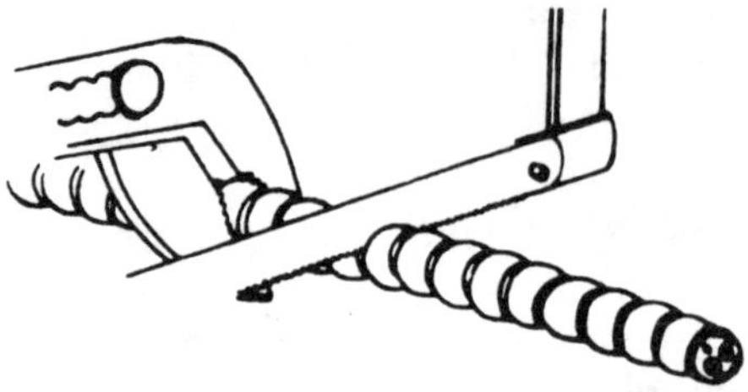

Figure 10–4
Cutting BX cable
with a hack saw.

plaster or other types of masonry and NM absolutely may not be. Where BX is so embedded, no fasteners are required.

The radius of bends in BX may be no less than five times the diameter of the cable (NEC 333-8). That measurement is to be made from the inside of the bend.

BX cable does not, like NM cable, contain a grounding conductor of the same size and material as the circuit conductors. It is grounded through its steel armor supplemented by a thin ***bonding strip*** often of aluminum.

After the fiber bushing is in place, the bonding strip is bent back to lie between the armor and the cable connector, making good electrical contact with both.

In a BX cable containing the usual copper conductors, the potential difficulties due to dissimilar metals are obviously considerable. In some jurisdictions the use of BX cable is limited or prohibited, either because of the proximity of dissimilar metals, or the ground is considered inadequate, or because they do not like the danger of damage to conductor insulation when cutting the cable, or a combination of these considerations. Consequently, unless you have seen BX currently in use in your immediate area, check with electrical inspection before using it to make certain it is permitted.

Flexible Metal Conduit—Greenfield

Boxes

With greenfield only metal boxes may be used. For mounting boxes to which greenfield is to be attached, use the same procedures already discussed. Locate them horizontally by scaling measurements from the plans, and set heights from the floor by customary practice. As always remove knockouts before nailing boxes in place.

The most commonly used greenfield box connector [Figure 5-10 (a)] threads inside the armor, and when in place appears as in Figure 5-9. After the boxes have been properly installed, run the flex from box to box as called for in plans. Cut with hacksaw across one loop just as is done with BX armor. Before cutting, be certain to leave enough slack for fastening, per NEC 350-18, which requires the same securing at 12" from each box, and then every 4½', that was required for both NM and BX cables.

Running Greenfield

Regarding bends the only limitation on NM or BX cables is the radius of any single bend. Not so with greenfield. In any one run from box to box, or between conduit bodies (Figure 10-6) which may also be used with greenfield, bends are limited to no more than 360° (NEC 350-16). Conforming to this limitation will, at times, require careful planning. It is because greenfield is an empty armor through which wiring must be pulled that this limitation is made. However, when there are several bends involved in a run, the easier way to proceed is to fit and cut the flex *without* stapling it in place. Then take it back out, lay it down, and pull in the wires. Proper use of the fish tape and techniques for pulling wires in conduit will be discussed in the section on EMT. After cutting the wires, leaving at least the required 6″ extra at each end, replace the greenfield, tighten the box connectors, and now staple it down per code.

When counting conductors for conformity with box limitations, and for conformity with Appendix A, Table 1, Chapter 9 covering limitations on number of conductors in a conduit, do not forget that flexible metal conduit is not approved as an adequate ground path when the length of the path is over 6'. This means that along with whatever circuit wires are required a ground wire will also have to be pulled through each box to box run. This will add one to the conductor count in every greenfield run, and in every box.

Liquidtight Flexible Metal Conduit

Except concerning exposure to moisture, the code requirements relating to the use of this material are similar to those for greenfield. However, liquidtight is mostly used for quite short final connecting runs to equipment in locations or applications where both vibration and water are liable to be encountered. Thus it will seldom require fastening. It is unlikely to have bends in excess of 360° in a run. Most runs will be under 6′ so it will not often need a grounding conductor. On few occasions will it ever carry anywhere close to its maximum conductor capacity.

In rough wiring using this material, cut it with a sharp, fine toothed hacksaw to avoid leaving a ragged edge. Then be careful to assemble the liquidtight connectors correctly, and be sure to cinch them down tight so that they will in fact be liquidtight, and so that they will be resistant to loosening due to vibration.

Electrical Metallic Tubing—EMT, Or Thinwall Conduit

Producing clean, accurate, professional work using EMT requires much more skill and practice than any of the other materials used in rough wiring. While its use is limited in residential applications, it is used extensively in both industrial and commercial wiring. Many industrial jobs are done entirely in EMT.

Box Mounting

Box locations are determined from a cable plan. Since one of the primary reasons for using EMT is that it provides good protection for exposed wiring, much of it is run exposed. Consequently, the boxes at which runs terminate are surface mounted as well. All boxes used with EMT are metal, and when surface mounted they are fastened through holes provided for this purpose in the backs of the boxes. These holes may be seen in boxes shown in Figures 6-1, 6-5, or 6-10.

When a box is to be mounted on drywall at a point where there happens to be a convenient wood stud it can be nailed. Most often, however, they must be placed on drywall or masonry at locations where nailing is impossible. Other types of fasteners are then required. These will be either toggle bolts, or plastic or lead expansion anchors (Figure 10-5). All of these fasteners will require drilling before inserting the fastener. A slow speed drill motor and carbide tipped masonry drills, described in Chapter 3, should be used for this purpose.

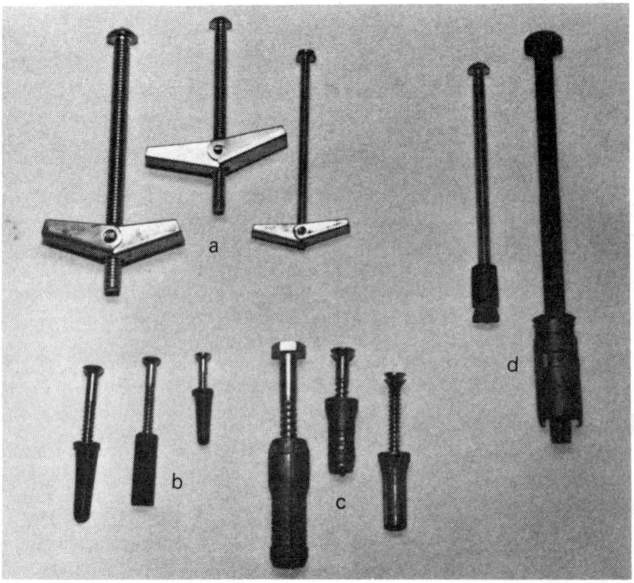

Figure 10–5
Various fasteners to masonry: (a) togglebolt; (b) plastic expansion anchor; (c) lead expansion anchor, wood screw type; (d) lead expansion anchor, machine screw type.

Cutting EMT

EMT, as was mentioned in Chapter 5, is supplied in 10′ pieces. It is easily cut to length with an ordinary hacksaw (Figure 3-4). Use a 32-tooth, or certainly nothing coarser than a 24-tooth, blade. Otherwise the saw will bind badly about half way through. ½″ EMT, the most commonly used size, can be cut very easily and cleanly with a tubing cutter (Figure 3-9 b). For larger sizes use the pipe cutter (Figure 3-9 a).

> ***IMMEDIATELY*** after cutting a piece of EMT ***ream it*** using a pipe reamer such as shown in Figure 3-8. Do not stop for coffee, stop for lunch or even go to the bathroom. Ream that pipe end before doing ***anything*** else. The inside edge of a pipe after cutting is very sharp. Anyone who fails to make it an absolute rule to ream every cut *at once*, sooner or later forgets one.

When wires are pulled through a conduit run containing an unreamed cut there is every likelihood that the insulation on one or more conductors will be damaged, if not completely torn, allowing wires to short out to each other or to the pipe. Only careful reaming of absolutely every cut will prevent this.

Bending

One of the primary advantages of EMT over rigid conduit is the ease with which the smaller trade sizes can be bent to follow irregular contours. The great majority of EMT work is done using ½″ and ¾″ trade sizes, which are the ones that can readily be bent using a tool known as a conduit bender (Figure 3-15). Regardless of manufacturer, any ½″ conduit bender will be shaped to a curve matching the code conduit bend radius given in Appendix Table 346-10 for ½″ conduit which is 4″. All ¾″ benders will match the code specified 5″ radius. ½″ conduit may be bent with a ¾″ bender, but ¾″ cannot be bent with a ½″ tool. One inch and even 1¼″ EMT can be bent by hand, but anything larger will require a power bender.

The radius of bends is specified in Table 346-10, depending on the trade size of the conduit. The amount of bending in a single continuous run between boxes or conduit bodies such as ells, or tees, or straights is limited to a total of 360°. Thus, there can be no more than four 90° bends between boxes, or eight 45° bends, or any combination of 90°, 45°, 30°, 60° angles, or whatever as long as the total number of degrees in bends between boxes does not exceed 360°. A run bent as shown in Figure 10-6 would be unacceptable. A 90° ell could be inserted at either A, B, or C to make the run meet code.

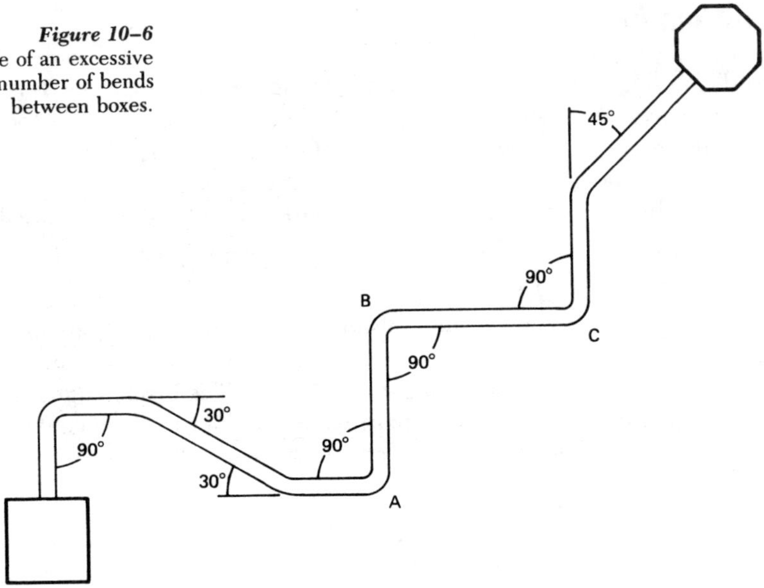

Figure 10–6
Example of an excessive number of bends between boxes.

The Conduit Bender

A conduit bender is shown in Figure 10-7. Since there are several different manufacturers of these tools, there are as many models differing in details as there are different manufacturers. The basic benders are all very similar. The major differences lie in the bending guides and aids provided to assist the electrician in making accurate bends. One very helpful system consists of an arrow close below the hook plus two additional guide marks along the channel rim (Figure 10-7). The arrow is fairly standard. The other two guide marks differ. To identify them read the instructions accompanying the tool.

One manufacturer provides a pair of built-in levels to precisely indicate 45° and 90° bends. Another provides a system of guide lines cast into the tool to indicate 10°, 22°, 30°, 45°, and 60° bends as well as right angles.

Right Angle Bend

a. First mark on the conduit the point at which the bend is to start.

b. Lay conduit on floor.

c. Slide bender over conduit until mark on conduit is aligned with start mark on bender (see bender instructions to identify start mark).

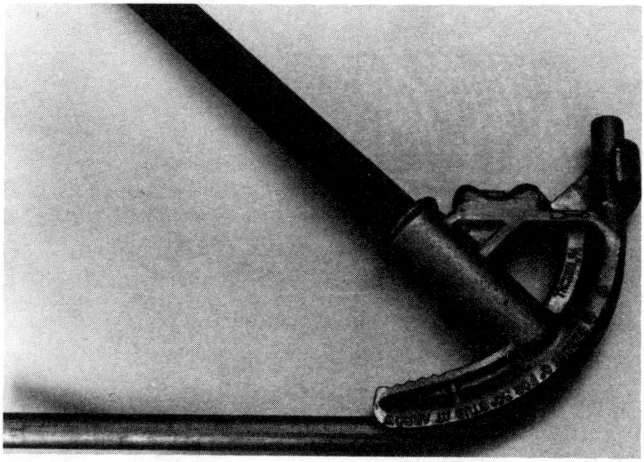

Figure 10–7
Bending with conduit bender.

d. Step firmly on foot pedal, and pull back on handle simultaneously until 90° guide point is reached (Figure 10-7).

e. Check bend with carpenter's square, and adjust with bender, if necessary.

Other Angles

To bend an angle for which the bender has a specific guide (such as a 45° level or an incised guide line) proceed as above stopping at the appropriate guide point. To make angles for which the bender has no guides, a common carpenter's tool called a T bevel can be set to the desired angle, and the bend then matched to the T bevel.

Offset Bends

Since EMT is required to be supported within 3′ of each box or fitting (NEC 348-13) and every 10′ thereafter, it is normally run tightly against the wall. However, when it comes to a box the knockout through which it will enter must be ¼″ (more or less) away from the wall. This requires an offset bend at the end of the conduit (Figure 10-8). For shallow offsets of this type, measure from the conduit end about 4″. Then make two marks 2½″ apart. Place the bender on the second mark and bend up only 10°. Now rotate the conduit 180° and place it on a 2 × 4 on the floor so the first bend is pointing down (Figure 10-9). Now move over to the first mark and bend up so the stub end is now parallel to the original run of the conduit but offset to the side.

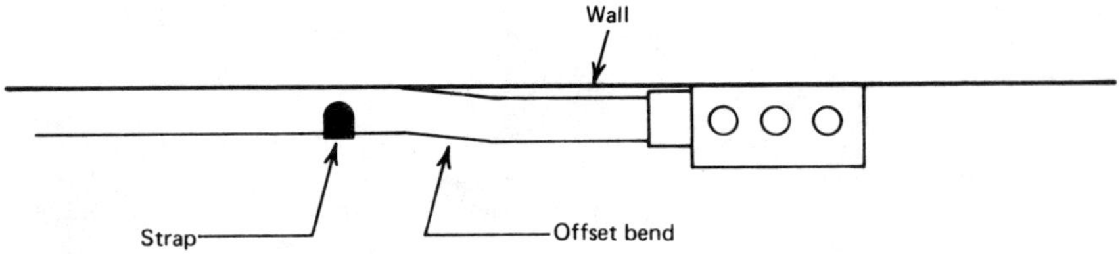

Figure 10–8
Offset into box.

Be careful to rotate the conduit exactly 180° between bends; otherwise the offset will not be straight. If the offset is either too shallow or too deep to align exactly with the knockout, adjust it by slightly increasing or decreasing the bends.

To pass around various obstructions, much deeper offsets may be required. This is accomplished by increasing both the angles that are bent as well as the distance between bends. To make offsets of various depths using 45° bends, following is a table of distances between bend marks:

Figure 10–9
Bending shallow offset: rotate and place on block so first bend is pointing down. Make second bend so offset end is parallel to conduit run.

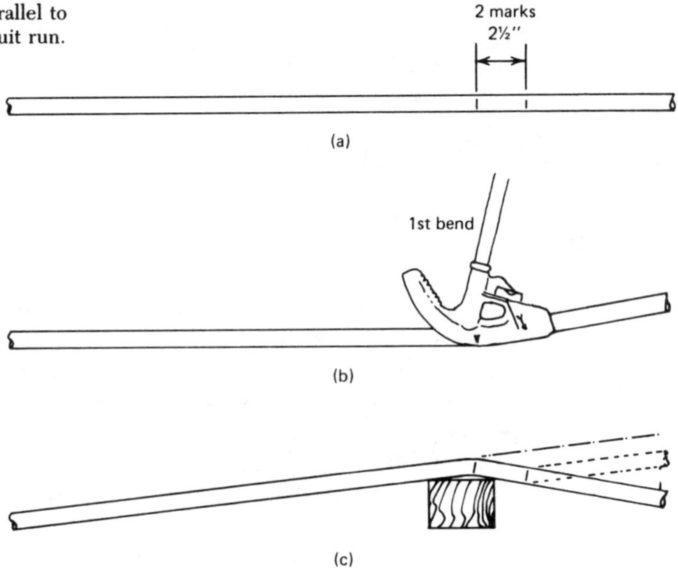

Offset Depth	*Mark Spacing*
3″	4¼″
4″	5½″
5″	7″
6″	8½″
7″	9¾″
8″	11¼″
9″	12½″
10″	14″
11″	15½″
12″	16¾″
13″	18¼″
14″	19¾″
15″	21″

For various other bend angles and offset depths a handy formula will produce the distances between bends:

$$\text{Offset Depth} \times \begin{array}{c}\text{Constant Multiplier}\\ \text{for Bend Angle}\end{array} = \text{Distance Between Bends}$$

The constant multipliers to be used in this formula for five common bend angles are:

Angle	*Multiplier*
10° × 10°	6
22½° × 22½°	2.6
30° × 30°	2
45° × 45°	1.4
60° × 60°	1.2

For an offset depth of 7″ using bend angles of 30°, the formula gives us the distance between bends as follows:

$$\text{Offset Depth} \times \begin{array}{c}\text{Constant Multiplier}\\ \text{for Bend Angle}\end{array} = \text{Distance Between Bends}$$

$$7'' \quad \times \quad 2 \quad = 14'' \text{ apart}$$

To pass an obstruction such as a beam or a column (Figure 10-10) a four bend ***saddle*** is required. This is simply two identical offsets of the type just discussed with a connecting section between. For an offset

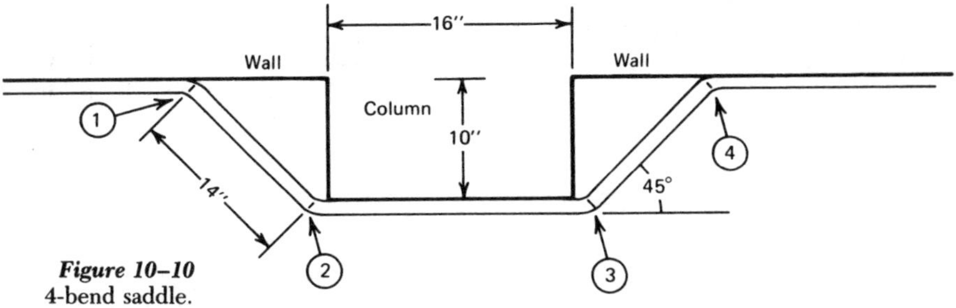

Figure 10–10
4-bend saddle.

depth of 10″ using 45° angles the table supplies the distance between bends as 14″. That takes us to the front of the column. The same 14″ with 45° bends will take us back to the wall on the far side of the column. The only information lacking to make the part is a distance between bends #2 and #3. After making bend #2, hold the conduit up to the column and measure between mark 2 and the column. That measurement doubled and added to 16″ is the distance between marks #2 and #3.

The most difficult part of making a four bend saddle is keeping the whole thing straight. After the first bend the conduit must be rotated ***exactly*** 180° three times to make the other bends; otherwise the piece comes out cockeyed. A pair of guide lines down opposite sides of the conduit will help immensely.

To go around a pipe, a three-bend saddle is used (Figure 10-11). This consists of a 45° at A, and two 22½° bends at B and C. Bends B and C are done using the usual start mark on the bender. However, bend A requires that the tool have a special alignment mark for the center of 3 bend saddles. After the 45° bend is made at A the identical distances to the 22½° bends at B and C are based on the saddle depth. For some common depths those distances are:

Saddle Depth	*Marks B and C from Center*
1″	2½″
2″	5″
3″	7½″
4″	10″
5″	12½″
6″	15″

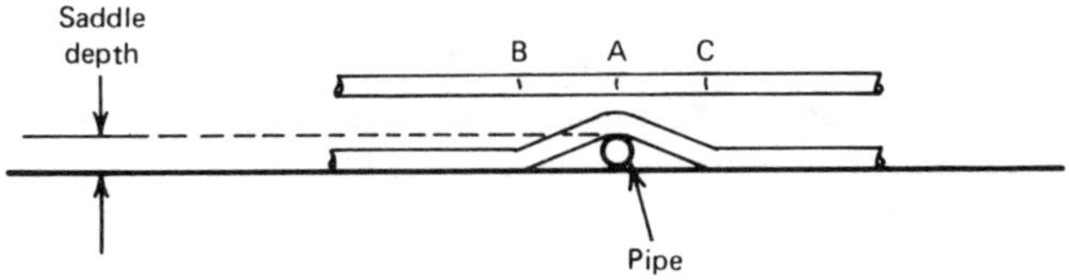

Figure 10–11
3-bend saddle.

Bends B and C are both made with the hook end of the bender pointing toward the center of the saddle. Again as with the four-bend saddle a pair of guide lines on opposite sides of the conduit will be a big help in keeping the piece straight through all three bends.

Pulling Conductors in Conduit

Conductors are pulled into conduits by the use of a fishtape. The tape is fed from its reel through the conduit from one box to another (Figure 10-12). There the required conductors are attached to the hook end, and pulled back. This operation is done most conveniently with two people. One person pulls on the fishtape while the second feeds the conductors into the pipe.

When, as is usually the case, several conductors are to be pulled at once only one of them is passed through the hook on the fishtape. It is then doubled back on itself by somewhere between 6″ and 1′. The other conductors to be pulled are now taped to the first one and to each other using plastic electrician's tape at intervals of about 3″ to 6″ for a distance of

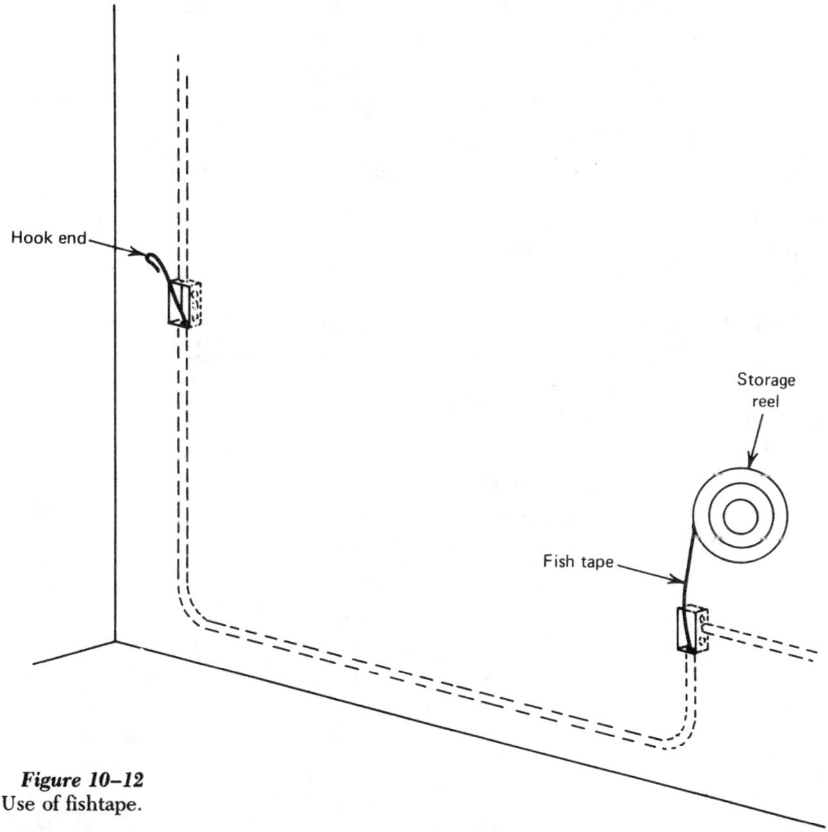

Figure 10–12
Use of fishtape.

1½′ to 2′. The more conductors there are the closer the tape intervals, and the longer the taping distance in order to insure that none of them slip out to get left behind part way through.

If it is to be a long pull, or there are a large number of conductors to be pulled, or both, before starting the pull unroll from the wire spools adequate lengths of all required conductors. Stretch them out together on the floor, and with one man at each end slap them hard against the floor several times. This will straighten them and take the kinks out, making them far easier to pull. Also when there are many conductors or many bends, or both, a ghastly yellow pulling grease is available with which the man feeding the wires lubricates them, making the job vastly easier for the fellow at the pulling end.

When making a long pull that is to pass through several boxes, or a combination of boxes and Ts or Ls, at each pull point whether it is a box a T or an L, drop out of the group of taped conductors any that are to terminate at that point. Then pull through lengths of each of the others sufficient to reach from here to their final destinations. Then pull the entire lot through to the next pull point and repeat the process.

At each box separate the conductors that are to connect to devices or splices from those that simply pass on through. Do not cut any conductors that are not required for connections. Leave a loop of each of them around the bottom of the box, and pass them on into the next conduit. Leave a foot of slack in each of the conductors that are to be cut so that after cutting each piece will be of the 6″ length required by NEC 300-14.

Surface Raceways—Wiremold

Where it is impossible, or undesirable, to conceal wiring, but where exposed EMT and its usual metal boxes would be visually unacceptable, a surface raceway may be used. This material (Figure 10-13) being of a basically flat design is far better looking than EMT. While Wiremold is certainly not presented as an aesthetic triumph it equally certainly is much less offensive than EMT for exposed wiring.

A very complete selection of boxes, Ts, inside elbows, outside elbows, flat elbows, twisted elbows, couplings, fastening straps, and adapter fittings for connection to standard concealed boxes, conduit, and other forms of conventional wiring is supplied (Figure 5-14).

Since the method of connecting this raceway and its various fittings to each other is by slide-in flanges, installation is best approached by starting at a box somewhere and attaching raceway lengths and fittings consecutively in a manner similar to the way a run of threaded pipe is connected.

Wiremold is cut with an ordinary hacksaw. After cutting, file the rough inside edges of every cut to prevent damage to insulation when

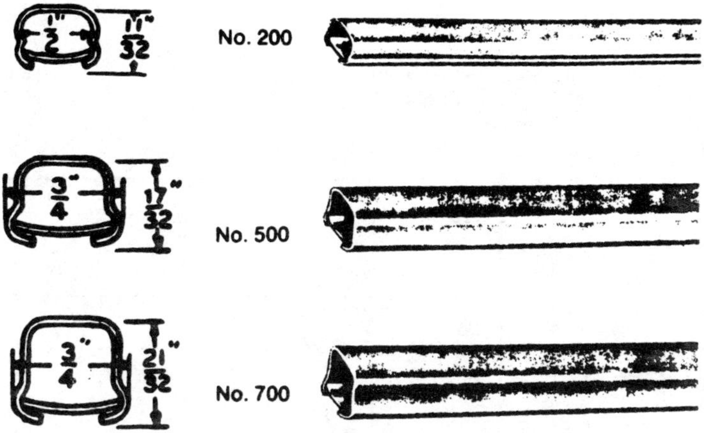

Figure 10–13
Wiremold: 200, 500, nd 700 sizes. *Courtesy of the Wiremold Company.*

wires are pulled. Each fitting consists of two parts. One is the backing plate which is flanged to slip into the raceway and contains holes for attachment to the wall. The other part is the finished cover. Except for long runs, or cases of irregularities in the walls, the backing plates only are attached to the walls, and the lengths of raceway are held in place by the backing plate flanges. Finished covers are screwed or snapped in place only after all wiring is pulled and connections completed.

Like EMT, and unlike greenfield, Wiremold is its own ground. No grounding conductors need be pulled through. As with all types of conduit, the number of conductors that may be pulled through Wiremold is limited by the size of the specific raceway, and the size and insulation type of the conductors being used. These limitations are illustrated in Figure 10-14 for reference.

Conductors are pulled through surface raceways using the same fishtape and in the same way as through EMT. No special equipment or procedures are required.

Identifying Rough Wiring

When rough wiring is completed there will be in each box at least 6″ lengths of each conductor (NEC 300-14). When the time comes to complete the finish wiring, most conductors in most boxes will be adequately identified simply by means of the color code used on the insulations.

However, in boxes containing several switches, particularly a combination of single pole and three-ways or four-ways, some confusion can develop as to which conductors go where. This mix-up is further compounded when, as often happens, the finish wiring is not done by the same person who does the rough. The man who does the rough can save the fellow who is going to do the finish a good deal of time and irritation, if

Figure 10–14
Conductors permitted in different sizes of wiremold.

Type of Raceway	Wire Size Gage No.	Number of Wires: Types RHH, RHW	Type THW	Type TW	Types THHN, THWN
No 200 	12		2	3	3
	14		2	3	3
No 500 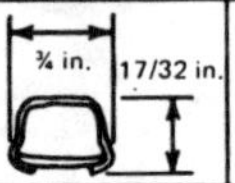	8			2	2
	10	2	2	3	4
	12	2	3	4	7
	14	2	4	6	9
No 700 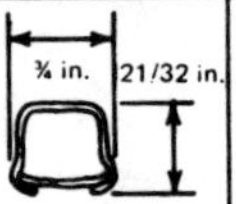	6				2
	8		2	2	3
	10	2	3	4	5
	12	2	4	6	8
	14	3	5	7	11
No 5700 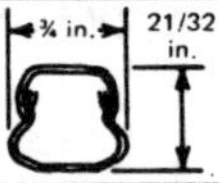	8		2	2	3
	10	2	3	5	6
	12	2	4	6	9
	14	3	5	8	12
No 1000 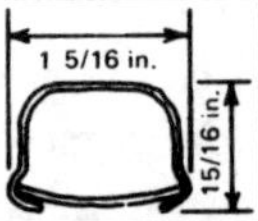	6	2	3	3	5
	8	4	5	7	9
	10	6	9	13	16
	12	7	12	17	25
	14	9	14	22	34
No 1500 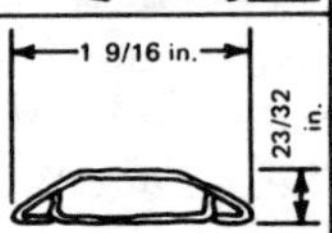	6				2
	8			2	3
	0	2	3	4	5
	12	2	3	5	7
	14	2	4	6	10
No 1900 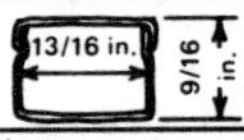	12			3	3
	14			3	3
No 2000† 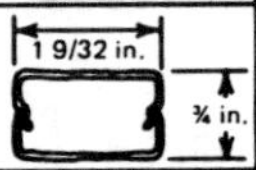	12			3* 3	3
	14			3* 3	3
No 2100† 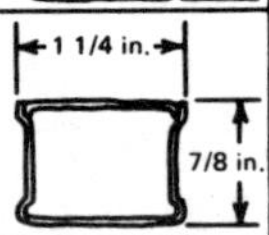	6	2	4	4	6
	8	4	6	8	10
	10	7	10	14	17
	12	8	13	19	28
	14	10	15	24	37
No 2200† 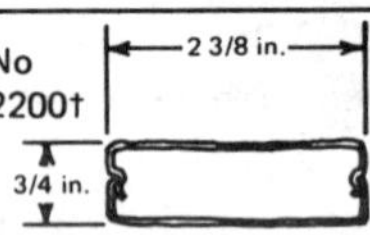	6	5	7	3* 7	11
	8	8	11	7* 14	19
	10	13	19	10* 26	32
	12	15	23	10* 34	51
	14	18	29	10* 44	69
No 2800 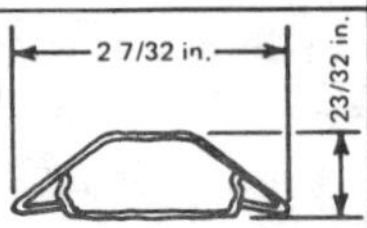	6	2	3	3	5
	8	4	5	7	9
	10	6	9	12	15
	12	7	11	16	24
	14	9	14	21	33

†Figures for Nos. 2000, 2100, and 2200 are *without receptacles,* except where noted.
*With receptacles.

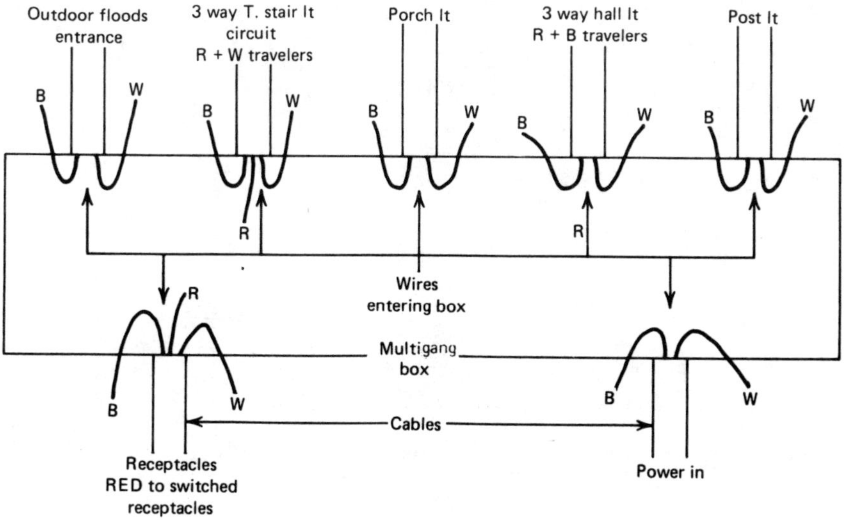

Figure 10–15
Sketch to record rough wiring of a large box.

he makes small sketches (Figure 10-15) of all boxes which might be confusing. Very simple sketches will do. Only a few lines are needed to show where on the box each cable or conduit enters, and a few words will take care of identifying where each goes. This drawing to go along with the cable plan will be a great help to the finisher. Be kind to the finisher—remember when the time comes the fellow they send back just might be ***you.***

CHAPTER 11

FINISH WIRING OF NEW WORK

Only after all boxes have been installed, and all box to box wiring completed, inspected, and approved, may finish wiring be done. First, however, the walls are closed and finished in whatever manner has been specified.

When the other trades are substantially finished, the electrician returns to install switches, fixtures, and outlets, and wire in any direct-wired fixed appliances. Obviously at this stage all electrical work must be done extremely carefully and neatly to avoid any damage to the now completed decorative finishes. So it is time to remove the drywall mud, paint globs, and other debris from our boxes, pull out and clean off the conductors, and prepare to complete the electrical system.

Receptacles

120 Volt

The usual 120V receptacle is shown in Figure 11-2. It consists of two three-hole sockets mounted on a metal strap called the ***yoke.*** Each socket will accommodate a standard 120V plug. This receptacle is termed a *duplex* since it will take two plugs. Also available, but not commonly used, are single and triplex receptacles.

To connect conductors to a receptacle first strip the insulation off each one for between ½″ and ⅝″ as shown in Chapter 3. At present most 120V duplex receptacles are furnished (Figure 11-1) with two screw terminals on the power side (brass heads), and two on the common side (silver heads). The two brass screws connect to the power slots of the two sockets, and to each other via a removable link.

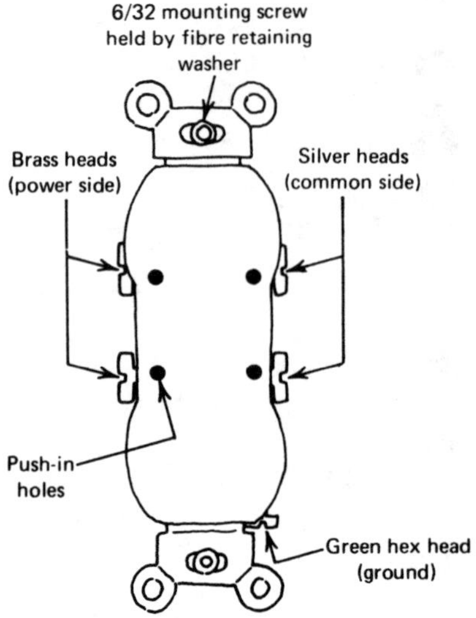

Figure 11–1
Back side of duplex receptacles.

The silver headed screws similarly connect to the common sides of each socket and to each other via another removable link. Thus one connection on each side ties both power slots and both common slots into the circuit. Any end of the line receptacle is connected in this way (Figure 11-2). Although the receptacle is furnished with four terminal screws ***only two need be used.*** The ground wire (bare or green) connects to the hexagonal green screw at one end of the receptacle. When the wiring has been done in EMT or wiremold, no ground wire is required as either one is approved per code as its own ground.

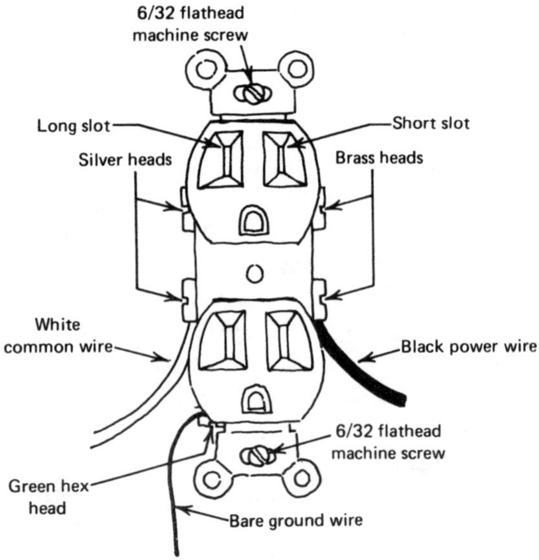

Figure 11–2
End-of-line receptacle.

A middle of the line receptacle is connected between one line coming from the direction of the power source, and another going on to another receptacle or other use point. Now the black power wires from both lines (Figure 11-3) are spliced to each other along with a short pigtail of black. The pigtail then connects to one of the brass power screws on the receptacle. The two white commons are similarly spliced to each other with a pigtail to one silver screw, and the two grounds splice with a pigtail to the green hex head screw.

Although there are two brass and two silver screws on the receptacle only one of each is being used. The reason is that by directly splicing the incoming and outgoing power and common lines with pigtails to the receptacle, subsequent damage to the receptacle or even complete removal will not interrupt the continuity of the rest of the circuit.

At the present most 120V receptacles are supplied with both terminal screws and holes for push-in connections (Figure 11-1). At the risk of being considered old fashioned, it is the opinion of this writer that the

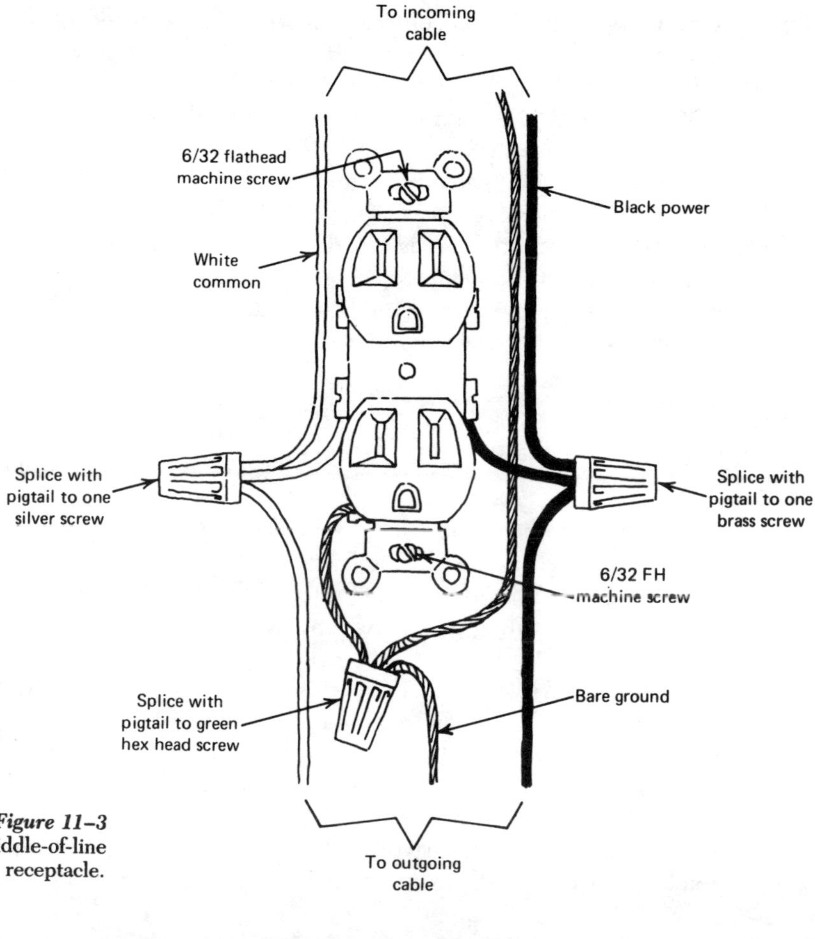

Figure 11–3 Middle-of-line receptacle.

screw terminal makes a significantly more positive connection than the push-in. While the push-in does appreciably reduce labor time, when installing receptacles it is electrically inferior to the traditional screw terminal.

When splices are completed and pigtails connected push the splices down to the bottom of the box to clear the space for the mounting of the receptacle. Get the grounds down first so they are below everything else. This will minimize any possibility of a stray loop of ground wire accidentally contacting a power terminal. Assuming that the proper 6″ length of each conductor was there prior to making the splices (NEC 300-14), there will be no difficulty separating the splices by pushing each wire nut down into a different corner leaving only the pigtails standing up and available for connection to the receptacle.

Each new receptacle is furnished with two flat head 6 × 32 machine screws which are held in the slots at the yoke ends by small square fiber washers. These screws fit threaded holes provided for them in device boxes. When attaching a receptacle to a ***metal box be sure to remove*** these fiber retaining washers. The receptacle yoke must make good positive metal to metal contact to insure proper grounding. In the case of a plastic box the washers can do no harm.

The slots in which the mounting screws are found are there to allow for alignment of the receptacle in case the box itself is somewhat cockeyed. After the receptacle is firmly tightened in place the installation is completed by adding the cover plate. That cover plate, incidentally, is by no means a decorative option. It is a code requirement (NEC 370-25).

When two or more duplex receptacles are mounted in a multi-gang box (Figure 11-4), they must be correctly aligned with each other or the required multi-gang cover plate will not fit.

240 Volt

The front faces of the 240 V receptacles used for ranges and dryers differ as seen in Figure 7-25, 7-26 and 7-27; however, the connections to be made at the backs of both are the same (Figure 11-5). In a normal installation the screw terminal marked "white" will be down. Connections are made by inserting the circuit wires straight into the holes below the screw heads, and simply tightening down. As long as the white common conductor goes to the "white" terminal either power wire may go to either of the other terminals. While 120 V duplex receptacles are supplied with appropriate mounting screws for box attachment the 240 V range and dryer receptacles are not. The electrician must furnish his own. The same cover plate requirement mentioned above applies to box mounted receptacles.

Where plans call for a weatherproof receptacle, WP on the drawing, the receptacle part itself is the same as one used in any other location.

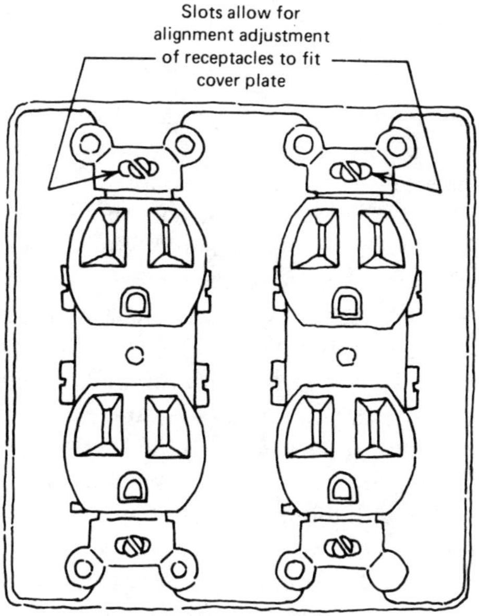

Figure 11–4
Mounting of two duplex receptacles in 2-gang box.

The weather-proofing is provided by the cover plate to which it is attached (Figure 6-16). This plate is provided with a neoprene gasket to seal the seam between it and the front edges of the box. The box also has spring loaded covers with neoprene gaskets that fit over the receptacle openings. A weatherproof outlet is normally mounted in a weatherproof box as in Figure 6-13; however, any standard non-waterproof box may be used for a waterproof outlet if that box is embedded in an exterior wall in such a manner that the waterproof cover when installed will adequately seal it.

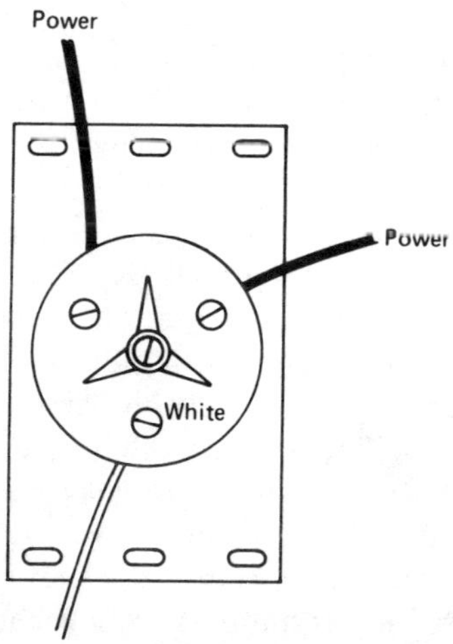

Figure 11–5
240 V range receptacle: rear view.

Switches

Single pole, three-way and four-way switches can all be obtained with either push-in or screw terminals, or both. In the opinion of the author push-ins are no more desirable on switches than they are on receptacles. However, regardless of how the conductors are physically connected to the switch terminals the switching circuits will be based on one or another of those shown in Chapter 7.

The first step toward finish wiring a switch box is to splice all grounds together, and if it is a metal box, clip a pigtail from that splice to the box to ground it out. If there are commons in the box, splice them to each other as well. Remember, commons never go to a switch. Just as in a receptacle box get those splices down in the bottom of the box and out of the way before connecting the switch or switches. Then make the connections in accordance with the wiring diagram. While one cannot go far wrong on a single pole switch, the same is not true of three-ways or four-ways.

Switches have the same type of slot at the yoke ends as found on receptacles, and for the same reason—to allow for adjustment of its position in the box. Like receptacles, most switches are furnished with 6 × 32 flat head mounting screws. Since there is no need to ground the switch yoke, the fiber retaining washer need not be removed for mounting in metal boxes as must be done with receptacles. When the switch is securely fastened in place, NEC 370-25 again states there will be a cover plate placed over it to complete the installation.

A weatherproof switch (Figure 7-6), whether single pole or three-way, is wired just as any other switch. However, it is supplied already mounted on its cover plate and accompanied by its weatherproofing neoprene gasket to seal the seams between that cover plate and the box on which it is mounted.

Photo-controls (Figure 7-7) are normally intended to be actuated by daylight, or the absence of it. Therefore, they are generally mounted on outdoor boxes where they are screwed into one of the threaded openings. Connections are made to the wire pigtails with which these controls are furnished.

Photo-controls are most commonly used to turn on various types of lighting at nightfall. Thus the primary concern when installing them is to insure that they are oriented or shielded, so that they will not be tripped out by other lighting during the night. Properly installed photo-controls are extremely reliable, and extremely durable as well.

The timer switch (Figure 7-9) uses screw terminal connections. These terminals are color-coded with brass heads for power, and silver heads for common. ***Be sure*** to check the diagram pasted inside the hinged cover to locate the correct input and output terminals. This would seem to be too simple to require any mention, but it is astonishing how many times these things are initially connected backwards. When this is done the first time the device cycles ***OFF***, it not only shuts the load off but also shuts off its own clock, stopping everything.

Appliances

Ovens-Ranges

Free-standing or drop-in type oven/ranges are usually connected to the power supply by a plug inserted in a range receptacle. However, when the equipment consists of a countermounted cooktop with a separate wall-mounted oven, the two units are often directly wired to a common junction box or load center, which is fed from the main power panel. The other way of handling separate cooktops and ovens is to run two feeds from the power panel—a 30 A with #10 wires to the oven, and a 40 A with #8 wires to the cooktop.

When separate supplies are furnished each goes back to an appropriate 240 V double breaker in the main panel. When the two are supplied by a single feed it splits in two directions and a junction box is used (Figure 11-6), the #6 or #8 feeders from the power panel are spliced in a 4 × 4 box to a set of #8s going on to the cook top and a set of #10s that connect to the oven. The overcurrent protection at the power panel will be either a 40 A or a 50 A 240 V double breaker.

The load center, if used, will be a small breaker box (Figure 11-7) containing two 240 V double breakers. One will be 30 A for the oven, and the other will be a 40 A for the range. In this case the subfeeders from the power panel feed the two breakers in the load center which then individually protect the two appliances. The advantage of this arrangement is that since each appliance has its own breaker, a difficulty in one of them does not knock them both out of service.

When the cooking appliances are gas fired do not forget to furnish a 120 V outlet to power accessories such as oven light, timer, or clock.

Dryer

A great many dryers plug into the typical dryer receptacle (Figure 11-5); however, some are directly wired. In either case the wire size will be #10, and the circuit will be 30 A at 240 V. EMT, flexible conduit (greenfield), or NM cable may be used to carry the supply. When EMT is used, a transition should be made either at a junction box or through a transition coupling so that the final connection to the dryer is through three or four feet of loose flex that will allow the dryer to be moved for cleaning or servicing without requiring disconnection. When the entire wire run

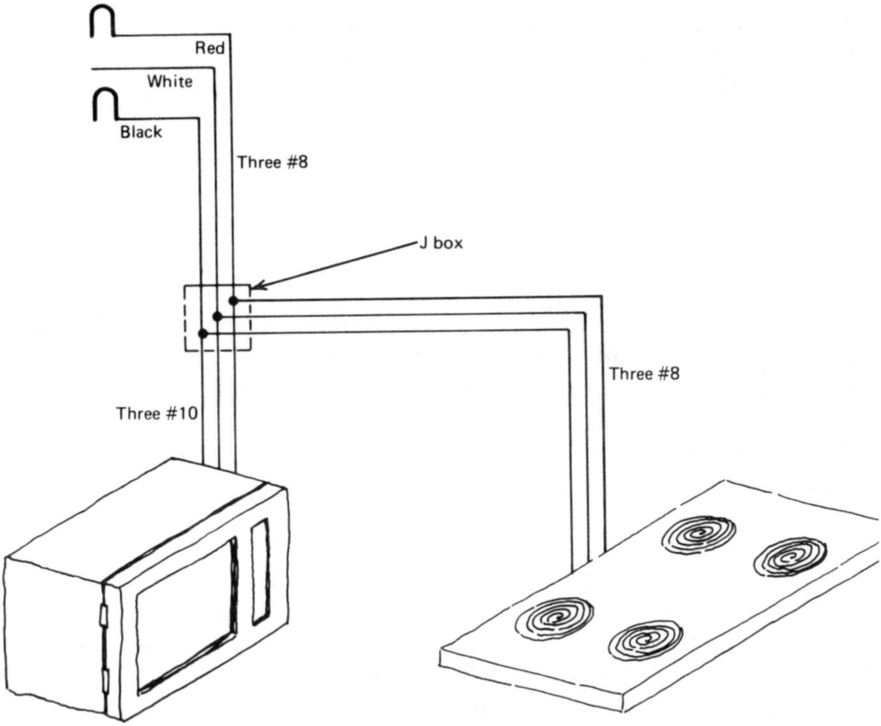

Figure 11–6 Wiring of separate wall-mounted oven and cooktop: single feed splitting at junction box.

from the power panel is made for a direct connection in cable or flexible conduit, a loose loop of a few feet should be left at the dryer for the same reason.

A gas dryer will need to plug into the same 120 V supply that operates the washer in order to power its timer. Also on the newer gas dryers the gas pilot lights have been replaced with a part called an ***intermittant ignition device*** (IID). This IID is used to ignite the gas electrically, and saves considerable energy, as the pilot is on only momentarily.

The small apartment size dryers are 120 V units. They can be plugged into any 120 V convenience outlet—the plug will fit. However, the heaters draw 10 or 12 amperes, and the motors a few more, so they will approach and in some cases even exceed the capacity of a 15 amp lighting circuit. To avoid difficulty connect them to 20 amp appliance circuits only. The required 120 V laundry circuit is fine, or any 20 amp small appliance circuit will do.

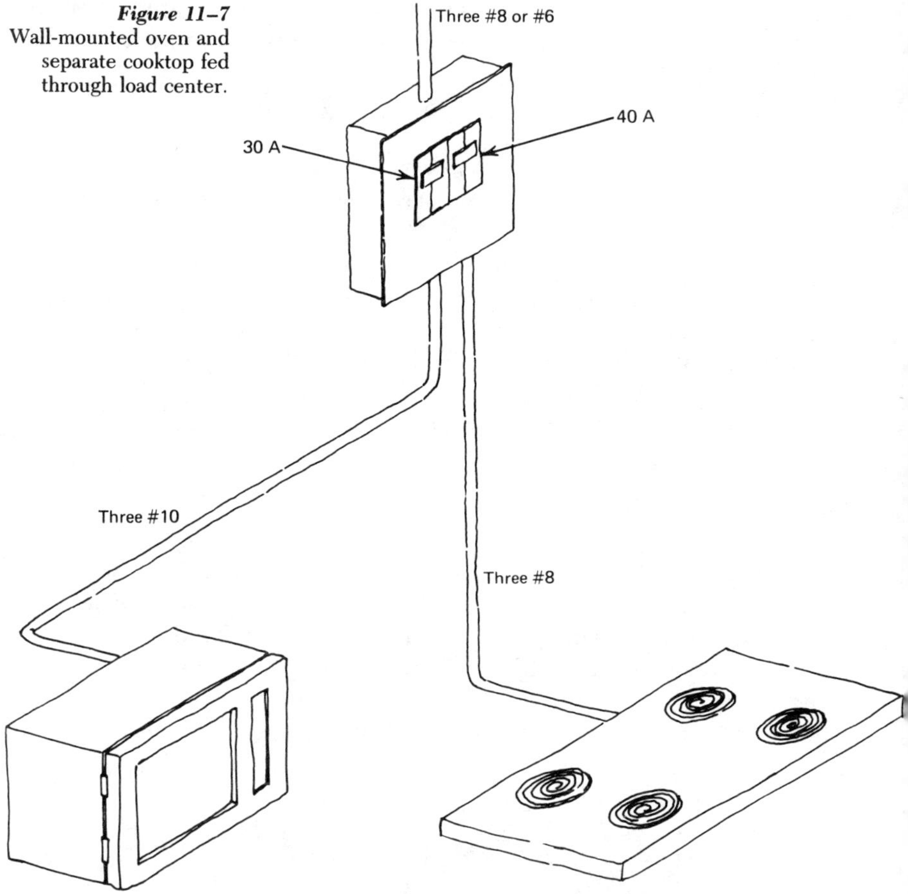

Figure 11–7
Wall-mounted oven and separate cooktop fed through load center.

Water Heater

An electric water heater will require another 240 V circuit. To determine wire and breaker size, check the nameplate on the tank. The power requirements for the heating elements will be given in wattage. There will be two heating elements, but the control is designed so that only one operates at a time; thus the maximum demand will be that of the larger of the heating elements, if they are not the same.

Suppose a water heater contains two elements rated at 2800 W and 3600 W respectively. The circuit will be designed for a demand of 3600 W at 240 V.

$$\text{Using Watt's Law:} \quad A = \frac{W}{V} = \frac{3600}{240} = 15 \text{ Amperes}$$

NEC 422-13 states that the water heater branch circuit rating must be no less than 125 percent of the nameplate rating of that heater, thus:

$$125\% \text{ of } 15 = 18.75$$

A 20 ampere 240 V circuit using #12 wire will be adequate for this heater.

Water heaters are direct wired not plugged. On the top (Figure 11-8) will be found a small removable metal plate held by one or two screws, which contains the stamping for a standard ½″ electrical knock-out. Under this plate are the wires to which incoming power is connected. NM cable, armored cable (BX), or flexible conduit may be brought through the knock-out, and secured to the plate with the usual box connector. If you use NM cable, inquire first at the local electrical inspection department. Some departments dislike this use of NM.

Disposal

This is a 120 V appliance, and may be either plugged or directly wired. While code directly states that the refrigerator may be connected to one of the two required small appliance receptacle circuits, the disposal is not so mentioned [NEC 210-52 (b)]. However, in some jurisdictions this disposal connection is permitted. Sometimes it is on a circuit by itself while at others it is doubled up with a dishwasher. In any event it is an

Figure 11–8
Top of water heater showing removable plate for electrical connections.

appliance; despite the fact that it draws little power, and does so very infrequently, it still goes on a 20 A circuit.

For plug-in connection, a box containing a receptacle is placed in the cabinet under the sink on which the disposal is mounted. When direct wired, a cable or a short piece of liquid tight conduit may be used. This cable or conduit will enter the appliance through a standard ½″ knock-out which will require the usual box connector.

Dishwasher

The only finish wiring required for the dishwasher will be the installation of the receptacle in the electrical box that was rough wired to the back of its compartment. If someone blundered and put the box on the side, move it to the back or the appliance installers will never be able to get the machine in place.

Trash Compactor

As with the dishwasher, the only finish electrical work required is the installation of a receptacle in the box provided for it when the rough wiring was done.

Garage Door Opener, Attic Exhaust Fan, Forced Air Furnace Fan

Assuming proper ***roughing in*** is complete, the only finish work required for any of these appliances is again the simple installation of a receptacle.

Light Fixtures

Incandescents Surface Mounted and Hanging Types

Where incandescent fixtures are surface mounted on a wall, mounted, or hung on a ceiling, octal metal or round plastic boxes were installed with the rough wiring. These boxes (Figure 11-9) have two mounting holes on the front threaded for 8-32 machine screws. To allow for variety in fixture design each fixture is supplied with an adapter plate (Figure 11-9). This plate is attached to the threaded holes in the box through the slots. The fixture is then secured to the adapter plate using its threaded holes into which fasteners furnished with the fixture are fitted.

The reason for the slots in the adapter plate is that although the

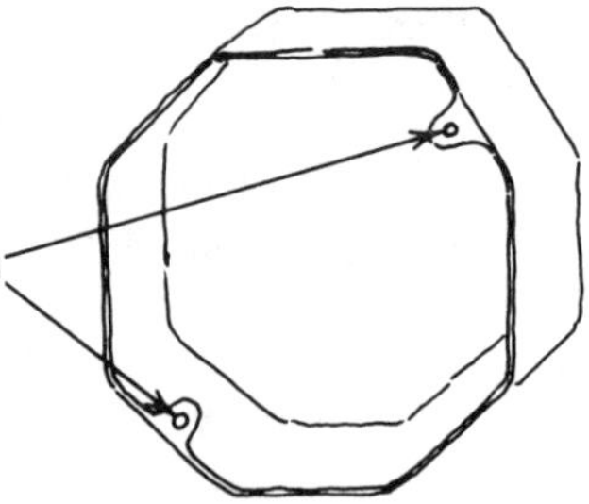

(a)

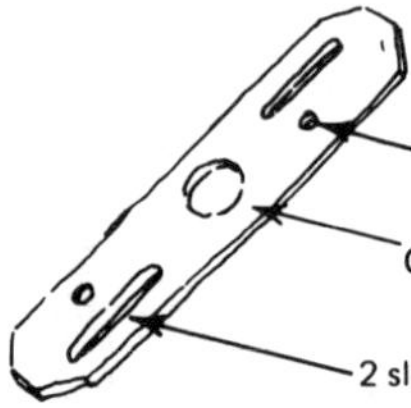

(b)

Figure 11–9 Octal box with typical fixture adapter plate attached using threaded mounting holes on flanges: (a) octal box; (b) adapter plate.

roughing-in is generally done using standard 4″ fixture boxes, from time to time 3″ fixture boxes may be used. The slots permit attachment of the adapter plate to either one. Adapter plates vary in other ways depending on the requirements of the fixture design. The spacing between the threaded holes on the adapter differs for different fixtures. In some cases the center hole is threaded for a ⅜″ nipple and the two small threaded holes are omitted. In other cases the two small threaded holes are replaced by one small one in the center.

Each fixture is accompanied by a drawing, or assembly instructions, or both. It is ***not*** a sign of weakness or stupidity to read and follow these instructions. There is a lively possibility that the fellows who made that fixture also know how to put it together and install it. Admittedly, instruction sheets are sometimes poorly written and confusing, but at least give them a try before striking out on your own!

Fixtures are often supplied with a pair of wire nuts considerably smaller than the smallest size used in the circuit wiring. This is because the fixture wires will be #16 or #18 stranded and must be securely attached to a #14 solid. The smaller wire nut is necessary in order to make a tight connection. In Chapter 3 it was noted that when splicing solid wires to each other with wire nuts it is not necessary to twist the wires together. When splicing stranded to solid as when installing fixtures, first twist the stranded around the solid, then cap with the small wire nut. Make sure they both turn in the same direction. Twist the stranded wire around the solid wire clockwise, then turn the wire nut down on the two of them clockwise as well.

Ceiling fixtures are normally not too heavy to be supported by a correctly installed fixture box. However, in the case of a large chandelier, the box support should be reenforced. Instead of using the usual offset bar hanger to hold the box, nail 2 × 6 across between the ceiling joists, then attach the box to that.

Recessed Incandescent Fixtures

No matter how elaborate or expensive it may be, one of these fixtures reduced to its essentials is simply a tin can facing down containing a very hot, and not very well ventilated, light source—an incandescent bulb. For that reason the installation of these fixtures must be done in accordance with code provisions intended primarily to reduce the danger of fires resulting from overheating.

As noted in Chapter 10 these fixtures contain their own junction boxes; thus no fixture boxes were provided for them in the rough wiring. When the plans call for power to feed from a switch to a single recessed fixture, as in the second floor baths in Figure 8-48, there are no special requirements relating to the fixture junction box. However, when the plan calls for the power to pass through several recessed lights in sequence, as in the family room in Figure 8-31, then the boxes on those lights must be approved for ***through*** wiring (NEC 410-11). A box that is approved for through wiring will be so marked. If it is not so marked ***do not*** use it for this purpose.

When a recessed fixture or surface mounted fluorescent is installed in a closet [NEC 410-8 (d)] only 6" horizontal clearance between it and storage space is necessary rather than the 12" required for a surface mounted fixture (Fig. 8-13).

In order to prevent the temperature of recessed fixtures from reaching dangerous levels, NEC 410-66 requires that the recessed parts be spaced at least ½" from any combustible material, such as wood ceiling joists. In addition there shall be 3" clearance between the fixture and any thermal insulation that is placed above the ceiling. This will prevent heat being trapped in the fixture.

While trimming thermal insulation 3" around each recessed fixture will encourage heat dissipation from the light when it is operating, the resulting gap in the insulation will also allow cold air to enter when the light is off. Responding to this fact and recognizing its effect on energy conservation efforts, a new provision, NEC 410-65, has been added to deal with this question. NEC 410-65 (c) requires recessed fixtures to have thermal protection and to be so marked. Also, of ***major*** importance, Exception No. 2 allows for the design and approval of recessed incandescent fixtures that may be in direct contact with thermal insulation making the insulation gap around them unnecessary. Such fixtures must also be clearly marked as suitable for this use. Any fixture ***not*** so identified will have to be installed in accordance with NEC 410-66.

Fluorescent Fixtures

In response to growing energy conservation efforts, a great deal of incandescent lighting is being replaced with fluorescents. Furthermore, a great many more initial installations are being done using fluorescents than in the past. While fluorescents generate a good deal less heat than incandescents, a fact often overlooked is that they are also vastly more temperature sensitive.

These fixtures contain a part called a ballast which is a type of transformer. The ballast will normally have a service life in the vicinity of 60,000 hours. However, if its operating temperature exceeds 90°C (194°F), that service life will be reduced. The higher the temperature, the greater the reduction.

At the other end of the spectrum the tubes are designed for optimum performance in an air temperature of about 25°C (77°F). While dropping off somewhat light output remains good down to about 10°C (50°F). Below that point light output decreases rapidly with decrease in temperature, and depending on variations in humidity, line voltage, fixture design, and other factors the lamp may not light at all unless special cold weather ballasts, or lamps, or both are used.

All of this means that when the finish wiring of fluorescents is done particular attention must be paid to the effect fixture mounting may have on operating temperature. Figure 11-10 shows the results of hanging a fixture at various distances down from a ceiling compared to mounting it directly on that ceiling. By hanging a fixture only 1″ below the ceiling the

	Ceiling	0.c
	½"	−1.5°c
	1"	−6°c
	1½"	−10°c
	2"	−14°c
	3"	−19°c
	4"	−20°c
	5"	−22°c
	6"	−22.5°c

Figure 11–10 Effect of distance from ceiling on fluorescent ballast temperature. The distance the light unit is suspended below the ceiling greatly affects the ballast operating temperature. Tests indicate a ballast case temperature variation of 22.5°C between a close-ceiling mounted lighting unit and a unit suspended 6 inches from the ceiling. Suspension distances greater than 6 inches have no appreciable effect on ballast case temperature. However, if a unit is suspended 1½ inches to 2 inches from the ceiling, the ballast case temperature will still operate 10°C to 14°C cooler than a surface-mounted unit. *Courtesy of Advance Transformer Company.*

ballast will run 6°C (10.8°F) cooler. At 2″ the temperature will be down by 14°C (25°F), and by the time it is 6″ down the ballast temperature has plummeted 22.5°C which is 40.5°F while the surrounding air temperature is unchanged. In cases where the ballast operating temperature is close to or exceeding the recommended maximum of 90°C, a small drop down from the ceiling can make a huge difference.

When the ambient temperature is too low, the solution is not as simple. Special cold weather ballasts that will start lamps at temperatures as low as −20°F can be obtained. These coupled with 800 M.A., 1000 M.A., VHO, or Power Groove tubes will often provide satisfactory low temperature service. Another factor that will help considerably is shielding around the tubes. Totally enclosing the lamp in a wrap-around plastic shield can raise the ballast operating temperature by as much as 14.5°C (26°F).

Although fluorescents, like recessed incandescents, need no junction boxes since the case is adequate there is a significant difference in that any fluorescent may be used for feedthrough wiring.

Breaker Box

Conductors entering a breaker box should be left far longer than the 6″ minimum required for junction or device boxes. Leave them 18″ long or more. A second look at Figure 9-11 shows why. The distance between the point where a conductor enters a breaker box and its point of connection in that box can be considerable. In addition it may have to go around a large block of breakers to get there. In the illustration the commons in the two conduits entering on the right side of the panel box must go down to the bottom of the box then across and all the way up the left side to reach the neutral bar where they are finally attached.

All commons and all grounds go to the neutral bar where they be attached to any available terminal screw. Power leads only (black, red, blue, etc.) go to the breakers. Never attach more than one conductor per breaker, and be certain that the conductor is large enough to handle the amperage that breaker will pass. #12 wire good for 20 A can be put on a 15 A breaker, but never put #14 good for only 15 A on a 20 A breaker. Keep all wiring neat and orderly so that if someone has to trace a problem, or make changes or additions, he will have no difficulty understanding the original installation.

> ***NOTE*** the exception: if the ground fault interrupter is included in one of the breakers, then and only then the branch circuit common wire (white or grey) goes to the terminal screw on the breaker marked with a white dot. The white wire furnished with the breaker then goes on to the neutral bar instead of the circuit common.

CHAPTER 12

ADDITIONS AND ALTERATIONS TO OLD WORK

When doing new work the electrician needs to know a good deal about the building structure in which he will work to determine how power will be distributed from the entry location to the various required use points. When doing new work the ***details*** of how a wood stud wall is framed, for example, are unimportant to the electrician. He simply comes in after the framing is up and runs his wiring through whatever is there, all of which is open and accessible. With old work, however, it is a different story. ***Old work*** refers to work in a building that has been finished. The term has nothing to do with the age of the building. The important difference is that the walls are now closed and finished. The framing along with any wiring or piping that may be inside are concealed and inaccessible. In order to add or change anything the electrician now needs to know exactly what is normally contained in that concealed structure.

The most common, and also the most complicated, internal structure is found in wood frame construction. However, due to the requirements of the Uniform Building Code, wood frame construction is also highly standardized. The electrician merely needs to familiarize himself with the basic characteristics of two framing systems. These are platform framing, also known as western framing, and balloon framing.

The basic parts are joists, studs, plates, headers, trimmers, and rafters are common to both systems. Most of the material used to make these parts is of nominal 2″ thickness (actual thickness 1½″ or 1⅝″), and of varying widths the majority falling between nominal 4″ and nominal 12″.

Platform Framing

Both framing systems start with bolting a piece, called the sill, to the top of the foundation wall (Figure 12-1-a). Resting directly on the sill are the floor joists which form the support for the sub-floor. The placement of a header across the joist ends completes the framing of the platform for which this system is named. The sub-floor is then nailed to the top of that frame.

The bottom frame member of the wall is the sole plate which rests on the sub-floor. On it stand the wall studs to which drywall, plaster, or paneling may be attached inside, and any of a variety of surfacings outside (Figure 12-2-b). Above the wall studs, a doubled top plate is placed. This doubled plate forms the support for the joists of the next platform or the ceiling, whichever comes next. In the case of another floor (Figure 12-1-b and 12-2-b), the sub-floor placed on those joists forms the base for the sole

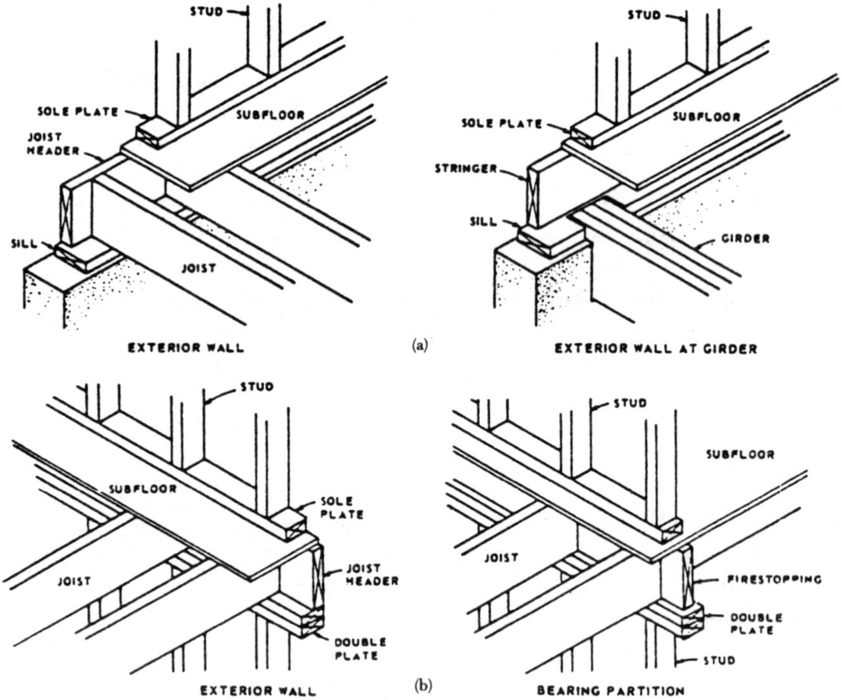

Figure 12–1 Platform framing: (a) first floor level; (b) second floor level. *Reprinted from* Modern Carpentry, *the Goodhart Willcox Company, Inc.*

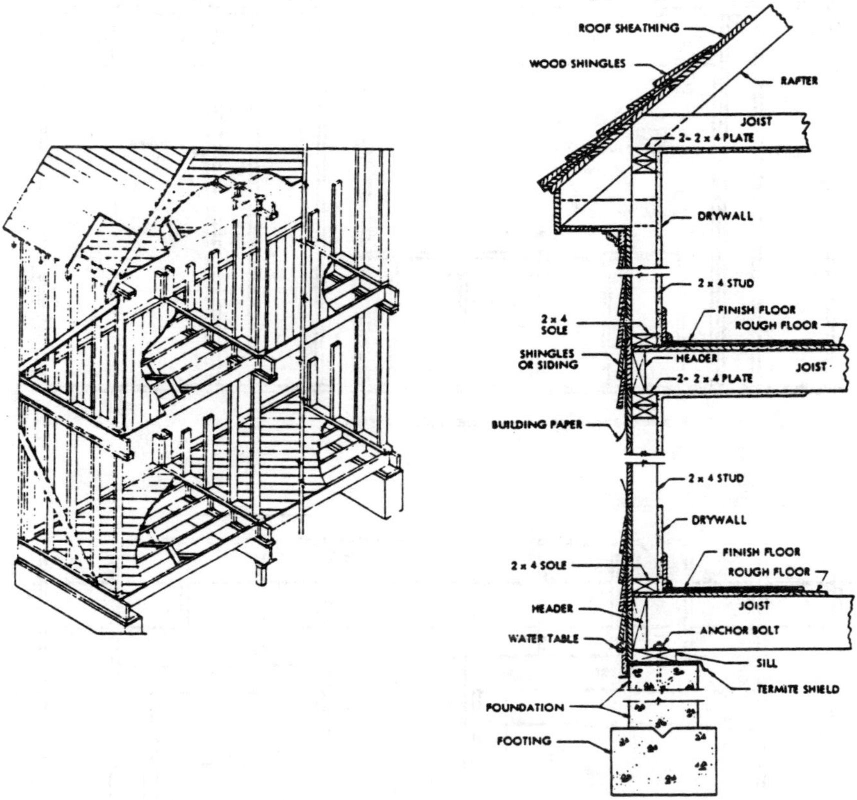

Figure 12–2 Platform framing: (a) perspective; (b) typical section.

plate and then the wall studs of that next floor. Each floor platform and wall frame is thus a repetition of the last one.

The basic wall frame (Figure 12-3-a) consists of regular 2 × 4 wall studs spaced 16″ apart on centers. Where a window or door opening is being framed, a regular stud on each side of the opening is doubled with trimmer stud on which the header forming the top of the opening is placed. In some cases the header is made thick enough to completely fill the space between the opening and the lower side of the top plates. This saves the labor of cutting and installing the short pieces called cripples that otherwise must go between the header and the top plates.

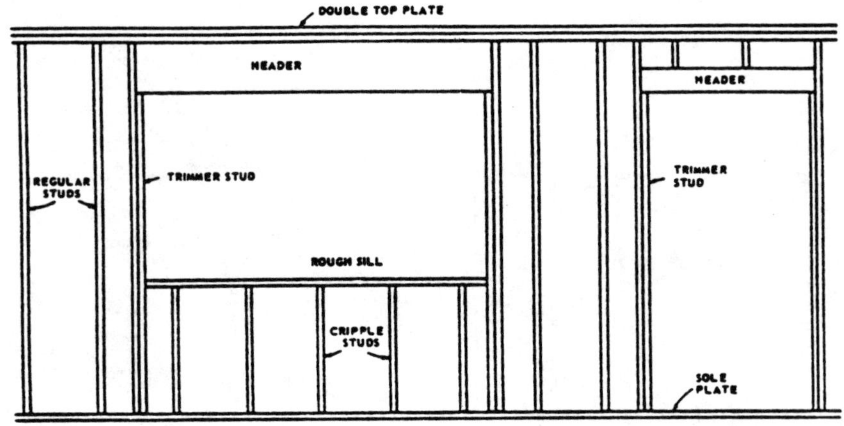

Framed openings in platform construction.

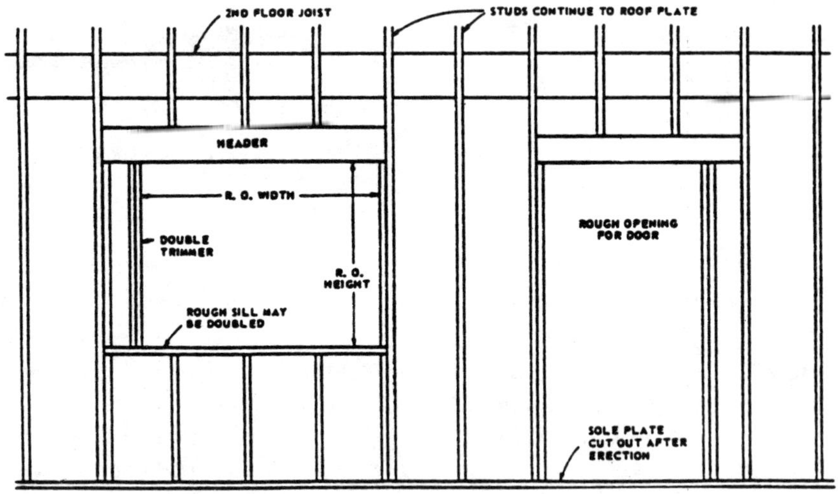

Framed openings in balloon construction.

Figure 12–3 Wall framing: (a) platform; (b) balloon.

To conserve energy the interior spaces between studs of exterior walls are filled with one of several types of insulation. The type and density used will vary depending on climatic conditions in the area where the building is located. Thus when planning additions or alterations it is important to remember that the spaces inside exterior walls are ***not*** going to be empty. This will complicate the job of wiring such walls.

Balloon Framing

As noted above balloon framing also starts with a sill bolted to the top of the foundation. The floor joists (Figure 12-4-a) again rest on that sill, but now the wall studs as well rest on that same sill. Joists and studs overlap and are nailed to each other in the case of a ***standard sill.*** In the case of a ***T***

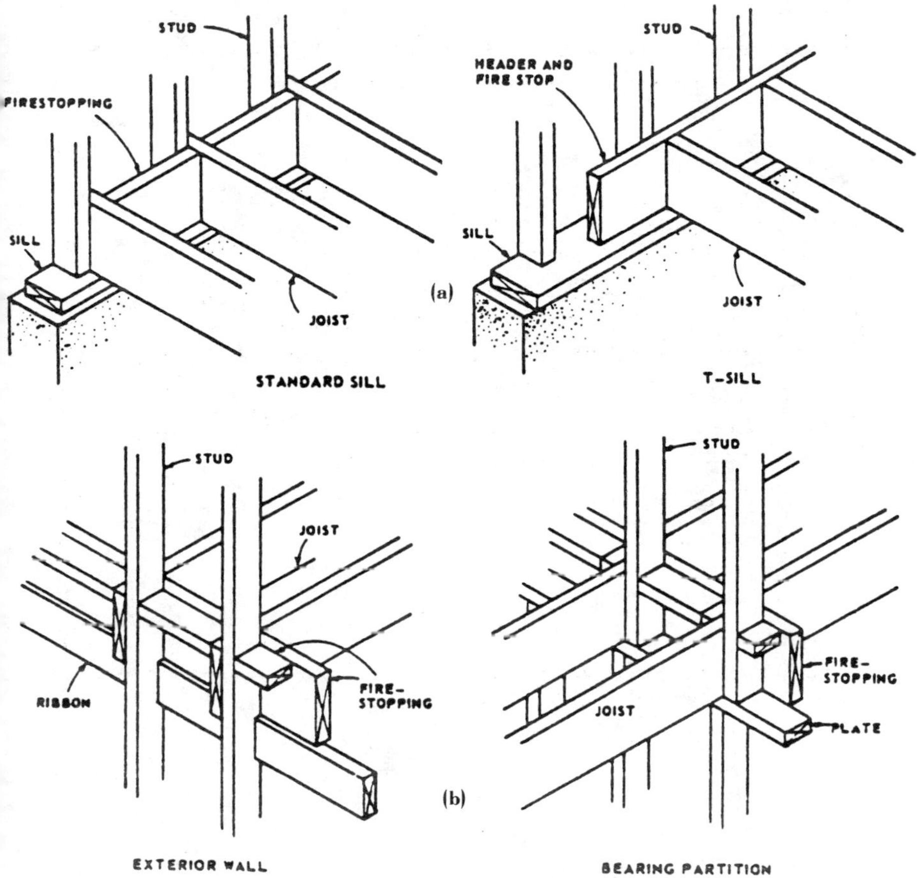

Figure 12–4 Balloon framing: (a) first floor level; (b) second floor level. *Reprinted from* Modern Carpentry, *the Goodhart Willcox Company, Inc.*

sill they abut. In either case the sub-floor ends at the studs instead of passing under them as it does in platform framing.

At second floor (Figures 12-4-b and 12-5-b) a piece called the ***ribbon*** is set into notches cut in the studs because the studs (Figure 12-5-a) continue uninterrupted on up to the roof plate. The second floor joists rest on the ribbon and are nailed to the studs. Note in Figure 12-4-b that firestops are inserted between both studs and floor joists because they will be in the way of fishing wiring between or under floors. Firestops are not needed in platform framing (Figure 12-1-b) except at bearing partitions because the wall top plates and joist headers between floors perform that function.

The spacing of normal studs in balloon framing is 16″ on centers exactly the same as in platform framing. However, the openings for doors and windows are handled differently (Figure 12-3-b). Instead of adding

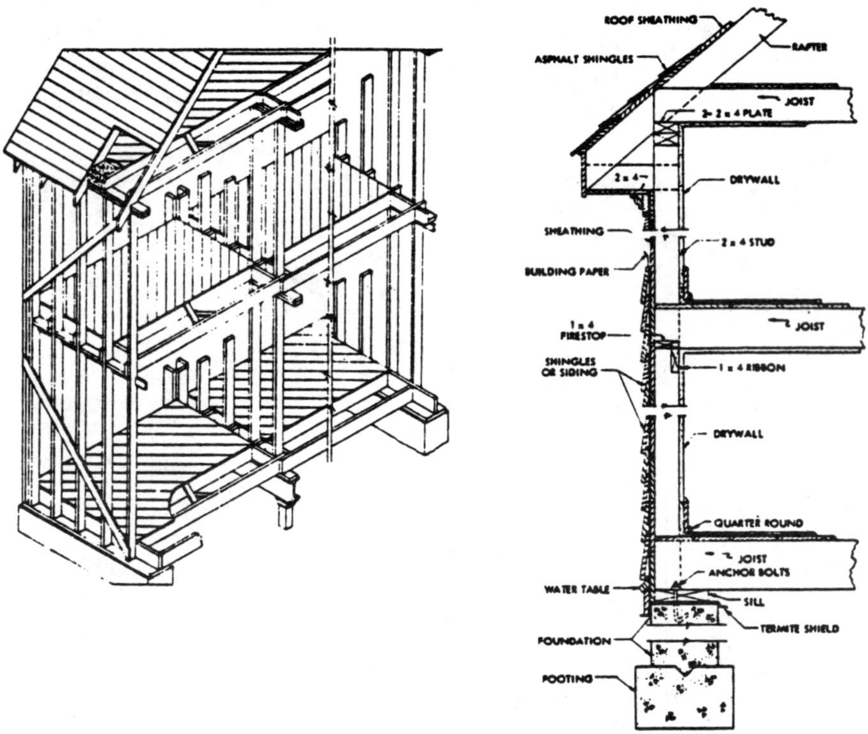

Figure 12–5 Balloon framing: (a) perspective; (b) typical section. *Reprinted from* Modern Carpentry, *the Goodhart Willcox Company, Inc.*

studs, trimmers, and a header cut to fit the required opening, the header extends beyond the required opening to the next standard spaced stud. The desired opening is then rough framed from the header down as shown.

The same types of both interior and exterior finishes are used with either platform or balloon framing. The same need for insulation in the exterior walls also applies to either framing method so the same insulation materials are used interchangeably.

Concealing Electrical Additions or Alterations

The most commonly desired minor changes to an electrical system, other than replacement of light fixtures, are the addition of receptacles, addition of switches and lights in new locations, or adding switching for pull-chain or self-switching fixtures. In utility areas exposed EMT will serve for these purposes, but in habitable parts of the building, its appearance is not satisfactory. It is preferable to conceal the new wiring in walls and ceilings or under floors in the same manner as the original wiring was hidden. Since the walls are now closed and finished, this requires fishing wiring through the inside voids, and calls often for the use of cut-in boxes (Figures 6-9 and 6-22).

Adding a Receptacle

To add a receptacle powered from an existing one (Figure 12-6), first make a hole for a cut-in box at the desired location. With a stiff wire feel back behind the wall for the first stud. Mark that distance on the wall to

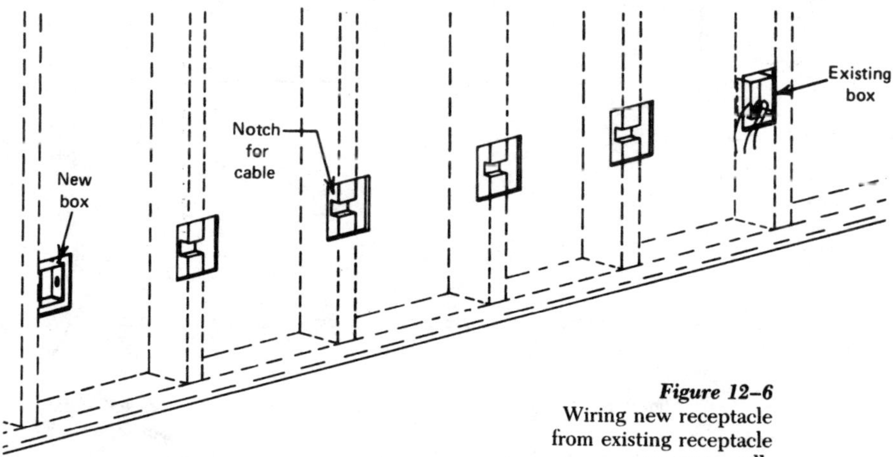

Figure 12–6
Wiring new receptacle from existing receptacle on same wall.

guide you in making a neat rectangular hole about 4″ wide and about 3″ high exposing the stud. At intervals of 16″ on centers there should be studs all the way back to the existing box. Make a neat rectangular hole to expose each of them. A notch cut in each stud will provide an open path for wiring from the existing box to the new one. Before inserting and securing the new box in its mounting hole, box connectors and cable, greenfield, or BX must be attached to it. Once it is in place only its interior is accessible. With the new wiring in place the holes over the studs are then patched and the wall repainted.

When the new box location is across the room from the existing one, wiring under the floor (Figure 12-7) may be a simpler way of running wires to the new one. This will require holes through the sole plates at the bottoms of the walls on both sides. These should be made from below. If the floor joists run in the same direction as the wire run, then the wiring can be secured to the side of a joist. Otherwise it will have to be secured to the bottoms of joists, or passed through a series of holes drilled through the joists. Where the wiring is passing over a usable basement area, taking the time to drill holes in the joists is probably worth while. If it is only a crawl space, certainly not.

If there is a doorway between the existing box location and the new it will be necessary to go over, or under it (Figure 12-8). When a basement or crawl space exists making it possible to drill up through the sole plates, this is the shortest and easiest way. Where a slab floor makes this impossible, the attic route is best. If both ways are closed the only alternative left is over the header and through the cripple or cripples above it. In case there is a full header all the way to the top plates (not an uncommon situation), go back to the drawing board, and start again; you cannot get through. The wise will tap above the doorway ***first*** to determine whether a path exists.

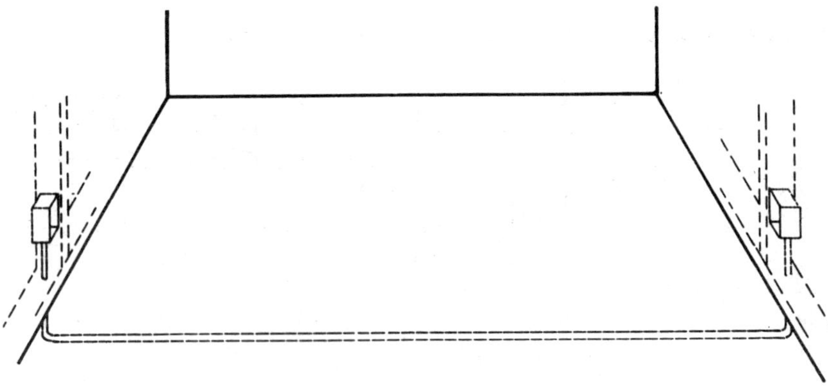

Figure 12–7 Power new receptacle from existing receptacle: under floor to opposite wall.

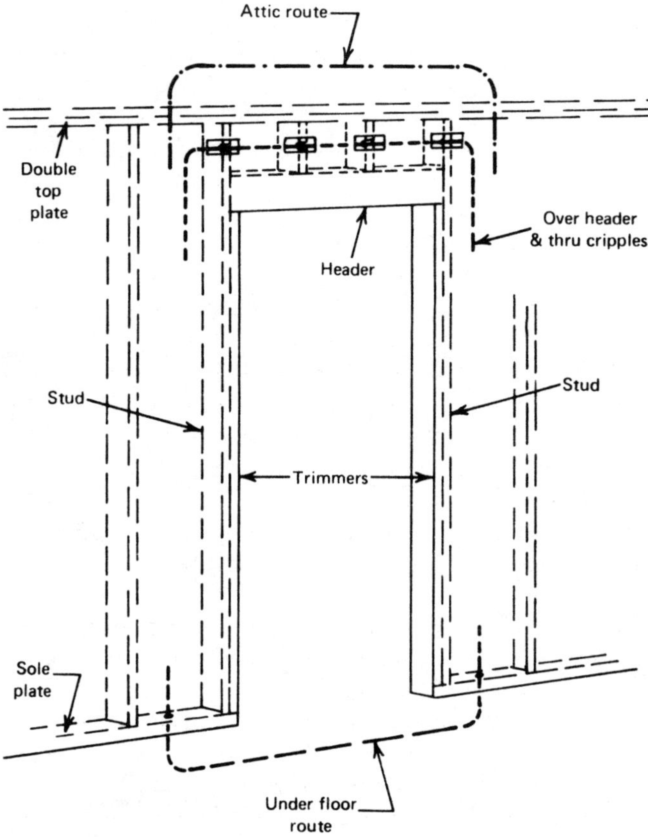

Figure 12–8 Wiring around doorway.

Adding Switch for Existing Fixture Location

In older buildings, particularly, light fixtures with self contained switches are often found. When modernizing these are generally replaced with fixtures that require wall switches. This is easily accomplished by running a switch loop (Figure 7-11) from the fixture box to a switch box (Figure 12-9) usually located on the latch side of a doorway. After the wiring has been dropped down into the wall from the hole in the top plates, it is pulled out and connected through the box hole beside the door. If it should be elusive a hook made from a wire coathanger will catch it.

If it is impossible to drill through the top plates from above, an access hole can be cut into the wall, and out on the ceiling (Figure 12-10). After the top plates have been notched, wiring is pulled to and from this access hole with the fish tape (Figure 3-16). The wiring is then secured in the notches and the holes patched in wall and ceiling.

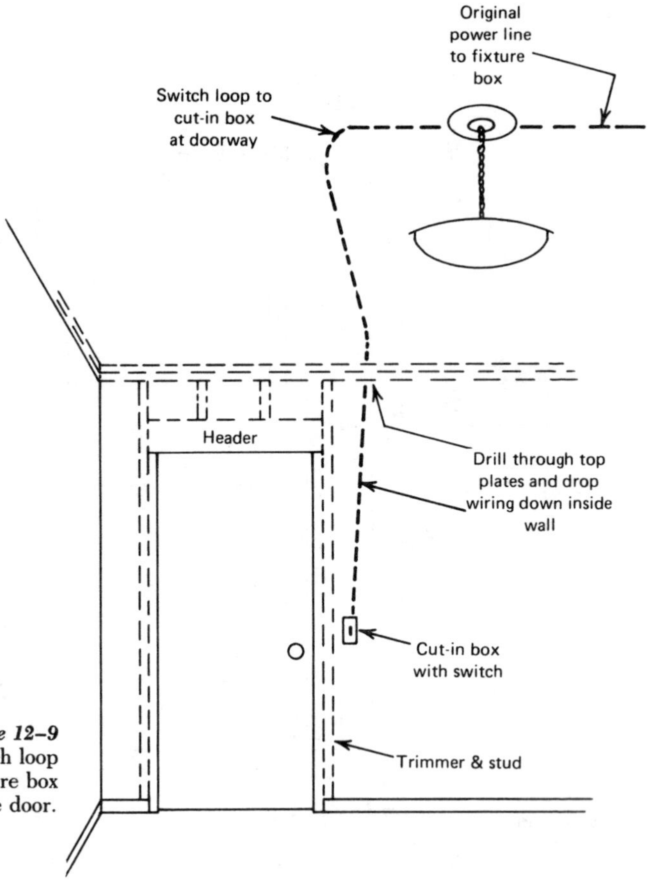

Figure 12–9
Running switch loop
from ceiling fixture box
to cut-in beside door.

Adding Switched Fixture

To add a switched fixture in a new location, a new box for the switch is needed. Next there must be a box on which to mount the fixture, ***unless*** it is a recessed fixture with its own junction box, or a fluorescent fixture, the case of which is acceptable as a junction box. Thirdly, a two-wire run has to connect the switch to the fixture, and finally a power input is needed at one end or the other.

A cut-in box will readily fit the switch. When the wiring is in NM cable, the fixture box may be a round plastic cut-in type as well. Where the wiring requires a metal box, octal cut-ins are not available. The metal

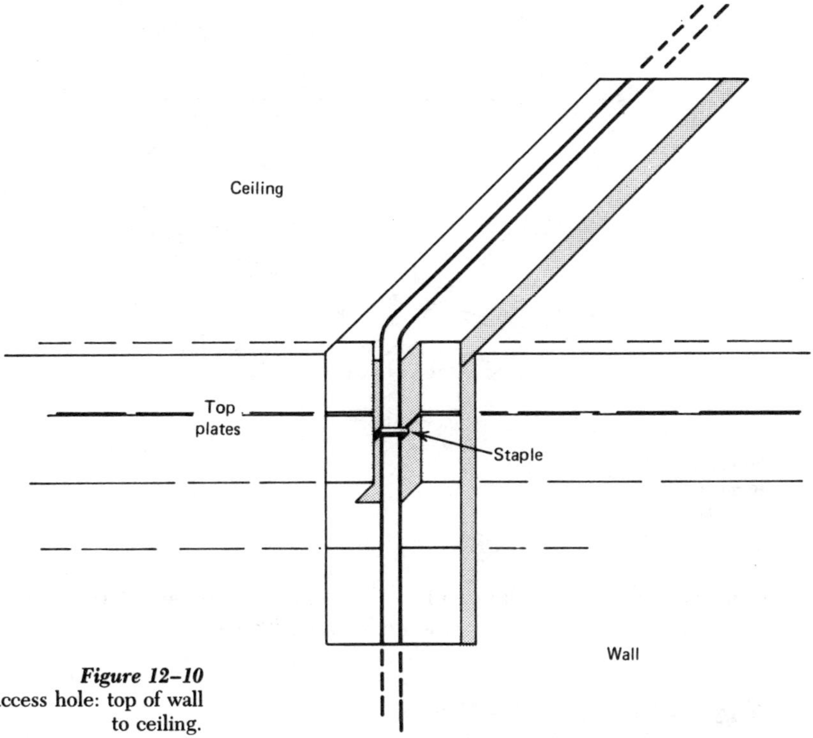

Figure 12–10
Access hole: top of wall to ceiling.

octal will have to be mounted from above in an attic. If attic mounting is impossible, make a hole for the box from below, then cut (Figure 12-11) slots in the ceiling about ¾″ wide from that hole to the joists on either side so that an offset bar hanger can be inserted to support the box. After the bar hanger and box are in place the slots are patched.

The wire run from fixture to switch box will be across the ceiling, then down inside the wall. When it is impossible to go through the wall top plates to get inside the wall, then the access hole technique (Figure 12-10) can be used.

With both switch and fixture boxes installed and the two-wire run connecting them in place, the power input is still lacking. If the fixture box and connecting wiring were installed from above, it may be that a junction box containing unswitched power can be located and a power input run from it to the fixture box. From there the job is completed by using the two-wire switch box connecting run as a switch loop.

If the fixture box had to be installed from below, or if no suitable power connection can be found in the attic, it may be necessary to pick up power from a receptacle box and go by whatever route is necessary into the switch box. If both boxes are on the same wall, notching studs as in Figure 12-6 may be required to get from one to the other. If they are on

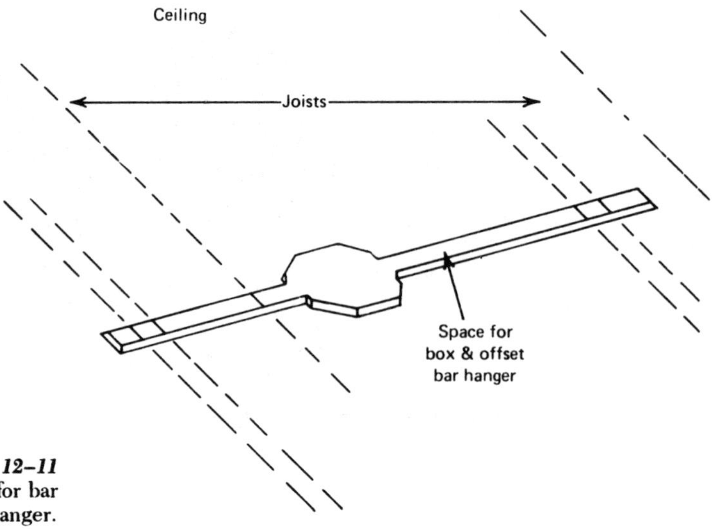

Figure 12–11
Ceiling cut for bar hanger.

different walls the under floor route (Figure 12-7) may be the best method, and if a doorway must be crossed see Figure 12-8.

Adding 240 V Circuit—Dryer, Heater, Airconditioner, or Other

240 V circuits are only needed for devices that require considerable power. Consequently, the first step before starting to install such a circuit is to determine whether the incoming service can handle the proposed additional load. This involves calculating the present loads on the system, adding the proposed load (or loads), and comparing the total obtained to the rated capacity of the service using the methods described in Chapter 8.

Assuming the service can handle the load, the next step is to find suitable breaker spaces in the distribution box. If you find a half dozen empty slots in the breaker box, the work will be greatly simplified. This seldom occurs. More often every slot in the breaker box is filled, and two more that must be next to each other are needed. The answer is going to be ***piggy-back*** breakers (Figure 12-12). These allow two breakers to be mounted in the space normally occupied by only one. Doubling up four circuits on two piggy-backs will open up enough room for the necessary 240V double breaker.

Both the rough and finish wiring of the circuit will be the same as for any other 240V circuit previously described, except that physically wiring and placing boxes is going to be considerably more awkward, since the work is being done in a completed building.

Figure 12–12
(a) standard breaker; (b) piggy-back breaker.

Adding a 120V Lighting Circuit

Remodeling work that includes room additions, space conversion, or changes in the sizes, shapes, and uses of existing rooms will usually require additional lighting and convenience outlet circuits. Sometimes added rooms can be supplied by using available capacity from one or more circuits supplying adjacent areas. More often this is not possible.

The design of such a new circuit will be the same as any of those discussed in Chapter 8. It will most likely be a 15 A circuit, wired in #14 wire, and will be limited to a computed connected load of no more than 80 percent of its capacity, or 1440 W. Both rough and finish wiring will be done in conventional manner. The only problem may be that there is no addition space in the distribution box. However, that is easily solved by using another piggy-back.

Exposed Additions and Alterations

Sometimes optional and other times necessary electrical system alterations or additions in habitable rooms must be done with exposed wiring. While EMT makes an appallingly ugly job when used for this purpose, another material, Wiremold, discussed in Chapter 5, is quite satisfactory.

Wiremold is a surface raceway furnished in 10′ lengths—the same as EMT—and in several different cross-sections. These are shown approximately ½ size in Figure 10-14. The most commonly used raceways are the 200, 500, and 700 sizes. The larger raceways (No. 1000 and higher) are used in office and commercial installations. The saddles (No. 1500 and No. 2600) are for floor runs in offices, showrooms, stores, etc. A very considerable variety of surface mounting fixture, device, and switch boxes is available adequate to meet most normal requirements. Unlike EMT, the Wiremold raceway sections are not bent. Changes in direction are made by cutting and joining raceways with *L*s and *T*s (Figure 12-13).

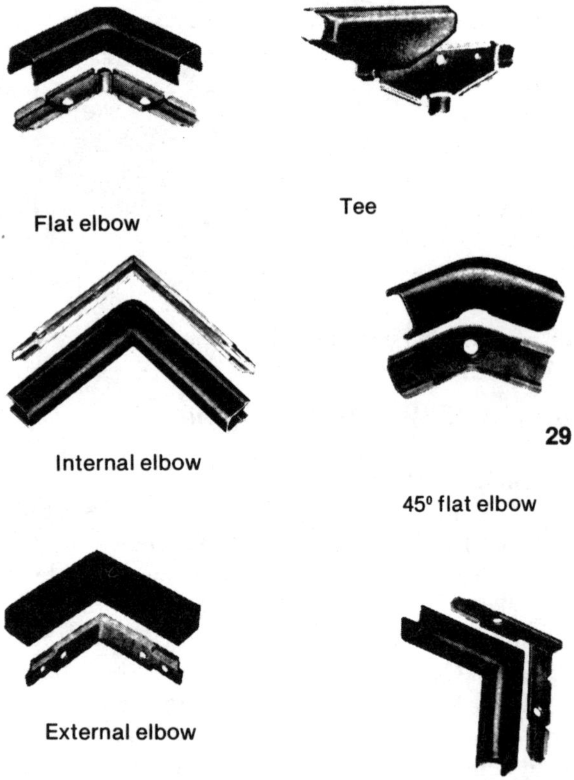

Figure 12–13
Wiremold connectors.
Courtesy of the Wiremold Company.

Both the raceways and the various boxes and connecting units each consist of two parts: a base of galvanized steel and a painted steel cover. On all boxes the covers are screwed to the base plates. On ***T***s and elbows the covers are snapped onto the base plates. The base plates of all fittings are punched for fasteners. Each length of raceway is supplied with a connector, which is punched for fasteners. The bases of all fittings (boxes, ***T***s, and elbows) are furnished with curved tongues, which fit behind the base strips of the raceways.

The parts of this system are assembled progressively. First the base plate of a fitting, a device or fixture box for example, is fastened to the wall. A precut length of raceway is then slipped onto one of its tongues. The baseplate of another fitting is slipped into the far end of that raceway, and is then also fastened to the wall. The raceway section is now securely held between the baseplates of the two fittings. Next another length of raceway is connected to the second baseplate. That raceway will be fixed in position by the baseplate of another fitting that will be placed at its far end. This progression is repeated as many times as necessary.

The raceways are cut with a fine tooth hacksaw. Since they cannot be reamed the cut ends should be cleaned with a small file to remove the burr from the sawing. A fiber bushing similar to the one required to the ends of BX cable is available for use in wiremold raceway ends. However, the author has seldom seen them used. Light filing is the normal procedure.

Those who have been paying attention have noticed that so far nothing has been said about putting the covers on any of the boxes or fittings. What is on the walls and/or ceiling at this stage is a number of baseplates for various fittings connected to each other by lengths of empty raceway. At this point, before any covers are installed, the wiring is pulled. The same fishtape used for EMT will do this job. Through many short raceway lengths the fishtape will be unnecessary as #14 wire, the smallest that may be used, is still stiff enough to slide through without it. As with other wiring methods runs are to be uninterrupted from box to box.

After the wiring is pulled leave the usual 6″ tails on all wires in all boxes for use in making connections. As when wiring in EMT or greenfield any conductors that merely pass through a box without making connections do not get cut. The finish wiring of switches and fixtures is the same as when the power wiring is concealed. Only when wiring the grounds on receptacles may the finish wiring differ.

A surface raceway wiring run correctly installed as described above provides continuous metal to metal contact from the tongues on the galvanized fitting baseplates to the galvanized backing strips in the lengths of raceway joining them. This continuous metal to metal contact is considered according to code as an adequate safety grounding path. Consequently, no grounding conductors need be pulled in surface raceways.

Receptacles may be installed on either of two types of boxes (Figure

(a)

Figure 12–14
Wiremold exploded receptacle boxes: (a) receptacle attached to flanges from base plate; (b) receptacle attached to painted box cover. *Courtesy of the Wiremold Company.*

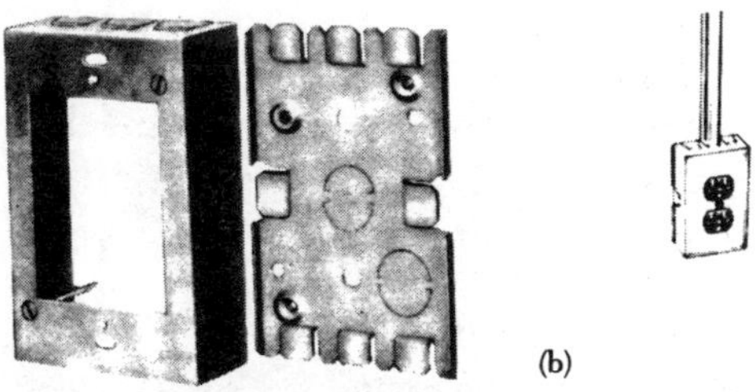

(b)

12-14). When the yoke of the receptacle is mounted on flanges (Figure 12-14-b) that are parts of the baseplate, removal of the fiber washers on the receptacle mounting screws will permit direct metal to metal contact forming a positive ground path from the yoke to the baseplate. However, when a receptacle is mounted on a painted metal cover plate (Figure 12-14-b) a wire pigtail from the ground screw on the receptacle to one of the fasteners holding the baseplate to the wall is desirable.

When installing switches or receptacles in boxes of the type shown in Figure 12-14, the box cover must go on before the device can be mounted. Machine screws for this purpose are provided with the box. Be careful that they are not lost in the process of doing other preliminary work as they are special screws, and therefore difficult to replace.

Adding Surface Wired Receptacles

To add surface wired receptacles powered from an existing box, start by removing the existing receptacle. The base plate of a surface mount adapter box (Figure 12-15) is now put in its place using the same mounting holes on the old box. To one or more of the connecting tongues on that baseplate attach raceways and whatever fittings are necessary to reach the location, or locations, of desired additional outlets. Fish new wiring through raceways and splice to existing wiring in the old box along with pigtails to reconnect the old receptacle. With raceways attached to box baseplates one or more knockouts will have to be removed from each box cover before those covers can be mounted. Remembering the code provision (NEC 370-18) that no unused openings may be left in a box, be careful to break out only the correct openings, for no plugs are available for these boxes.

The difference between the adapter box and the standard mount device box is the large rectangular opening in the baseplate of the adapter box. For the reason just mentioned the adapter may only be used when attached to an existing box.

Figure 12–15 Wiremold device adapter box from concealed to surface wiring. *Courtesy of the Wiremold Company.*

Adding Switch for Existing Fixture Location

Surface mount fixture boxes, like surface mount device boxes, are available in both standard type with full baseplate and adapter type with a large round hole in its baseplate. To add a switch to an unswitched fixture box start with the round adapter.

Working from a connecting tongue on the adapter baseplate attach raceways and fittings to reach the desired switch location. Pull a pair of wires for the switch loop, and complete finish wiring in the usual manner. If additional wall receptacles are desired as well, pull a third conductor for the common down to the switch box. Connect raceways from there to the receptacle locations and complete as above.

Adding Switched Fixture at New Location

This time use the fixture box with the complete baseplate. If power will be brought into the circuit at the fixture box, that baseplate has a standard ½″ knockout through which it may enter. Raceways from fixture box to switch box are connected as previously described.

If power will enter at the switch box, it probably will have to come from a nearby receptacle. In this case move the receptacle forward onto an adapter box. Raceways can now be run from there to the switch location, and from there to the fixture location.

CHAPTER 13

TROUBLESHOOTING AND REPAIRS

Common electrical difficulties will fall into one of two general categories: either some part of the permanent building electrical system has failed or some piece of equipment using power from that system has broken down. Our concern here is with the first category.

Consequently, we eliminate discussion of the servicing of all of the small plug-in appliances, except to the extent that they cause malfunctions in the electrical supply system. The same is true for the larger appliances such as ranges, dryers, water heaters, ovens, microwave ovens, disposals, air conditioners, and baseboard or radiant heating systems. We are concerned with failures in these appliances only when they cause difficulties with the building electrical system. The electrician who installs or maintains the power wiring in buildings is not generally an appliance repairman. Appliance repair is another independent, and very extensive, field of activity—far too extensive to be addressed as only part of a chapter in a book on wiring buildings. Discussion of appliance repairs here will be deliberately very limited.

Shorts

In accordance with code, all branch circuits must be safeguarded by overcurrent protective devices which may be either breakers or fuses (NEC 210-20). As was discussed in Chapter 9, these devices shall have a current rating no greater than the ampacity of the wiring in the circuit. Thus since for example a copper wire size #12 AWG with THW as insulation is rated for 20 A, the breaker or fuse protecting that wire may be rated at no more than the same 20 A. If the breaker were smaller, that would be permissible, but it may never be larger.

In the event any piece of electrical equipment fails to function, the first step in troubleshooting is to verify that power is being delivered. While it may sound absurd, it is also astonishing how much time the author has seen wasted puzzling over equipment connected to dead power supplies. What makes this doubly absurd is that testing the power supply is so simple. This can be done using the AC scales on a voltmeter, or simpler yet a small pocket test light.

Where the appliance is plugged into a receptacle, pull the plug out and insert test probes in its place (Figure 13-1). At times the test light or meter probes fail to make a good connection with the interior contacts in the receptacle. Such cases result in a false indication that the receptacle is dead. In case a receptacle appears dead, make absolutely certain by removing the cover plate in order to test the terminal screws directly (Figure 13-2). If they too produce no reading, move on to the next step.

Shorts in Appliances

When a receptacle is dead, move to the breaker box and look for a tripped breaker (Figure 13-3). When one is found, switch it to the ***OFF*** position, and try resetting to ***ON***. If it will not stay, but drops back immediately to

Figure 13–1
Checking receptacle with test light.

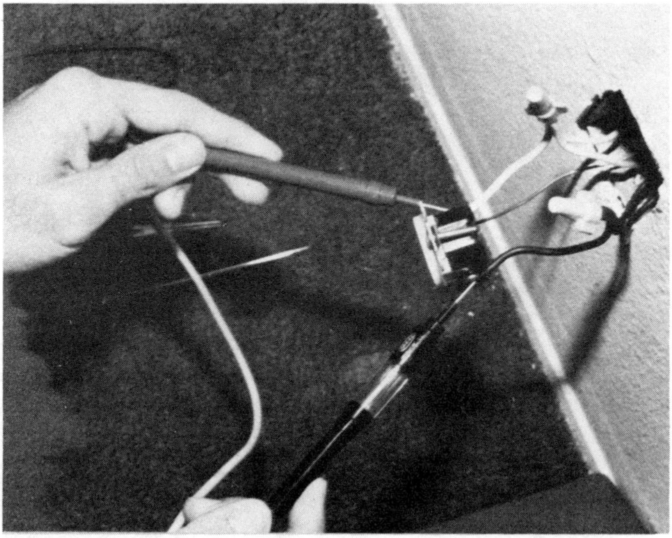

Figure 13–2
Testing receptacle at terminal screws with test light.

the ***tripped*** position, *do not attempt to hold it at* **ON**—do not *attempt again to reset it*. Either attempt will be useless, and holding it in could be dangerous as well. Somewhere the circuit is shorted. This means that at some point the ungrounded *hot* side of the line is in direct contact with either ground or common or both.

If you are working in an old building it may be that the overcurrent protective devices in use are fuses. In residential structures, these will almost always be plug fuses (Figure 13-4). When the fuse is blown, the discoloration of that transparent top will be obvious. The standard old Edison base plug fuse is made in three sizes: 15, 20, and 30 amperes. This is a potentially dangerous situation because of the ease with which a 15 amp circuit can be over fused with a 20 or 30 amp fuse. To prevent overfusing, or under fusing for that matter, all standard plug fuses should be replaced with *S* type plug fuses (Figure 13-4). Only the fusable inner part is removable, and each adapter will accept only one fuse size. Thus once a 15 amp adapter is in place that circuit can no longer, either

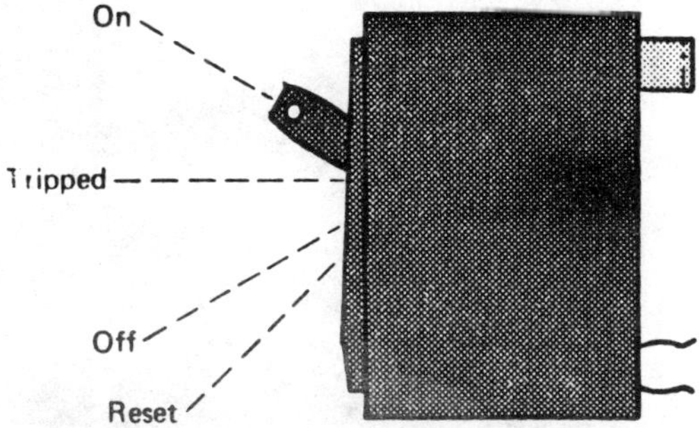

Figure 13–3
Toggle positions on a "Square D" breaker.

(a) (b) (c)

Figure 13–4
S type plug fuses: (a) 0–30 amperes; (b) adapter; (c) type delay (dual-element).

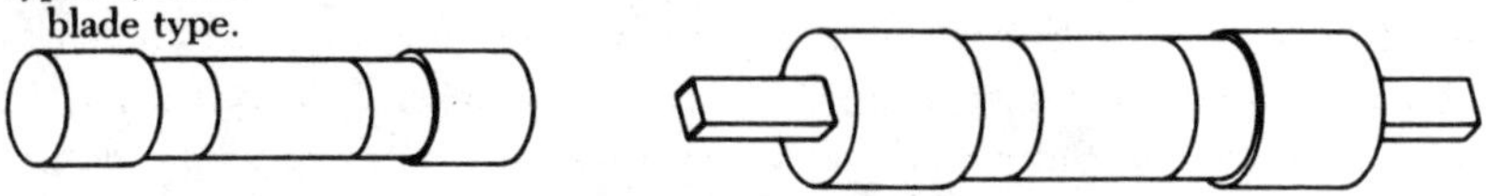

(a) (b)

Figure 13–5
Cartridge fuses: (a) ferrule type; B) knife blade type.

accidentally or on purpose, be overfused with a 20 or a 30. Many jurisdictions are now requiring all standard plug fuses to be replaced with *S* type.

In fused disconnects, or if you prefer fused switches, the fuses are generally cartridge type (Figure 13-5) with either ferrule contacts or knife blade contacts. With these fuses there is no visible indication whether a fuse is good. They can be tested in place with power on using a voltmeter, or they can be tested separately with an ohmmeter. In place with power on, if the fuse is good the voltmeter will read 0 volts across the two ends. A blown fuse will show 120V difference between the two ends. A good fuse tested by itself with the ohmmeter will show continuity (0 ohms resistance) between the two ends, a bad one will show no continuity (infinite resistance).

Figure 13–6
Testing for short in power lead.

When it appears that a short exists the next step is to try to isolate it as either in some piece of equipment, or in the building wiring. Unplug or disconnect all readily removable equipment from the circuit, and turn all building switches *on,* and try the breaker again. If it now holds in the *on* position the short must be in one of the units that was just disconnected. If it still will not stay on, the short has to be in the circuit wiring, or in one of the permanently wired-in devices such as a receptacle, a switch, or a fixture.

If, instead of a breaker, the overcurrent protective device is a plug fuse the circuit can be tested by replacing the blown fuse with a new one after disconnecting all using equipment as just described. Of course if the short is in the building wiring, the second fuse will promptly blow and follow the first into the trash barrel. Blowing out a second fuse to no good purpose could be avoided by shutting off the main disconnect to the fuse box and testing the circuit with an ohmmeter for continuity between the power wire coming from the fuse, and the ground buss. If the meter reads a very low resistance a light bulb or two are still in place somewhere. If it reads absolutely no resistance there is a short at some point in the circuit.

When, after the removal of the equipment from the circuit, the breaker holds at *on,* or if it is a fused circuit, the fuse holds, the difficulty is a matter of appliance or equipment repair that is beyond the scope of this book.

Building Circuit Shorts

Given a circuit protected by a breaker that will not hold in the *on* position after the using equipment has been disconnected, one last check remains to be made before we will believe absolutely that a short exists in that circuit. Disconnect the power wire from the output terminal of the breaker (Figure 13-6), and test with the ohmmeter from it to the ground buss in the breaker box. A zero reading eliminates breaker failure from consideration, and demonstrates the certain existance of a short somewhere out in the circuit.

Wiring problems in the run between boxes inside a wall, under a floor, or across an attic are possible, however, unlikely. The vast majority of shorts will occur inside an electrical box either in the wiring that has been stuffed in, or in a wiring device or fixture mounted there. Tracing down the short requires going through the circuit from box to box, opening each one. After examining the box carefully for any indications of defects, test every entering cable for shorts with the ohmmeter. To accomplish this all splices must be opened so each cable can be checked completely independent of the others. Most commonly when testing cables in this way, if there are two or three or more cables entering a box, one of them will show zero resistance between its conductors (indicating

Figure 13–7
Disassembled duplex receptacle.

the short), and the others will indicate there is no problem. If the sequence of boxes in the circuit is known, or can be determined, tracing the short is a simple matter of following the shorted cable to its next box, and its next after that if necessary, until the short is found.

When the circuit sequence is unknown and cannot be accurately reconstructed, tracing a short can become a rather time consuming and frustrating exercise. You must proceed in hit or miss fashion until luck allows you to hit the trouble. This is contrary to the way one normally does electrical work.

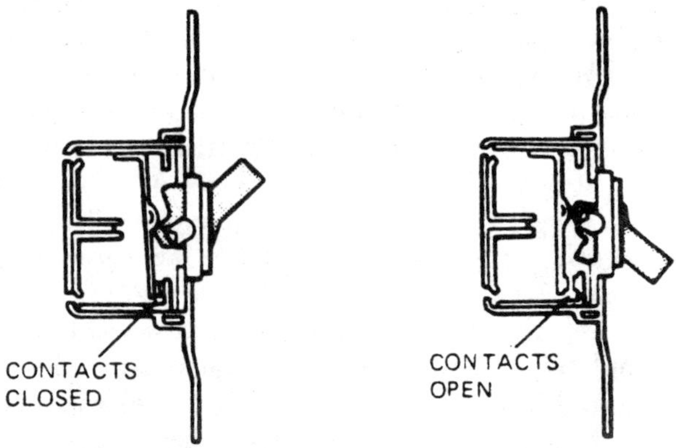

Figure 13–8
Inside single pole switch showing contact points. *Reprinted from* Electrical Wiring Fundamentals, *McGraw-Hill Publishing Company.*

Overloads

The same overcurrent protective devices (breakers or fuses) that are activated by short circuits are also activated by overloads. However, the symptoms of overload look rather different from a short. In case of a short the tripped breaker absolutely will not reset until the short has been removed from the circuit. In the event of an overload that trips the breaker, it ***will*** reset after a few minutes cooling time. How long it will hold before it trips again varies depending on the magnitude of the overload, and the make of the breaker. It might last only a minute or two, it might hold for 5 to 10 minutes. It is certain that as long as the overload remains connected to the circuit, the breaker will trip again, but there will be a perceived time lag. The immediate and obvious difference, then, between a short and an overload is that although the breaker will not hold the ***ON*** position in either case, it can temporarily be reset when the circuit is overloaded, but not when there is a short in the line.

The same time difference will be observed in a fuse protected circuit. A short will instantly blow a fuse even as it is being inserted. An overload will permit a time lag, even if it is only quite short.

What happens during an overload is that the circuit is required to deliver more power than the wires and the breaker are rated to supply. Breakers, and some fuses, will permit temporary overloads such as the starting surge of tungsten lights, but when the overload persists the breaker trips, or the fuse blows. Increasing the capacity of the breaker or fuse is ***not*** the proper solution.

Replacing a 15 ampere breaker, or fuse, with perhaps a 30 ampere is a deceptively simple ***wrong*** answer. While there is every likelihood that the breaker, or fuse, will hold the circuit ***ON***, the wiring and its insulation are now operating under overloaded conditions which will overheat the wires. Assuredly the wires will not get hot enough to melt the metal, but just as assuredly they can get hot enough to melt the insulation.

If that happens arcing can occur between a now inadequately insulated power lead and a common wire or ground. Another possibility is that the melting of the insulation may allow the hot wire to directly touch the common or ground and thus short out the circuit. Neither of these consequences is desirable, but of the two the arcing is the more dangerous. A dead short will activate the circuit overcurrent protective device, which will shut off the power. In a situation involving arcing the overcurrent protective device might not be activated, permitting the arcing to continue for some time. This sort of thing has produced some very lively and destructive fires.

The only correct cure for an overload is to reduce the power demand on the circuit until the current being drawn falls within the amount that circuit is rated to supply. This means one or more items must be disconnected from the overloaded circuit, and transferred.

In order to decide what load, or loads, to transfer, it is necessary first to know the magnitude of the overload. This can be determined directly with a meter, if one is available; if not it can be done by calculation. If an Amprobe, or other clamp-on field sensing ammeter is available, clamp it around the power lead coming from the breaker, switch the breaker on, and quickly read the amperage being drawn. The breaker will usually hold long enough for such a reading to be made. With the meter reading available, inspect the area served by that circuit, and list each load individually. Total the listed loads, then compare that total with the meter reading. If the sum of loads is significantly lower than the meter (0.5 amps or more) a load has been missed. If the sum is significantly high, obviously some load belonging on another circuit has mistakenly been included.

With the total actual load on an overloaded circuit known, and also broken down into its component parts, the difficulty may easily be corrected by disconnecting enough individual items (going by either wattage or amperage) to bring that total down within required limits. For example, suppose a 15 ampere breaker repeatedly trips, will reset, and trips again. Since this appears to be an overloaded circuit, a meter reading is taken, as described above, revealing a current of 17.7 amperes being drawn. Direct examination and listing of loads produce the following:

Loads	*Wattage*	*Amperage*
1 @ 200 W	*200 W*	*1.67 A*
1 @ 350 W	*350 W*	*2.92 A*
4 @ 100 W	*400 W*	*3.33 A*
1 @ 250 W	*250 W*	*2.08 A*
1 @ 900 W	*900 W*	*7.50 A*
	2100 W	*17.50 A*

The difference between the meter reading of 17.7 A, and the calculated total of the loads found on the circuit of 17.5 A is only 0.2 A, an acceptable error. A 25 W light bulb or some other small overlooked item would account for an error of that size.

To get down to the maximum load that this 15 A circuit will carry, at least 2.7 amperes will have to be removed. Cutting out the one load of 350 W would do it; however, that will not really be adequate. We found in Chapter 8 that NEC 210-20(a) requires that a circuit must be rated no less than 125% (5/4ths) of its load, or working as we are now with an established circuit capacity, the load may be no more than 80% (4/5ths) of it.

This means that the total load should be reduced not merely by 2.7 amps to 15A, but actually by 5.7 amps to 12 A (80 percent of 15 = 12). To accomplish this not only the 350 W load, but the four 100 W loads as well will have to go. Depending on where and how things are located it might be simpler to change, or run a separate circuit to, the one 900 W item rather than have to deal with five different items.

As was mentioned earlier, if a suitable meter is not available overload situations can be handled by calculation alone. Since the initial step of reading the actual load cannot be done, proceed at once to the second step of listing each of the individual loads that are connected to the circuit that appears overloaded. At the outset it makes no difference whether the size of each load is noted in watts or in amperes because using Watt's formula (see Chapter 1), either can easily be converted to the other:

Since Volts × Amperes = Watts, therefore, Watts / Volts = Amperes

To arrive at a proper total, obviously, all the individual loads must be converted to a common unit of measurement, either watts or amperes. When that total exceeds the rated maximum for the circuit the existance of the overload is clearly demonstrated. For example: 15 A × 120 V = 1800 W; thus if the figures are in watts anything over 1800 will overload a 15A circuit. Since 20 A × 120 V = 2400 W, a load of 2700W is certainly too much for a 20 A circuit. 2700 W divided by 120 V = 22.5 A, a clear overload of 2.5 A.

The transparently obvious disadvantage of determining the amount of an overload by calculation alone is the uncertainty that all the loads have been counted. As long as the count does at least show an overload even if something is left out, the chances are it will be dealt with adequately by reducing the known load to 80 percent of the circuit capacity.

When the total of the known loads connected to the circuit fails to show an overload, something has been overlooked. This will occur most often in an old building that has been remodeled, and in which the remodeling has included changes in circuitry. If thorough investigation of the system discovers absolutely no unidentified load or loads that could be tied to the troublesome circuit, the answer could be a defective breaker.

Defective Breakers

Breaker failure is an extremely uncommon occurrence, so uncommon that it is often misdiagnosed at first when it does happen. The author has never encountered a case of a breaker that failed in the *on* position, and could not be shut off, although since it is a mechanical device such a failure is conceivable. However, breakers that were impossible to get on have been encountered several times, and also breakers that would snap into the *on* position but not hold under load have been encountered as well.

A breaker makes its internal connection through a pair of contact points similar to, but larger than, the contact points in a car. Not only when a short occurs, but also any time a breaker is switched off or on there is arcing across those points whenever any appreciable load exists on the circuit. This arcing, sometimes gradually, sometimes suddenly, ruins the points creating a high resistance spot through which the rated current will not pass or where enough heat is generated to trip the breaker quickly. Thus a breaker that repeatedly trips, although measurements and/or calculations show it is not overloaded, is probably bad. If a breaker keeps tripping when no overloads can be found, try replacing it. If the new breaker holds in, the old one was bad. If the new one still will not stay on, start looking again for that overload.

A breaker, as we saw in Chapter 9, is activated to trip *off* by the heating and subsequent bending of a bi-metal strip. When one has been operating with a current close to its maximum capacity for some time that bi-metal strip may have been running hot and become weak. If this happens the breaker may refuse to stay on, and must be replaced.

Receptacles

The vast majority of receptacles installed are the common duplex grounded 15 A, 120 V type. Today most of these are furnished with both terminal screws and holes for push-in connections. As may be seen from Figure 13-7 the outside receptacle case consists of two parts: a back shell and a front plate slotted to receive the standard three-prong grounded plug. Both parts are molded of a hard thermosetting plastic. Inside the case are metal contact strips for power and common, plus ground which connects to the mounting yoke.

The outside plastic case while hard is also quite brittle. It is easily cracked on impact. The customary convenience outlet mounting height of 1′ off the floor was intended to minimize damage to receptacles by cleaning implements such as brooms, mops, or vacuum cleaning equipment. While this type of damage has been reduced it is by no means eliminated. A receptacle cracked by any kind of accidental impact is dangerous and should not be used. Replace it as soon as possible.

Other than breakage, difficulty with receptacles can result from poor contact between plug and receptacle due to dirt in the slots, or from an oxide coating forming on the metal of the internal contact strips. In either case replacement is probably easier than attempting to clean the contacts. Occasionally a short in an appliance or a loose wire in a plug may cause arcing that severely burns both plug and receptacle. Again replacement is necessary.

Aluminum wire is another source of difficulties. Very few 120 V duplex outlets are made for use with either copper or aluminum. Those that are marked for copper only or for copper or copper-clad aluminum will not handle bare aluminum. Also those that will take bare aluminum

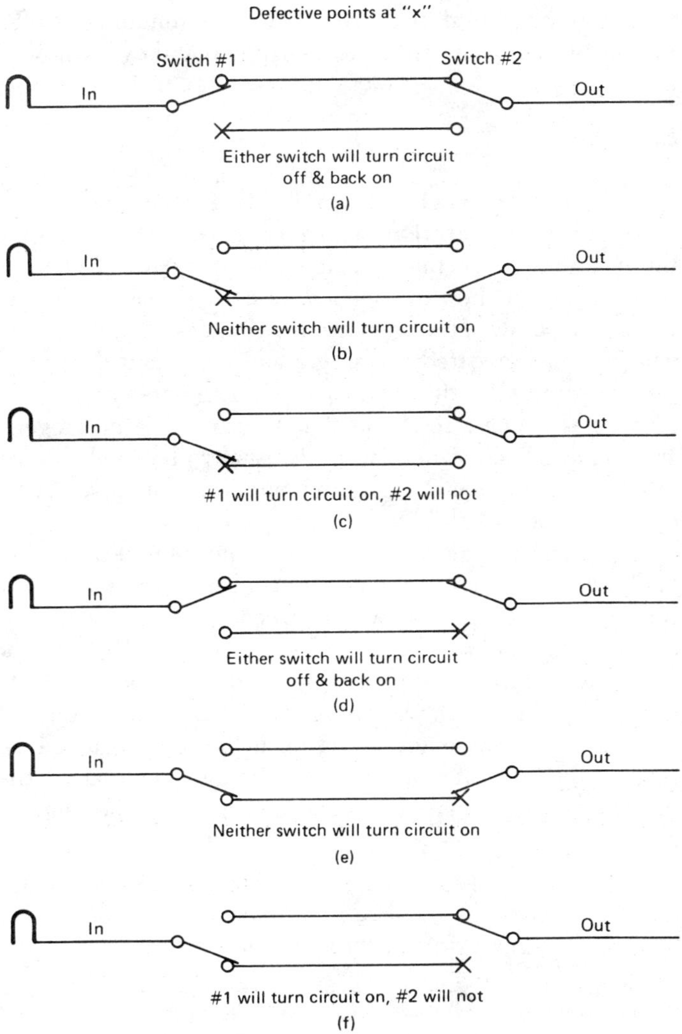

Figure 13–9 switch defects.

must have attachments made only at screw terminals—***never*** use push-in contacts with aluminum wire.

The larger amperage receptacles, 30A and 50A, are generally marked *Al-Cu* meaning bare aluminum may be used with them; however, a great many cases of arcing and burned receptacles in these sizes occur daily for reasons discussed in Chapter 4. If aluminum wire is used to feed such receptacles, usually to supply power to dryers or ranges, the receptacle should be completely removed from its box about once a year, the aluminum wires disconnected, treated with antioxidant, and replaced. If this is not done regularly the chances of trouble with such receptacles are quite high. Unfortunately, in recent years copper's high price has encouraged aluminum wiring in range and dryer outlets while the rest of the residence was wired in copper. These aluminum lines often remain unknown to the owner until the unhappy day they burn out.

Switches

The single pole switches generally used are toggle type that internally make or break connections with a pair of contact points (Figure 13-8). It is those contact points that are the vulnerable parts of such switches. Every time these switches are opened or closed a slight electrical arc occurs across the points. The larger the load being switched, the stronger the arc will be. The repeated arcing gradually pits and deforms them. Finally they become so badly misshapen, and the area of contact between the points has become so small, that a current can no longer pass, and the switch fails. Total switch failure is usually preceded by a period of intermittant malfunction. During this period it may take several attempts at operating the switch before it actually functions.

Switches normally fail in the *off* position because the points are so badly damaged that adequate contact across them is no longer possible, but occasionally they appear to freeze in the *on* position. Moving the toggle back and forth has no effect. When this happens the reason they appear to be frozen at *on* is simply because they ***are*** frozen there. At some time when the switch was being closed a strong arc occured that actually welded the points together. Even if flicking the toggle back and forth repeatedly does succeed in breaking the points apart that switch will never operate properly and must be replaced. The same thing is true of

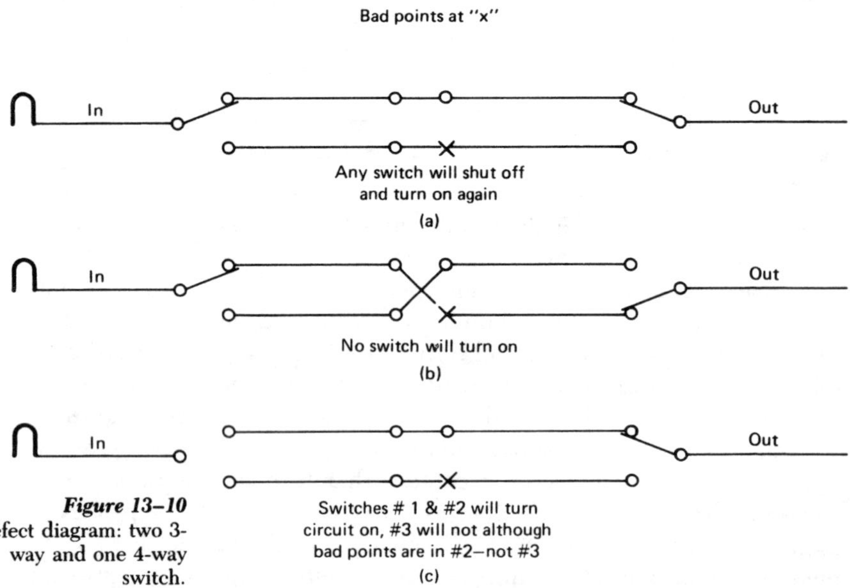

Figure 13–10
Defect diagram: two 3-way and one 4-way switch.

switches that fail in the *off* position—they cannot be repaired. The only possible course is replacement.

Like single pole switches, three-way and four-way switches also make and break connections through contact points. These points are also subject to arcing, pitting, and eventual failure. When a single set of points in any three-way or four-way switch fails, that entire multiple switch circuit is disrupted. In addition it is difficult to determine which switch has gone bad. Figure 13-9 shows why. In example a, the defect is in switch #1, in d switch #2 is bad, but in both instances with the switches in the positions shown the circuit will be on and either switch will turn it off and back on as though nothing were wrong.

With the defects in the same places, but switch positions changed as in examples b and e, neither switch will operate the circuit, but in each case only one switch is bad. Finally, in examples c and f switch #1 will operate the circuit while #2 will not. Replacing #2 will help in the f situation, but certainly not in the case of c.

With two three-ways and one four-way all sorts of deceptive combinations exist. A few are shown in Figure 13-10. In example a, any one switch will shut the circuit off and turn it back on again although the four-way is bad. In example b none of them will turn it on, but only one is bad. In case c switches #1 and #2 will turn the circuit on and off, and #3 will not even though there is nothing wrong with it.

Since defective multiple switch circuits show seemingly inconsistent malfunctions, the only way to isolate which switch is bad is to test each one individually to see whether it will pass power in both positions of the toggle. Another sometimes simpler way of dealing with trouble in a multiple switch circuit is the substitution method. With a three-way switch circuit replace either one of the switches. There is a 50 percent chance you got the right one. If luck was looking the other way and the circuit still does not operate properly, use the switch that was removed to replace the second one. The logic behind the substitution approach is that while it is not known which switch is bad, you do know that when the smoke clears one surely will have to be replaced. By simply starting off with a replacement in 50 percent of cases you will be right the first time, while at worst five minutes or so were wasted changing the wrong switch. With two three-ways and a four-way, if changing both three-ways fails to solve the difficulty, only the four-way is left. The odds are 2 to 1 it will be one of the three-ways, right?

Incandescent Light Fixtures

Incandescent fixtures are basically of three types: surface mounted, hanging, and recessed. All three types are comparatively trouble free equipment. When an incandescent fixture fails to light, the trouble is seldom

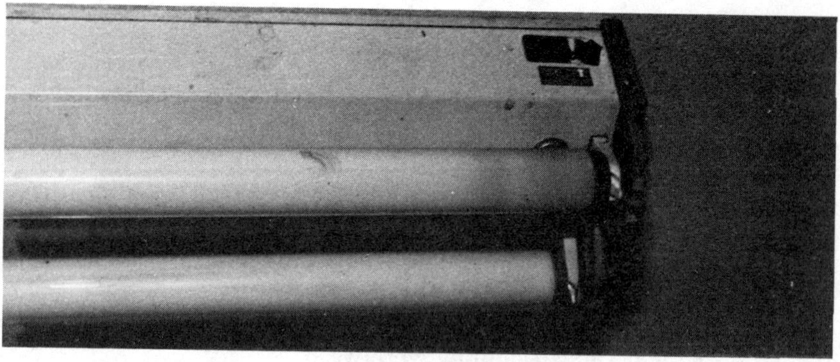

Figure 13–11
Pre-heat fluorescent showing starters under tube ends at right.

anything more than a burnt out bulb. Of course in the case of fixtures with internal switches the switch part of the fixture is subject to the same kind of failure, and for the same reason, as any of the wall switches discussed above.

In light of the currently growing interest in energy conservation insulation is being installed in a great many formerly uninsulated attics, and existing insulation is being increased in thickness in many other attics. When this is done without proper regard for the 3″ clearance around recessed fixtures required per NEC 410-66, and already discussed in Chapter 11, difficulties with recessed lights are very likely. At the very least the fixture will burn out bulbs at an alarming rate. Also operating at excessively high temperatures may scorch the inside finish, substantially reducing light output, and the lenses in fixtures with plastic lenses may be damaged. While the insulation on the wiring in such fixtures is rated to survive temperatures well above anything likely to be encountered in normal operation, a large bulb burning base up in a fixture surrounded by insulation and not dissipating its heat adequately can damage the fixture wiring.

Another common cause of overheating, and consequent damage to recessed incandescent light fixtures, is the use of bulbs larger than the maximum wattage for which the fixture is rated. As was also mentioned in Chapter 11, never use a bulb of a size larger than the wattage stamped on the inside of the fixture case. Code requires (NEC 410-70) that this number be stamped *in letters at least ¼″ high*, where it can be seen when changing bulbs. Be sure to look for it.

Fluorescent Fixtures

The operation of fluorescents is considerably more complex than that of incandescents. Consequently they are subject to a good many more kinds of malfunctions. Some of these, as was noted in the fluorescent installation discussion in Chapter 11, can be caused by improper installation while others result from unfavorable operating conditions. Fluorescent ballasts and lights are far more sensitive to temperature extremes and voltage variations than incandescents. Ballasts are particularly subject to premature failure if operated at excessively high temperatures (above 105°C) either due to inadequate ventilation or excessive line voltage (above 120V). The light output of fluorescent lamps decreases at temperatures below 50°F and depending on humidity, line voltage, and lamp design at varying temperatures below that, the point is reached when they will fail to light altogether.

Fluorescent fixtures are of three basic types: preheat, rapid start, and instant start. Some parts of the troubleshooting drill are the same for all three. When a fixture fails to light, the first step in all cases is to change tubes. In most instances that will be all that is necessary. In most two tube fixtures one tube will not operate alone; thus if one goes bad they both go dead. In four-tube fixtures, most tubes are wired in pairs so that if one tube is bad its mate will go out, but the other two will continue to operate. When changing tubes, before installing a replacement tube examine both sockets to assure that all lamp pins are making good contact. If the fixture is a pre-heat type (Figure 13-11), it will have one or more starters. With these fixtures if changing tubes did not help try changing starters. Actually with the pre-heat type reverse the order and try changing starters first before changing tubes. Starters are a lot smaller and easier to handle. Incidentally, the only preheat fixtures in service are old ones. They are being replaced with either rapid start or instant start.

The majority of both new and existing fluorescent installations today are of the rapid start type. For proper operation it is most important that these fixtures be securely grounded. If not, random starting or non-starting will commonly result. If the tubes are good, and the fixture is properly grounded, but it still will not start, the cathode voltage to the lamps may be inadequate. Blue-blue, yellow-yellow, and red-red leads (Figure 13-12) are to the filament windings in the ballast, and supply between 3.4 V and 4 V to the lamp cathodes. If that voltage cannot be measured at the lamp sockets the lamps are not properly heated and may not start, or if they start may not operate correctly. If the voltage is inadequate the cause could be an improperly connected lead wire from the ballast, low power supply voltage to the ballast, or the ballast winding has gone bad.

Assuming the cathode voltage is satisfactory the trouble may be

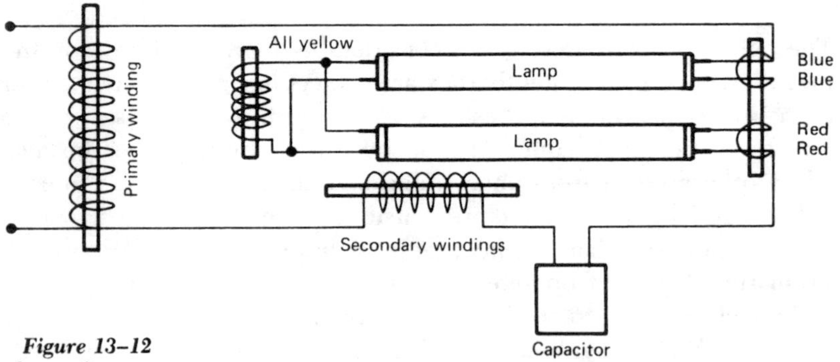

Figure 13–12
Rapid start fixture.

Rapid start- 430 MA lamp type	*Starting Voltage (Minimum @ 50°F)* Single lamp	Two lamp	*Filament voltage*
F14T12	108	162	7.5-9.0
F15T8	108	162	7.5-9.0
F15T12	108	162	7.5-9.0
F20T12	108	162	7.5-9.0
F22T9	185	–	3.4-3.9
F30T12/RS	150	215	3.4-3.9
F32T10	205	–	3.4-3.9
F40T12/RS	205	256	3.4-3.9
22-32 (circline comb.)	–	235	3.4-3.9
32-40 (circline comb.)	–	295	3.4-3.9

Rapid Start— 800 & 1000 MA lamp type	*Starting Voltage (Minimum)* Single lamp 50°F	0°F	-20°F	Two lamp 50°F	0°F	-20°F
F24T12/HO	85	110	140	145	195	225
F36T12/HO	115	155	190	195	235	260
F48T12/HO	165	215	255	256	290	310
F60T12/HO	210	240	290	325	350	365
F64T12/HO	240	270	330	345	370	388
F72T12/HO	275	300	360	395	410	420
F84T12/HO	285	325	370	430	445	455
F96T12/HO	280	330	360	465	480	490

Note: Filament voltage 3.4-4.3

Rapid Start— 1500 MA lamp type	*Starting Voltage (Minimum)* Single lamp 50°F	0°F	-20°F	Two lamp 50°F	0°F	-20°F
F48PG, VHO, SHO	160	205	240	250	265	300
F72PG, VHO, SHO	225	270	310	350	360	400
F96PG, VHO, SHO	300	355	400	470	470	500

Note: Filament voltage 3.4-4.3
Courtesy of Advance Transformer Company.

Figure 13-13
Lamp types and starting voltages: rapid starts.

inadequate starting voltage. To read starting voltage, connect the meter from the highest reading red lead to blue. Correct starting voltage varies with lamp type and size, temperature, and number of lamps (Figure 13-13).

If under high humidity rapid start lamps start slowly, or fail to start when cathode heater voltage and starting voltage are correct, the cause may be dirt on the lamps affecting the silicon coating, or it could be a poor job of silicon coating. If the installation has been in service for some time try washing the lamps to get the dirt off. Small droplets of water should then form on the tubes similar to the droplets that will form on a newly waxed car.

When it is a new installation, but random starting is occuring when the humidity is high, generally the line voltage is low, or there is a poor silicon coating on the lamps. If droplets do not form when water is run on the tubes the silicon coating is poor.

Occasionally with a two lamp rapid start fixture one tube will light properly, but the other will not. See Figure 13-12: if the tube that lights is between the red and the yellow leads, look for a damaged yellow lead. However, if the lamp between the red and yellow leads will not light, but the other does, the ballast is probably shorted and must be discarded.

Depending on the frequency with which it is switched on and off a fluorescent tube can be expected to last between 15,000 and 20,000 hours in service. As a tube reaches the end of its service life the ends will gradually blacken before it fails entirely. Premature blackening and tube failure will result from inadequate cathode heating which can result from inadequate voltage to the cathodes from the ballast as mentioned above. However, inadequate heating can also occur when the ballast is fine. Other causes could be poor seating of the lamp in the socket, a broken socket, a broken lamp pin, a socket loose in the fixture causing socket spacing to be too great, or perhaps an internally damaged lamp cathode.

A ballast, again depending on frequency of switching, can be expected to give around 60,000 to 100,000 hours of service. Properly installed so they are not running overheated, they are extremely reliable parts. However, they too will eventually fail requiring replacement. When the ballast finally fails in a simple utility type fixture consisting only of ballast and lamp sockets mounted on a light-weight sheet metal case it will often cost more for labor and material to replace a ballast then to replace the complete fixture.

The tubes used in both preheat and rapid start fixtures are of the familiar bi-pin type. Instant starts will be single pin, and are most commonly found in the 6′ and 8′ long fixtures. Starting voltages for various sizes of instant-starts are given in Figure 13-14. These voltages are considerably higher than for comparable sized rapid-starts.

The effects of aging in service on instant-start (or slimline) tubes is similar to what one sees with rapid-starts (Figure 13-15). Dark bands near

To determine **Starting Voltage**, the lamp must be removed and voltmeter connected between the respective primary and secondary leads of each lamp as designated on ballast label. For series-sequence ballasts, the red lead must be in position while measuring the starting voltage of the remaining lamp.

LAMP TYPE	*STARTING VOLTAGE (MINIMUM)
F24T12	270
F36T12	315
F40T12	385
F40T17	385
F42T6	405
F48T12	385
F64T6	540
F72T8	540
F72T12	475
F96T8	675
F96T12	565

Figure 13–14 Starting voltage table: instant starts. *Courtesy of Advance Transformer Company.*

***For Single Lamp**, measure voltage between Red & White Leads.
For Two Lamp (Series Sequence), measure voltage between Red & White, insert lamp in Red & White position, then read voltage between Blue & Black.
For Two Lamp (Lead Lag), measure voltage between Red & White and Blue & White leads.

Slimline Lamp

End band develops gradually during normal lamp life.

Heavy end discoloration occurs at the end of normal lamp life.

Rapid Start Lamp

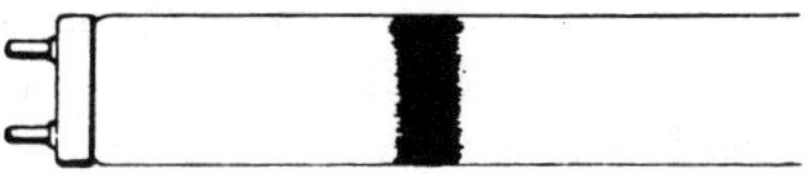

End banding may occur gradually during normal lamp life.

Excessive end blackening could be caused by insufficient cathode heating which will result in shortened lamp life.

Bi-pin for 30 or 40 watt lamps

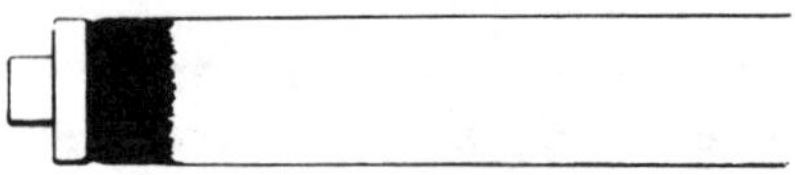

1500 M.A. lamps (VHO, SHO and PG) may discolor at end of lamps during operation. This is normal for this type of lamp.

Recessed bi-pin for 800 M.A. or 1500 M.A. lamps.

Caution: 1500 M.A. T12 and 800 M.A. lamps are the same size. Be sure to use the correct lamp ballast combination.

Figure 13–15 Drawing of normal darkening through age of slimline and rapid starts.

tube ends may develop during normal service. Approaching end of tube life is signalled by gradual development of heavy end blackening.

Remembering that normal tube life lies between 10,000 hours and 20,000 hours; end blackening after only a few thousand hours, or less, of operation is a clear sign that something is very wrong. Line voltage to the fixture may be low, lamp to socket contact may be poor, or the fixture could be miswired. Check all three. The correct wiring diagram for the fixture is normally found on the ballast.

With slimlines or instant starts, in many instances one tube of a two tube fixture will light when the other tube has a defective filament. The defective one will flicker and blacken at one end very rapidly. ***Do not continue*** to operate the fixture with this bad tube. The ballast will overheat and fail.

The tube sockets at both ends of either preheat or rapid-start, bi-pin tubes are exactly the same. Tubes are installed by slipping the pins into the end slot in the socket, then rotating the tube by 90° to seat the pins in the contacts (Figure 13-16). Single-pin, slimline tubes are suspended between two sockets that are not the same. One end is fixed in position while the other has a movable spring loaded contact. The tube is installed by depressing the spring loaded end to allow space to seat the single pins at both ends.

In the opinion of this writer the spring loaded single pin mounting system is by no means adequately engineered. The spring loaded socket has been seen repeatedly to overheat resulting in a sudden burn-out accompanied by dropping and shattering of the tube. This burn-out occurs uniformly at the spring loaded end not the fixed end.

Check-out of New Wiring

After the cable to a new circuit is brought back to the breaker panel and the power lead connected to its assigned breaker, as a routine procedure test the circuit for shorts. First make sure that no equipment is connected or plugged into any outlet, and that there are no light bulbs in any fixtures. Now with the breaker off, using the ohmmeter, test from the output terminal on the breaker to the ground buss in the box. The meter should read *infinity* (∞). If it reads a low resistance—2 or 3 ohms—a light bulb or something else that should not be there is in the circuit. If it reads O it is probably shorted.

To trace down the short, follow the procedure outlined above under *Building Circuit Shorts*. In the case of new wiring this should be quite easy because the sequence in which the various outlets, switches, and fixtures were wired will be known. This will make it possible to follow the power distribution from box to box through the circuit in an orderly manner.

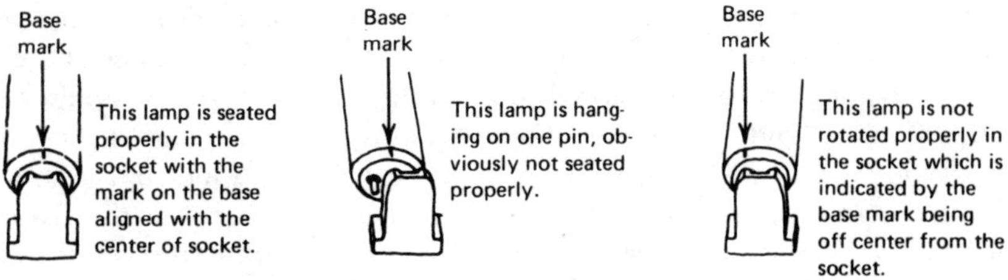

Figure 13–16 Seating bi-pin tube properly in socket. Check for proper rotation and fit of the lamps in the sockets. The lamps must be well seated in the sockets with spacing between the socket and lamp small enough to assure proper contact but free enough to prevent binding. Check for loose or broken sockets and dirty socket contacts.

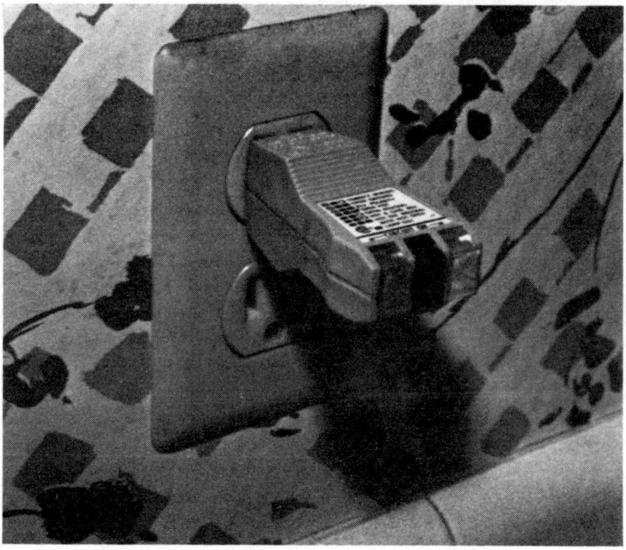

Figure 13–17 Using 120V outlet tester.

Assuming there was no short, or if there was one it has been located and corrected, the next step is to turn on the power and check all switches, fixtures, and outlets. At this point place bulbs in all light fixtures to test both fixtures and switches. If any fixture fails to light with the switch on, remove the switch from its box and with a pocket test light check from both input and output terminals to ground to make certain power is getting into and out of the switch. If power is not getting into the switch work back toward the source. Somewhere there is probably a poor splice. If power is getting into but not out of the switch replace the switch. It is defective.

If power is going through the switch, test at the fixture. If power is getting to the fixture, and it is certain that the bulb is good, the trouble has to be in the common return wiring—either a bad terminal connection or a bad splice.

With lights and switches all operating properly all receptacles should be tested for proper operation. A special outlet tester (Figure 13-17) can be used for this purpose, or it can be done with a voltmeter or even a little pocket test light. The outlet tester is simply plugged into each receptacle, and the way it lights up, or fails to light as the case may be, indicates the presence of any difficulty that may exist such as open ground, reversed polarity, open common, or open power lead.

Figure 13-18-a is a normal 120 V receptacle. To test this with a voltmeter first test across the slots (1). The reading should be approximately 120 V. From the shorter slot to the U shaped ground hole (2) should also read 120 V, and from the ground hole to the long slot (3) should read O V. Possible defective connections could be as follows:

1	*2*	*3*	*Defect*
120	0	120	Power and common reversed—polarity backwards
120	0	0	Open ground—ground continuity broken somewhere
0	120	0	Open common—common continuity broken
0	120	120	Power and ground reversed—ground is hot
0	0	120	Common is hot, and power is open
0	0	0	Power is open—no power

The same tests just described can be done with the pocket test light. In every case where the meter would show 120V, the test light will light. The only thing the test light does not show is whether the voltage itself is correct. If the voltage is running either too high or too low the test light does not show it.

To test dryer or range receptacles (Figure 13-18-b and c) the voltmeter should be used. Across the two top angled slots (1) the reading should be 240 V. With one test lead in the bottom vertical slot the reading to each of the angled slots (2 and 3) will be 120 V.
Defects could be as follows:

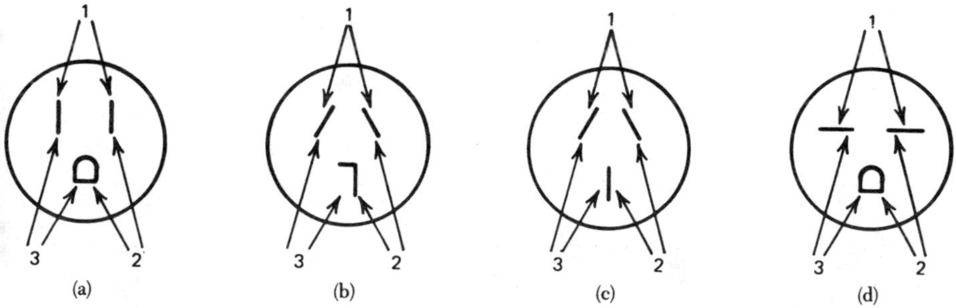

Figure 13–18 120 V and 240 V outlets showing test sequence.

1	*2*	*3*	*Defect*
120	120	0	Left side power line open
120	0	120	Right side power line open
240	0	0	Common open
120	240	120	Common and left side power line reversed
120	120	240	Common and right side power line reversed

Figure 13-18-d is a type of 240 V receptacle often used for small room air conditioners or space heaters. Both the tests and the defects applicable to Figure 13-18-b and c will also apply to it as well.

CHAPTER 14

SUPPLEMENTARY SYSTEMS

Three basic categories of supplementary electrical systems will be discussed: signaling and warning systems, communication systems, and entertainment systems. Signaling and warning systems include doorbells, chimes and buzzers as well as smoke detectors and burglar alarms. Communication systems include telephone for outside communications and internal communication via voice intercom systems. The entertainment area includes radio, TV, and stereo where the electrician may need to deal with wiring runs for antennae or remoted speakers. The people who do power wiring are not required to get involved in electronic installations.

Most of the supplementary systems are low voltage meaning that they operate on anywhere from 6 V to 24 V instead of the 120 V and 240 V circuits with which we have been dealing. Doorbells, chimes and buzzers are uniformly powered by step-down transformers attached to the 120 V supply. Smoke detectors and burglar alarms are usually directly connected to the 120 V line, their step-down transformers being internal. In many cases they are separately powered by batteries, or have battery back-ups that take over in case of failure of the 120 V line.

When the same emergency that caused a fire also cuts off the electrical power a battery powered smoke detector, or a battery back-up, can be litterally a life saver. A battery powered, or battery backed up, burglar alarm will continue to operate in case a thief shuts off the main electrical disconnect. Thus battery powered warning systems independent of the building electrical system have distinct advantages.

The batteries used in these systems have a usual service life of about a year. Their very durability is in a way a disadvantage since it is all too easy to forget when they should be replaced. Therefore, occupants should be strongly advised to follow manufacturer's instructions regarding routine testing of these units and should be encouraged to institute a rigid schedule of battery replacement.

Figure 14–1
Bell transformer with box mounting bracket.

Signaling and Warning Systems

Doorbells, Chimes, and Buzzers

The doorbell circuit starts with the bell transformer (Figure 14-1). This is a step-down transformer that reduces the 120 line voltage down to somewhere between 10 and 24 volts. The primary is furnished with two wire pigtails, a black and a white, that are attached to the 120V line. Since all 120V connections must be made in an approved electrical box (NEC 300-15), many transformers are supplied with a special bracket that fits a standard ½″ metal box knock-out. Using this bracket the transformer is mounted on any standard metal box having ½″ knock-outs so that the primary pigtails are inside ready for connection leaving the secondary screw terminals outside and available for attachment of the low voltage bell wires that will go on to the bell button and the bell.

Although bell transformers with secondary voltages as high as 24 V are available a 10 V secondary is adequate for most purposes. Some transformers have multiple taps providing two or three voltages such as 8 V, 16 V, and 24 V. When all else has failed consult the instructions accompanying the bell or chime. They should indicate the minimum voltage required by the device. There is no danger in erring on the high side. If the bell wants 8 V, a 10 V, 12 V, or even 16 V supply will cause no problem. The reverse might not work out. Where the device wants 16 V it may not operate on 8V.

While connection of the transformer primary to the 120 V supply must be made inside a box, the secondary terminals are outside because low voltage connections do not require the protection of a box. Also the wire size used is vastly smaller than the #14 minimum required for the 120 V power circuits. Bell wires range from #18 down to #22. Splices may also be made at any point no box being necessary.

Typical bell wire consists of two individually insulated conductors (Figure 14-2) furnished in two colors as a twisted pair. The colors commonly used are red and white. Different colors are used solely to help the electrician identify which is which. They have no other significance.

From the transformer forward the circuit consists simply of a switch (the bell button) and a load (the bell). As always, one side of the line goes through the switch. Both sides go to the load. Depending on the physical locations of the parts the wiring sequence may be (Figure 14-3) transformer to button, to bell, or it may be transformer to bell with switch loop to bell button. These two circuits are exactly the same as the ones shown in Figures 7-10-c and 7-11-b. They merely connect different components using much lighter wire.

Very commonly two tone chimes are being used to indicate by a difference in sound which of two entrances has been approached. These are normally wired as shown in Figure 14-4. A low voltage pair runs from the transformer to the chime where one wire connects to the terminal marked either *trans* or *common*. The other splices to one side of each of two switch loops going out to bell buttons. The return side of one goes to *front* the other to *rear* allowing each button to activate a different chime tone.

Figure 14–2
Bell wire.

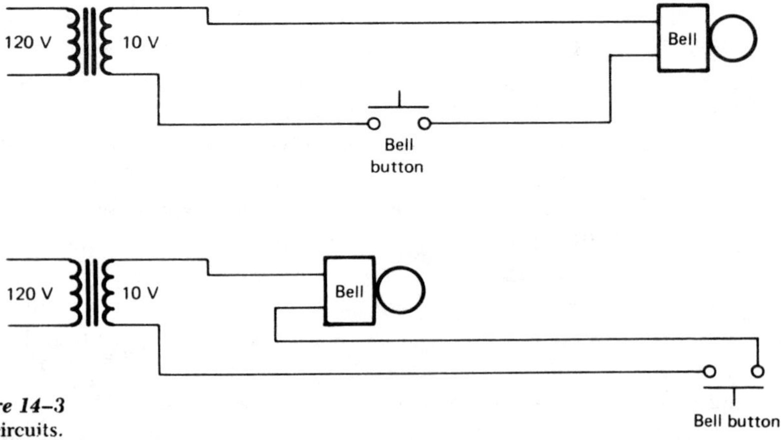

Figure 14–3
Bell circuits.

Because of the distances and shapes of halls and rooms, a single bell or chime is often inadequate. If so, repeaters are easily connected (Figure 14-5) to the initial unit. In order to power adequately the repeaters, it may be necessary to increase the secondary voltage of the bell transformer, if it is supplying only 10 V or 12 V.

With all due respect to our colleagues in manufacturing, doorbell buttons are generally not very well made. In addition they are placed outdoors exposed to the weather. Consequently more often than not difficulty with a doorbell circuit is caused by failure of that bell button. If a bell or chime fails to operate, start by pulling out the bell button. Then take off one wire and short it to the other wire. Generally the bell will then ring merrily indicating that the button is bad. Merely replace it.

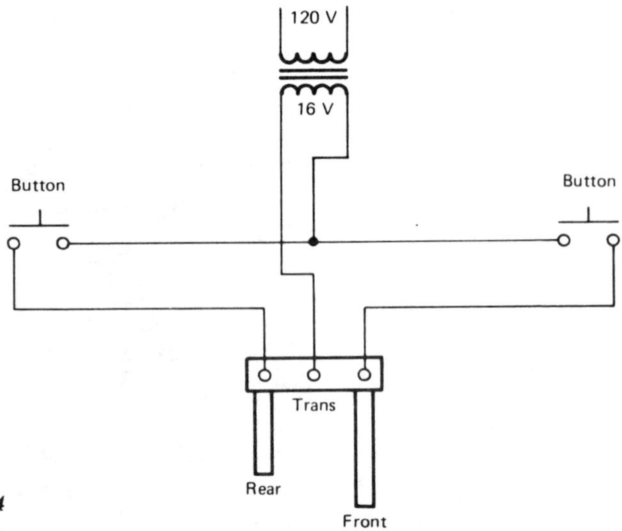

Figure 14–4
Two-tone chime circuit.

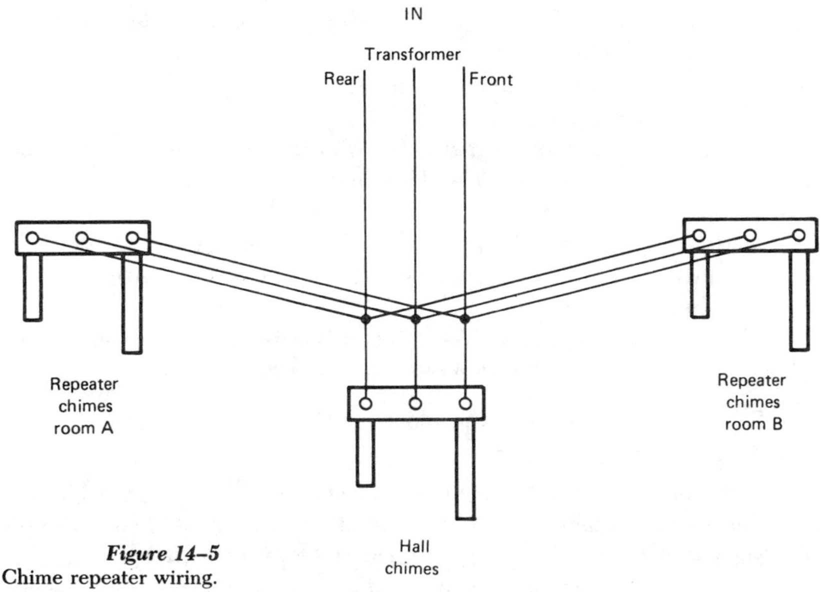

Figure 14–5
Chime repeater wiring.

If the bell did not ring, measure across the wires at the bell with a voltmeter. When the transformer is all right a reading somewhere between 10 and 24 volts will be found. If the transformer is bad, no voltage will be found. As a general rule the bell itself is the last component to go bad. Far more often it is the transformer.

Smoke Detectors

Two types of detectors are used commonly: photoelectric and ionization types. The photoelectric unit contains a light sensor that responds to a constant light in a small chamber. Smoke entering that chamber obstructs that light. The sensor detects the change and activates an alarm. This type of detector is not highly sensitive to fires produced by comparatively clean burning materials such as gasoline or other highly refined liquids.

Ionization detectors contain a small amount of radioactive material that causes ionization of the air in a test chamber, fitted with two oppositely charged plates. Due to the ionization an extremely tiny current flows between those plates. The entrance of smoke into this chamber reduces the flow. When that reduction is detected the alarm is sounded. This type detector does much better with clean burning fires than the photoelectric type.

Some smoke detectors are furnished with black and white wire pigtails for direct connection to the 120 V line, and must therefore be mounted directly on a box. Others are equipped with a cord and plug for connection to a normal wall receptacle. Be certain it is an unswitched receptacle, and also be sure to place the clip on the cord under the screw that holds the receptacle faceplate in position. This is a safety device to prevent the plug from accidentally being pulled out of the socket thus disabling the detector.

Still other detectors are battery powered. These units require no supply wiring. This gives the installer a most desirable freedom in terms of locating them in the most strategic possible places. Good quality batteries will last approximately a year, but they should be tested regularly in accordance with the manufacturer's instructions to insure that the device is operating properly.

The exact location of a smoke detector is critical to proper operation. Following are some *dos* and *don'ts* in this regard:

Smoke from a fire rises; therefore install the detector on the ceiling if possible.

Install it as close to the center of a room or hallway as possible. If a detector is installed off to one side of a room a lively fire could get started at the other side before the detector picks it up.

In an unfinished attic install the detector on the bottom edge of a rafter or joist. The space in between is considered a dead air space.

In an open stairway install the detector at the top (Figure 14-6). Do not put it in the dead air space that could be closed off by a door.

When the detector must be placed on a wall, its top can be no higher than 4″ from the ceiling (above that is dead air), and no lower than 12″ from the ceiling.

When it must be placed off to one side on a ceiling it may be no closer than 4″ from the wall (again dead air).

Do not install in bathrooms, laundry areas or other places where visible water vapor may be created.

Mount away from forced air registers (either heating or airconditioning). Moving air may keep smoke from entering the detection chamber.

Do not install in kitchens where cooking smoke could trip a false alarm.

Burglar Alarms

The burglar system consists of a number of small, inconspicuous low voltage sensors placed so as to be activated by the movement of a door or a window. When activated the sensor trips a relay which sounds the alarm. Some of the more expensive and elaborate systems not only sound an alarm at the premises where the intrusion is occurring, but also a signal at the local police station or at a commercial guard service.

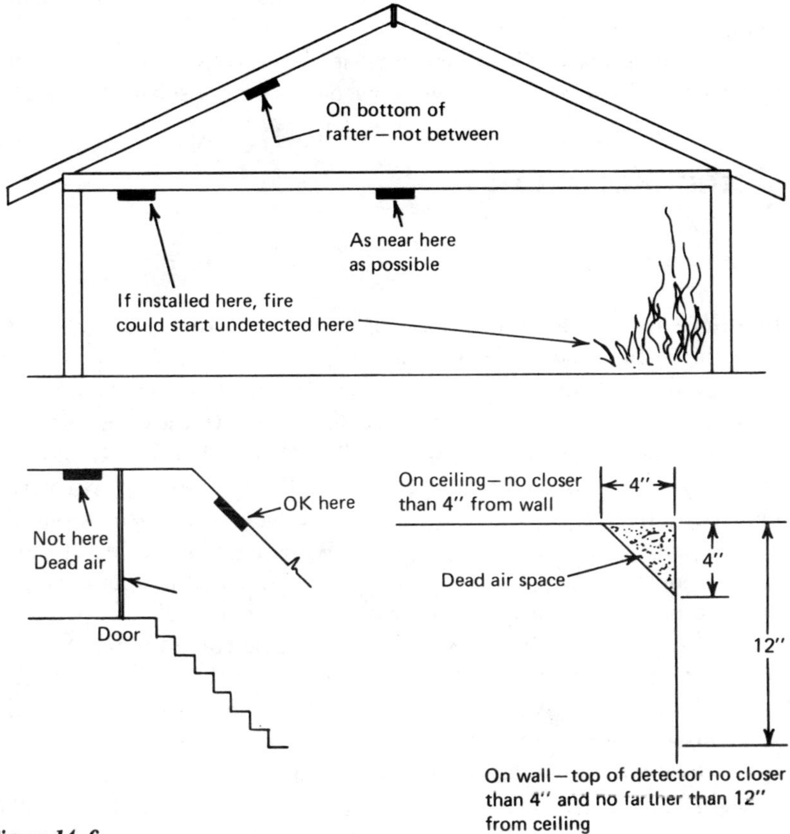

Figure 14–6
Smoke detector location.

A common type of sensor consists of two metal contacts, one mounted on the door or window, the other on its frame. In closed position the contacts touch completing the circuit. If the door or window is opened the circuit is opened as well, tripping the alarm. On a window the sensor may be combined with metallic tape so that if the glass is cut or broken the circuit will also be opened.

A burglar alarm system that operates on the main building power supply generally is equipped with a battery back-up in case that supply either fails or is deliberately interrupted. The battery supply is normally maintained in the *off* position by a relay that is held open by the main power supply. When that fails the relay is released switching the battery circuit on.

Once a sensor has been opened, and the alarm has tripped on, merely closing that sensor will not shut the alarm off. The relay that activates the alarm latches in the alarm position and can only be turned off manually.

As with smoke detectors burglar alarm systems should be tested regularly and back-up batteries replaced on a rigid schedule. Again battery shelf life can be estimated at about a year.

Communications

Telephones

In a majority of cases the electrician wiring a building has nothing to do with the installation of the telephone system; however, in a minority of cases he does. When he does it is to provide positions for the new modular type sockets and routes for concealed wiring to those positions.

The actual telephone wire as well as all associated parts and fittings are supplied by the telephone company. It also pulls its own wire and installs all parts of the telephone system. However, the electrician may be called upon to provide boxes and fish wires running either up into the attic or down to a basement or crawl space to enable the telephone people to get their wiring through the insides of finished walls.

In commercial installations the telephone company will frequently require that a conduit system be furnished in which it will run its cabling. Occasionally it may want conduit for a residential system as well. This is particularly likely in the case of masonry walls.

Since telephones are another low voltage system, standard electrical boxes may be used, but are not required. In some cases a plain plaster cover plate may be used without the box, and in others just a hole in the wall with a fishwire in it will do. Consult ahead of time with both the building owner and the telephone company to determine whether the electrician will be called upon to make any provisions for telephone lines, and if so what is wanted.

Intercommunication Systems

Several different types of intercom systems are available, each providing different capabilities. One simple answer is to have the telephone company supply extension phones equipped with switches and buzzers to permit communication through that existing system from room to room. No special provision needs to be made for this in the building wiring.

Another method is an independent system of telephone handsets powered and wired separately from the outside telephone system. Such a system will require a power supply in the form of a step-down transformer connected to the 120 V line and delivering 10 V. This transformer is exactly the same type as is used for the doorbell system, and is mounted on any junction box in the same way. Although both circuits use the same type of transformer do not put them both on one. Give them each a separate transformer.

A third type of intercom system is considerably more versatile than the telephone handset type. It consists of a master control station connected to a number of remote stations any of which can be called from the master, and any of the remotes can call in to the master as well. The master generally includes an AM-FM radio receiver that can be piped through the system when it is not in use for communication. This is possible because each remote station contains a small loudspeaker through which the messages from the master station are received, and which also serves as the microphone for transmissions back to the master. Another important feature of these systems is that a remote station can be switched to the transmit position and left that way allowing a person on watch at the master location to monitor that remote station. This feature is particularly useful in cases of serious illness or chronic incapacity, and is helpful for babysitting small children as well.

These systems are again low voltage, and usually require a step-down transformer connected to the building power although some plug directly into the 120V line when they contain their own internal step-downs. When drawings indicate that an intercom is to be installed in a building the electrician should inquire as to what measures he should take to provide for it.

Entertainment Systems

Television, FM Radio

Television is a complex and highly technical specialty that lies outside the normal area of expertise of the power electrician. However, the electrician may be called upon to leave fish wires or even run the antenna cables so they can be concealed inside the walls at the same time as the rest of the building wiring.

Normally a single roof antenna will be erected to supply the signals for several TV sets. When this is done the lead from the antenna goes first to a signal splitter, then separate leads go from that splitter to the individual sets. It is the lines from the splitter to the various set locations that often need to be concealed in the walls. At the set locations no boxes are necessary. A cover plate for the sake of appearance is enough.

TV signals are transmitted using the same principle as is used for FM radio; consequently many TV antennas also incorporate a section that covers the FM radio band as well. In such a case additional antennas for FM radio receivers are unnecessary.

In many instances TV antennas are equipped with rotators so they can be directionally adjusted to get the best possible signal from each broadcast station. The rotator control is plugged into the 120 V supply. It is also wired directly up to a small electric motor on the antenna mast; however, the power to that motor is only 24 V having passed through a step-down transformer in the control box. This rotator power cable is another line that is preferably concealed. It can generally be passed upward following the same route that the signal lead from the antenna follows coming down.

Inside a building, antennae and antenna lead-ins [NEC 810-18 (b)] shall not be run closer than 2" away from other wiring systems.

Stereo and Remoted Speaker Systems

To get a convincing stereo effect the speakers must be placed fairly far apart, and generally fairly far from the control center as well. With a bit of thought the wires connecting them can be concealed very effectively. For example, instead of running a wire around the room so that it lies on the carpet close to the baseboard and in plain sight, take the baseboard off. Put the wire behind it and replace the baseboard.

Simpler yet, when the lead has to go to another side of the room instead of following the walls, go under the carpet. Rather than staple wiring around a door frame, take the trim off, put the wiring behind, then replace the trim.

For placing speakers remotely in other rooms, any of the procedures for concealing wiring that have been previously discussed will do. As with other low voltage wiring keep speaker wires at least 2″ away from any power wiring encountered along the way.

APPENDIX A

NATIONAL ELECTRICAL CODE TABLES

The tables in Appendix A are reprinted with permission from NFPA 70-1999, *National Electrical Code*, Copyright ©1998, National Fire Protection Association, Quincy, MA 02269. This reprinted material is not the complete and official position of the National Fire Protection Association on the referenced subject, which is represented only by the standard in its entirety.

NOTE: The table numbers refer specifically to their positions within the NEC.

Table 1. Percent of Cross Section of Conduit and Tubing for Conductors

Number of Conductors	All Conductor Types
1	53
2	31
Over 2	40

FPN No. 1: Table 1 is based on common conditions of proper cabling and alignment of conductors where the length of the pull and the number of bends are within reasonable limits. It should be recognized that, for certain conditions, a larger size conduit or a lesser conduit fill should be considered.

FPN No. 2: When pulling three conductors or cables into a raceway, if the ratio of the raceway (inside diameter) to the conductor or cable (outside diameter) is between 2.8 and 3.2, jamming can occur. While jamming can occur when pulling four or more conductors or cables into a raceway, the probability is very low.

Notes to Tables

1. See Appendix C for the maximum number of conductors and fixture wires, all of the same size (total cross-sectional area including insulation) permitted in trade sizes of the applicable conduit or tubing.

2. Table 1 applies only to complete conduit or tubing systems and is not intended to apply to sections of conduit or tubing used to protect exposed wiring from physical damage.

3. Equipment grounding or bonding conductors, where installed, shall be included when calculating conduit or tubing fill. The actual dimensions of the equipment grounding or bonding conductor (insulated or bare) shall be used in the calculation.

4. Where conduit or tubing nipples having a maximum length not to exceed 24 in. (610 mm) are installed between boxes, cabinets, and similar enclosures, the nipples shall be permitted to be filled to 60 percent of their total cross-sectional area, and Section 310-15(b)(2)(a) adjustment factors need not apply to this condition.

5. For conductors not included in Chapter 9, such as multiconductor cables, the actual dimensions shall be used.

6. For combinations of conductors of different sizes, use Tables 5 and 5A for dimensions of conductors and Table 4 for the applicable conduit or tubing dimensions.

7. When calculating the maximum number of conductors permitted in a conduit or tubing, all of the same size (total cross-sectional area including insulation), the next higher whole number shall be used to determine the maximum number of conductors permitted when the calculation results in a decimal of 0.8 or larger.

8. Where bare conductors are permitted by other sections of this *Code*, the dimensions for bare conductors in Table 8 shall be permitted.

9. A multiconductor cable of two or more conductors shall be treated as a single conductor for calculating percentage conduit fill area. For cables that have elliptical cross sections, the cross-sectional area calculation shall be based on using the major diameter of the ellipse as a circle diameter.

Table 220-3(a). General Lighting Loads by Occupancies

Type of Occupancy	Unit Load per Square Foot (Volt-Amperes)
Armories and auditoriums	1
Banks	3½[b]
Barber shops and beauty parlors	3
Churches	1
Clubs	2
Court rooms	2
Dwelling units[a]	3
Garages — commercial (storage)	½
Hospitals	2
Hotels and motels, including apartment houses without provision for cooking by tenants[a]	2
Industrial commercial (loft) buildings	2
Lodge rooms	1½
Office buildings	3½[b]
Restaurants	2
Schools	3
Stores	3
Warehouses (storage)	¼
In any of the above occupancies except one-family dwellings and individual dwelling units of two-family and multifamily dwellings:	
Assembly halls and auditoriums	1
Halls, corridors, closets, stairways	½
Storage spaces	¼

Note: For SI units, 1 ft^2 = 0.093 m^2.

[a]See Section 220-3(b)(10).

[b]In addition, a unit load of 1 volt-ampere per square foot shall be included for general-purpose receptacle outlets where the actual number of general-purpose receptacle outlets is unknown.

Table 220-11. Lighting Load Demand Factors

Type of Occupancy	Portion of Lighting Load to Which Demand Factor Applies (Volt-Amperes)	Demand Factor (Percent)
Dwelling units	First 3000 or less at	100
	From 3001 to 120,000 at	35
	Remainder over 120,000 at	25
Hospitals *	First 50,000 or less at	40
	Remainder over 50,000 at	20
Hotels and motels, including apartment houses without provision for cooking by tenants *	First 20,000 or less at	50
	From 20,001 to 100,000 at	40
	Remainder over 100,000 at	30
Warehouses (storage)	First 12,500 or less at	100
	Remainder over 12,500 at	50
All others	Total volt-amperes	100

* The demand factors of this table shall not apply to the computed load of feeders or services supplying areas in hospitals, hotels, and motels where the entire lighting is likely to be used at one time, as in operating rooms, ballrooms, or dining rooms.

220-18. Electric Clothes Dryers - Dwelling Unit(s)

The load for household electric clothes dryers in a dwelling unit(s) shall be 5000 watts (volt-amperes) or the nameplate rating, whichever is larger, for each dryer served. The use of the demand factors in Table 220-18 shall be permitted. Where two or more single-phase dryers are supplied by a 3-phase, 4-wire feeder or service, the total load shall be computed on the basis of twice the maximum number connected between any two phases.

Table 220-18. Demand Factors for Household Electric Clothes Dryers

Number of Dryers	Demand Factor (Percent)
1	100
2	100
3	100
4	100
5	80
6	70
7	65
8	60
9	55
10	50
11-13	45
14-19	40
20-24	35
25-29	32.5
30-34	30
35-39	27.5
40 and over	25

220-19. Electric Ranges and Other Cooking Appliances - Dwelling Unit(s)

The demand load for household electric ranges, wall-mounted ovens, counter-mounted cooking units, and other household cooking appliances individually rated in excess of 1¾ kW shall be permitted to be computed in accordance with Table 220-19. Where two or more single-phase ranges are supplied by a 3-phase, 4-wire feeder or service, the total load shall be computed on the basis of twice the maximum number connected between any two phases. Kilovolt-amperes (kVA) shall be considered equivalent to kilowatts (kW) for loads computed under this section.
FPN: See Example D5(a) in Appendix D.

Table 220-19. Demand Loads for Household Electric Ranges, Wall-Mounted Ovens, Counter-Mounted Cooking Units, and Other Household Cooking Appliances over 1¾ kW Rating
(Column A to be used in all cases except as otherwise permitted in Note 3.)

	Maximum Demand (kW) (See Notes)	Demand Factor (Percent) (See Note 3)	
Number of Appliances	Column A (Not over 12kW Rating)	Column B (Less than 3½kW Rating)	Column C (3½kW to 8¾ kW Rating)
1	8	80	80
2	11	75	65
3	14	70	55
4	17	66	50
5	20	62	45
6	21	59	43
7	22	56	40
8	23	53	36
9	24	51	35
10	25	49	34
11	26	47	32
12	27	45	32
13	28	43	32
14	29	41	32
15	30	40	32
16	31	39	28
17	32	38	28
18	33	37	28
19	34	36	28
20	35	35	28
21	36	34	26
22	37	33	26
23	38	32	26
24	39	31	26
25	40	30	26
26-30	15 kW + 1 kW for	30	24
31-40	each range	30	22
41-50	25 kW + ¾ kW for	30	20
51-60	each range	30	18
61 and over		30	16

Notes: 1. Over 12 kW through 27 kW ranges all of same rating. For ranges individually rated more than 12 kW but not more than 27 kW, the maximum demand in Column A shall be increased 5 percent for each additional kilowatt of rating or major fraction thereof by which the rating of individual ranges exceeds 12 kW.

2. Over 8¾ kW through 27 kW ranges of unequal ratings. For ranges individually rated more than 8¾ kW and of different ratings, but none exceeding 27 kW, an average value of rating shall be computed by adding together the ratings of all ranges to obtain the total connected load (using 12 kW for any range rated less than 12 kW) and dividing by the total number of ranges. Then the maximum demand in Column A shall be increased 5 percent for each kilowatt or major fraction thereof by which this average value exceeds 12 kW.

3. Over 1¾ kW through 8¾ kW. In lieu of the method provided in Column A, it shall be permissible to add the nameplate ratings of all household cooking appliances rated more than 1¾ kW but not more than 8¾ kW and multiply the sum by the demand factors specified in Column B or C for the given number of appliances. Where the rating of cooking appliances falls under both Column B and Column C, the demand factors for each column shall be applied to the appliances for that column, and the results added together.

4. Branch-Circuit Load. It shall be permissible to compute the branch-circuit load for one range in accordance with Table 220-19. The branch-circuit load for one wall-mounted oven or one counter-mounted cooking unit shall be the nameplate rating of the appliance. The branch-circuit load for a counter-mounted cooking unit and not more than two wall-mounted ovens, all supplied from a single branch circuit and located in the same room, shall be computed by adding the nameplate rating of the individual appliances and treating this total as equivalent to one range.

5. This table also applies to household cooking appliances rated over 1¾ kW and used in instructional programs.

FPN No. 1: See Table 220-20 for commercial cooking equipment.
FPN No. 2: See the examples in Appendix D.

220-31. Optional Calculation for Additional Loads in Existing Dwelling Unit

For an existing dwelling unit presently being served by an existing 120/240-volt or 208Y/120-volt, 3-wire service, it shall be permissible to compute load calculations as follows:

Load (kVa)	Percent of Load
First 8 kVA of load at	100
Remainder of load at	40

Load calculations shall include lighting at 3 volt-amperes/ft^2 (0.093 m^2); 1500 volt-amperes for each 2-wire, small-appliance branch circuit and each laundry branch circuit as specified in Section 220-16; range or wall-mounted oven and counter-mounted cooking unit; and other appliances that are permanently connected or fastened in place, at nameplate rating.

If air-conditioning equipment or electric space-heating equipment is to be installed, the following formula shall be applied to determine if the existing service is of sufficient size.

Air-conditioning equipment*	100%
Central electric space heating*	100%
Less than four separately controlled space-heating units*	100%
First 8 kVA of all other loads	100%
Remainder of all other loads	40%

Other loads shall include the following:

1. 1500 volt-amperes for each 20-ampere appliance circuit
2. Lighting and portable appliances at 3 volt-amperes/ft^2 (0.093 m^2)
3. Household range or wall-mounted oven and counter-mounted cooking unit
4. All other appliances fastened in place, including four or more separately controlled space-heating units, at nameplate rating

*Use larger connected load of air conditioning and space heating, but not both.

Table 250-66. Grounding Electrode Conductor for Alternating-Current Systems

Size of Largest Service-Entrance Conductor or Equivalent Area for Parallel Conductors[1]		Size of Grounding Electrode Conductor	
Copper	**Aluminum or Copper-Clad Aluminum**	**Copper**	**Aluminum or Copper-Clad Aluminum[2]**
2 or smaller	1/2 or smaller	8	6
1 or 1/0	2/0 or 3/0	6	4
2/0 or 3/0	4/0 or 250 kcmil	4	2
Over 3/0 through 350 kcmil	Over 250 kcmil through 500 kcmil	2	1/0
Over 350 kcmil through 600 kcmil	Over 500 kcmil through 900 kcmil	1/0	3/0
Over 600 kcmil through 1100 kcmil	Over 900 kcmil through 1750 kcmil	2/0	4/0
Over 1100 kcmil	Over 1750 kcmil	3/0	250 kcmil

Notes:

1. Where multiple sets of service-entrance conductors are used as permitted in Section 230-40, Exception No. 2, the equivalent size of the largest service-entrance conductor shall be determined by the largest sum of the areas of the corresponding conductors of each set.
2. Where there are no service-entrance conductors, the grounding electrode conductor size shall be determined by the equivalent size of the largest service-entrance conductor required for the load to be served.

[1]This table also applies to the derived conductors of separately derived ac systems.

[2]See installation restrictions in Section 250-64(a).

FPN: See Section 250-24(b) for size of ac system conductor brought to service equipment.

Table 310-5. Minimum Size of Conductors

Conductor Voltage Rating (Volts)	Minimum Conductor Size (AWG)	
	Copper	Aluminum or Copper-Clad Aluminum
0 – 2000	14	12
2001 – 8000	8	8
8001 – 15,000	2	2
15,001 – 28,000	1	1
28,001 – 35,000	1/0	1/0

240-3(d) Small Conductors.

Unless specifically permitted in (e) through (g), the overcurrent protection shall not exceed 15 amperes for No. 14, 20 amperes for No. 12, and 30 amperes for No. 10 copper; or 15 amperes for No. 12 and 25 amperes for No. 10 aluminum and copper-clad aluminum after any correction factors for ambient temperature and number of conductors have been applied.

Table 300-5. Minimum Cover Requirements, 0 to 600 Volts, Nominal, Burial in Inches (Cover is defined as the shortest distance in inches measured between a point on the top surface of any direct-buried conductor, cable, conduit, or other raceway and the top surface of finished grade, concrete, or similar cover.)

	Type of Wiring Method or Circuit				
Location of Wiring Method or Circuit	Column 1 Direct Burial Cables or Conductors	Column 2 Rigid Metal Conduit or Intermediate Metal Conduit	Column 3 Nonmetallic Raceways Listed for Direct Burial Without Concrete Encasement or Other Approved Raceways	Column 4 Residential Branch Circuits Rated 120 Volts or Less with GFCI Protection and Maximum Overcurrent Protection of 20 Amperes	Column 5 Circuits for Control of Irrigation and Landscape Lighting Limited to Not More than 30 Volts and Installed with Type UF or in Other Identified Cable or Raceway
All locations not specified below	24	6	18	12	6
In trench below 2-in. thick concrete or equivalent	18	6	12	6	6
Under a building	0 (in raceway only)	0	0	0 (in raceway only)	0 (in raceway only)
Under minimum of 4-in. thick concrete exterior slab with no vehicular traffic and the slab extending not less than 6 in. beyond the underground installation	18	4	4	6 (direct burial) 4 (in raceway)	6 (direct burial) 4 (in raceway)

Table 300-5 (continued). Minimum Cover Requirements, 0 to 600 Volts, Nominal, Burial in Inches (Cover is defined as the shortest distance in inches measured between a point on the top surface of any direct-buried conductor, cable, conduit, or other raceway and the top surface of finished grade, concrete, or similar cover.)

	Type of Wiring Method or Circuit				
Location of Wiring Method or Circuit	**Column 1 Direct Burial Cables or Conductors**	**Column 2 Rigid Metal Conduit or Intermediate Metal Conduit**	**Column 3 Nonmetallic Raceways Listed for Direct Burial Without Concrete Encasement or Other Approved Raceways**	**Column 4 Residential Branch Circuits Rated 120 Volts or Less with GFCI Protection and Maximum Overcurrent Protection of 20 Amperes**	**Column 5 Circuits for Control of Irrigation and Landscape Lighting Limited to Not More than 30 Volts and Installed with Type UF or in Other Identified Cable or Raceway**
Under streets, highways, roads, alleys, driveways, and parking lots	24	24	24	24	24
One- and two-family dwelling driveways and outdoor parking areas, and used only for dwelling-related purposes	18	18	18	12	18
In or under airport runways, including adjacent areas where trespassing prohibited	18	18	18	18	18

Notes:

1. For SI units, 1 in. = 25.4 mm.
2. Raceways approved for burial only where concrete encased shall require concrete envelope not less than 2 in. thick.
3. Lesser depths shall be permitted where cables and conductors rise for terminations or splices or where access is otherwise required.
4. Where one of the wiring method types listed in Columns 1-3 is used for one of the circuit types in Columns 4 and 5, the shallower depth of burial shall be permitted.
5. Where solid rock prevents compliance with the cover depths specified in this table, the wiring shall be installed in metal or nonmetallic raceway permitted for direct burial. The raceways shall be covered by a minimum of 2 in. of concrete extending down to rock.

Table 310-13. Conductor Application and Insulations

Trade Name	Type Letter	Maximum Operating Temperature	Application Provisions	Insulation	Thickness of Insulation: AWG or kcmil	Thickness of Insulation: Mils	Outer Covering[1]
Fluorinated ethylene propylene	FEP or	90ºC 194ºF	Dry and damp locations	Fluorinated ethylene propylene	14-10 8-2	20 30	None
	FEPB	200ºC 392ºF	Dry locations - special applications[2]	Fluorinated ethylene propylene	14-8	14	Glass braid
					6-2	14	Asbestos or other suitable braid material
Mineral insulation (metal sheathed)	MI	90ºC 194ºF 250ºC 482ºF	Dry and wet locations For special applications[2]	Magnesium oxide	18-163 16-10 9-4 3-500	23 36 50 55	Copper or alloy steel
Moisture-, heat-, and oil-resistant thermoplastic	MTW	60ºC 140ºF 90ºC 194ºF	Machine tool wiring in wet locations as permitted in NFPA 79 (see Article 670) Machine tool wiring in dry locations as permitted in NFPA 79 (see Article 670)	Flame-retardant moisture-, heat-, and oil-resistant thermoplastic	 22-12 10 8 6 4-2 1-4/0 213-5000 501-1000	(A) (B) 30 15 30 20 45 30 60 30 60 40 80 50 95 60 110 70	(A) None (B) Nylon jacket or equivalent
Paper		85ºC 185ºF	For underground service conductors, or by special permission	Paper			Lead sheath
Perfluoroalkoxy	PFA	90ºC 194ºF 200ºC 392ºF	Dry and damp locations Dry locations - special applications[2]	Perfluoroalkoxy	14-10 8-2 1-4/0	20 30 45	None

[1]Some insulations do not require an outer covering.
[2]Where design conditions require maximum conductor operating temperatures above 90ºC (194ºF).
[3]For signaling circuits permitting 300-volt insulation.

Table 310-13 (continued). Conductor Application and Insulations

Trade Name	Type Letter	Maximum Operating Temperature	Application Provisions	Insulation	Thickness of Insulation: AWG or kcmil	Thickness of Insulation: Mils	Outer Covering[1]
Perfluoroalkoxy	PFAH	250ºC 482ºF	Dry locations only. Only for leads within apparatus or within raceways connected to apparatus (Nickel or nickel-coated copper only.)	Perfluoroalkoxy	14-10 8-2 1-4/0	20 30 45	None
Thermoset Thermoset	RH RHH	75ºC 167ºF 90ºC 194ºF	Dry and damp locations Dry and damp locations	Flame-retardant thermoset	14-12[4] 10 8-2 1-4/0 213-500 501-1000 1001-2000 For 601-2000, see Table 310-62	30 45 60 80 95 110 125	Moisture-resistant, flame-retardant, nonmetallic covering[1]
Moisture-resistant thermoset	RHW[5]	75ºC 167ºF	Dry and wet locations Where over 2000 volts insulation, shall be ozone-resistant	Flame- retardant, moisture-resistant thermoset	14-10 8-2 1-4/0 213-500 501-1000 1001-2000 For 601-2000 volts, see Table 310-62	45 60 80 95 110 125	Moisture-resistant, flame-retardant, nonmetallic covering[6]

[1]Some insulations do not require an outer covering.
[4]For size Nos. 14-12, RHH insulation shall be 45 mils thickness.
[5]Listed wire types designated with the suffix "-2," such as RHW-2, shall be permitted to be used at a continuous 90ºC (194ºF) operating temperature, wet or dry.
[6]Some rubber insulations do not require an outer covering.

Table 310-13 (continued). Conductor Application and Insulations

Trade Name	Type Letter	Maximum Operating Temperature	Application Provisions	Insulation	Thickness of Insulation		Outer Covering[1]
					AWG or kcmil	Mils	
Moisture-resistant thermoset	RHW-2	90ºC 194ºF	Dry and wet locations	Flame- retardant moisture-resistant thermoset	14-10 8-2 1-4/0 213-500 501-1000 1001-2000 For 601-2000 volts, see Table 310-62	45 60 80 95 110 125	Moisture-resistant, flame-retardant, nonmetallic covering[6]
Silicone	SA	90ºC 194ºF 200ºC 392ºF	Dry and damp locations For special application[2]	 Silicone rubber	14-10 8-2 1-4/0 213-500 501-1000 1001-2000	45 60 80 95 110 125	Glass or other suitable braid material
Thermoset	SIS	90ºC 194ºF	Switchboard wiring only	Flame-retardant thermoset	14-10 8-2 1-4/0	30 45 95	None
Thermoplastic and fibrous outer braid	TBS	90ºC 194ºF	Switchboard wiring only	Thermoplastic	14-10 8 6-2 1-4/0	30 45 60 80	Flame-retardant, nonmetallic covering
Extended polytetra-fluoroethylene	TFE	250ºC 482ºF	Dry locations only. Only for leads within apparatus or within raceways connected to apparatus, or as open wiring (Nickel or nickel-coated copper only)	Extruded polytetrafluoroethylene	14-10 8-2 1-4/0	20 30 45	None

[1]Some insulations do not require an outer covering.
[2]Where design conditions require maximum conductor operating temperatures above 90ºC (194ºF).
[6]Some rubber insulations do not require an outer covering.

Table 310-13 (continued). Conductor Application and Insulations

Trade Name	Type Letter	Maximum Operating Temperature	Application Provisions	Insulation	Thickness of Insulation: AWG or kcmil	Thickness of Insulation: Mils	Outer Covering[1]
Heat-resistant thermoplastic	THHN	90ºC 194ºF	Dry and damp locations	Flame- retardant, heat-resistant thermoplastic	14-12 10 8-6 4-2 1-4/0 250-500 501-1000	15 20 30 40 50 60 70	Nylon jacket or equivalent
Moisture- and heat-resistant thermoplastic	THHW	75ºC 167ºF 90ºC 194ºF	Wet location Dry location	Flame-retardant, moisture- and heat-resistant thermoplastic	14-10 8 6-2 1-4/0 213-500 501-1000	30 45 60 80 95 110	None
Moisture- and heat-resistant thermoplastic	THW[5]	75ºC 167ºF 90ºC 194ºF	Dry and wet locations Special applications within electric discharge lighting equipment. Limited to 1000 open-circuit volts or less. (Size 14-8 only as permitted in Section 410-31)	Flame-retardant, moisture- and heat-resistant thermoplastic	14-10 8 6-2 1-4/0 213-500 501-1000 1001-2000	30 45 60 80 95 110 125	None

[1]Some insulations do not require an outer covering.
[5]Listed wire types designated with the suffix "-2," such as RHW-2, shall be permitted to be used at a continuous 90ºC (194ºF) operating temperature, wet or dry.

Table 310-13 (continued). Conductor Application and Insulations

Trade Name	Type Letter	Maximum Operating Temperature	Application Provisions	Insulation	Thickness of Insulation: AWG or kcmil	Thickness of Insulation: Mils	Outer Covering[1]
Moisture- and heat-resistant thermoplastic	THWN[5]	75ºC 167ºF	Dry and wet locations	Flame-retardant, moisture- and heat-resistant thermoplastic	14-12 10 8-6 4-2 1-4/0 250-500 501-1000	15 20 30 40 50 60 70	Nylon jacket or equivalent
Moisture- resistant thermoplastic	TW	60ºC 140ºF	Dry and wet locations	Flame- retardant, moisture-resistant thermoplastic	14-10 8 6-2 1-4/0 213-500 501-1000 1001-2000	30 45 60 80 95 110 125	None
Underground feeder and branch-circuit cable - single conductor	UF	60º 140º	See Article 339	Moisture- resistant	14-10 8-2 1-4/0	60[7] 80[7] 95[7]	Integral with insulation
(For Type UF cable employing more than one conductor, see Article 339)		75ºC 167ºF[8]		Moisture- and heat-resistant			

[1]Some insulations do not require an outer covering.
[7]Includes integral jacket.
[8]For ampacity limitation, see Section 339-5.

Table 310-13 (continued). Conductor Application and Insulations

Trade Name	Type Letter	Maximum Operating Temperature	Application Provisions	Insulation	Thickness of Insulation: AWG or kcmil	Thickness of Insulation: Mils	Outer Covering[1]
Underground service-entrance cable - single conductor (For Type USE cable employing more than one conductor, see Article 338)	USE[5]	75ºC 167ºF	See Article 338	Heat- and moisture-resistant	14-10 8-2 1-4/0 213-500 501-1000 1001-2000	45 60 80 95[9] 110 125	Moisture-resistant nonmetallic covering [see Section 338-1(b)]
Thermoset	XHH	90ºC 194ºF	Dry and damp locations	Flame- retardant thermoset	14-10 8-2 1-4/0 213-500 501-1000 1001-2000	30 45 55 65 80 95	None
Moisture-resistant thermoset	XHHW[5]	90ºC 194ºF 75ºC 167ºF	Dry and damp locations Wet locations	Flame- retardant, moisture-resistant thermoset	14-10 8-2 1-4/0 213-500 501-1000 1001-2000	30 45 55 65 80 95	None

[1]Some insulations do not require an outer covering.
[5]Listed wire types designated with the suffix "-2," such as RHW-2, shall be permitted to be used at a continuous 90ºC (194ºF) operating temperature, wet or dry.
[9]Insulation thickness shall be permitted to be 80 mils for listed Type USE conductors that have been subjected to special investigations. The nonmetallic covering over individual rubber-covered conductors of aluminum-sheathed cable and of lead-sheathed or multiconductor cable shall not be required to be flame retardant. For Type MC cable, see Section 334-20. For nonmetallic-sheathed cable, see Section 336-30. For Type UF cable, see Section 339-1.

Table 310-13 (continued). Conductor Application and Insulations

Trade Name	Type Letter	Maximum Operating Temperature	Application Provisions	Insulation	Thickness of Insulation: AWG or kcmil	Thickness of Insulation: Mils	Outer Covering[1]
Moisture-resistant thermoset	XHHW-2	90ºC 194ºF	Dry and wet locations	Flame- retardant, moisture-resistant thermoset	14-10 8-2 1-4/0 213-500 501-1000 1001-2000	30 45 55 65 80 95	None
Modified ethylene tetrafluoroethylene	Z	90ºC 194ºF 150ºC 302ºF	Dry and damp locations Dry locations - special applications[2]	Modified ethylene tetrafluoroethylene	14-12 10 8-4 3-1 1/0-4/0	15 20 25 35 45	None
Modified ethylene tetrafluoroethylene	ZW[5]	75ºC 167ºF 90ºC 194ºF 150ºC 302ºF	Wet locations Dry and damp locations Dry locations - special applications[2]	Modified ethylene tetrafluoroethylene	14-10 8-2	30 45	None

[1]Some insulations do not require an outer covering.
[2]Where design conditions require maximum conductor operating temperatures above 90ºC (194ºF).
[5]Listed wire types designated with the suffix "-2," such as RHW-2, shall be permitted to be used at a continuous 90ºC (194ºF) operating temperature, wet or dry.

From the *National Electrical Code* ©1998 National Fire Protection Association

Table 310-15(b)(6). Conductor Types and Sizes for 120/240-Volt, 3-Wire, Single-Phase Dwelling Services and Feeders. Conductor Types RH, RHH, RHW, RHW-2, THHN, THHW, THW, THW-2, THWN, THWN-2, XHHW, XHHW-2, SE, USE, USE-2

Conductor (AWG or kcmil)		
Copper	Aluminum or Copper-Clad Aluminum	Service or Feeder Rating (Amperes)
4	2	100
3	1	110
2	1/0	125
1	2/0	150
1/0	3/0	175
2/0	4/0	200
3/0	250	225
4/0	300	250
250	350	300
350	500	350
400	600	400

From the *National Electrical Code* ©1998 National Fire Protection Association

Table 310-16. Allowable Ampacities of Insulated Conductors Rated 0 through 2000 Volts, 60ºC through 90ºC (140ºF through 194ºF) Not More than Three Current-Carrying Conductors in Raceway, Cable, or Earth (Directly Buried), Based on Ambient Temperature of 30ºC (86ºF)

Size	Temperature Rating of Conductor (See Table 310-13)						Size
	60ºC (140ºF)	75ºC (167ºF)	90ºC (194ºF)	60ºC (140ºF)	75ºC (167ºF)	90ºC (194ºF)	
AWG or kcmil	Types TW, UF	Types FEPW, RH, RHW, THHW, THW, THWN, XHHW, USE, ZW	Types TBS, SA, SIS, FEP, FEPB, MI, RHH, RHW-2, THHN, THHW, THW-2, THWN-2, USE-2, XHH, XHHW, XHHW-2, ZW-2	Types TW, UF	Types RH, RHW, THHW, THW, THWN, XHHW, USE	Types TBS, SA, SIS, THHN, THHW, THW-2, THWN-2, RHH, RHW-2, USE-2, XHH, XHHW, XHHW-2, ZW-2	AWG or kcmil
	COPPER			ALUMINUM OR COPPER-CLAD ALUMINUM			
18	—	—	14	—	—	—	—
16	—	—	18	—	—	—	—
14*	20	20	25	—	—	—	—
12*	25	25	30	20	20	25	12*
10*	30	35	40	25	30	35	10*
8	40	50	55	30	40	45	8
6	55	65	75	40	50	60	6
4	70	85	95	55	65	75	4
3	85	100	110	65	75	85	3
2	95	115	130	75	90	100	2
1	110	130	150	85	100	115	1
1/0	125	150	170	100	120	135	1/0
2/0	145	175	195	115	135	150	2/0
3/0	165	200	225	130	155	175	3/0
4/0	195	230	260	150	180	205	4/0
250	215	255	290	170	205	230	250
300	240	285	320	190	230	255	300
350	260	310	350	210	250	280	350
400	280	335	380	225	270	305	400
500	320	380	430	260	310	350	500

Table 310-16 (continued). Allowable Ampacities of Insulated Conductors Rated 0 through 2000 Volts, 60ºC through 90ºC (140ºF through 194ºF) Not More than Three Current-Carrying Conductors in Raceway, Cable, or Earth (Directly Buried), Based on Ambient Temperature of 30ºC (86ºF)

Size	Temperature Rating of Conductor (See Table 310-13)						Size
	60ºC (140ºF)	75ºC (167ºF)	90ºC (194ºF)	60ºC (140ºF)	75ºC (167ºF)	90ºC (194ºF)	
AWG or kcmil	Types TW, UF	Types FEPW, RH, RHW, THHW, THW, THWN, XHHW, USE, ZW	Types TBS, SA, SIS, FEP, FEPB, MI, RHH, RHW-2, THHN, THHW, THW-2, THWN-2, USE-2, XHH, XHHW, XHHW-2, ZW-2	Types TW, UF	Types RH, RHW, THHW, THW, THWN, XHHW, USE	Types TBS, SA, SIS, THHN, THHW, THW-2, THWN-2, RHH, RHW-2, USE-2, XHH, XHHW, XHHW-2, ZW-2	AWG or kcmil
	COPPER			ALUMINUM OR COPPER-CLAD ALUMINUM			
600	355	420	475	285	340	385	600
700	385	460	520	310	375	420	700
750	400	475	535	320	385	435	750
800	410	490	555	330	395	450	800
900	435	520	585	355	425	480	900
1000	455	545	615	375	445	500	1000
1250	495	590	665	405	485	545	1250
1500	520	625	705	435	520	585	1500
1750	545	650	735	455	545	615	1750
2000	560	665	750	470	560	630	2000
	CORRECTION FACTORS						
Ambient Temp. (ºC)	For ambient temperatures other than 30ºC (86ºF), multiply the allowable ampacities shown above by the appropriate factor shown below.						Ambient Temp. (ºF)
21-25	1.08	1.05	1.04	1.08	1.05	1.04	70-77
26-30	1.00	1.00	1.00	1.00	1.00	1.00	78-86
31-35	0.91	0.94	0.96	0.91	0.94	0.96	87-95
36-40	0.82	0.88	0.91	0.82	0.88	0.91	96-104
41-45	0.71	0.82	0.87	0.71	0.82	0.87	105-113
46-50	0.58	0.75	0.82	0.58	0.75	0.82	114-122
51-55	0.41	0.67	0.76	0.41	0.67	0.76	123-131
56-60	—	0.58	0.71	—	0.58	0.71	132-140
61-70	—	0.33	0.58	—	0.33	0.58	141-158
71-80	—	—	0.41	—	—	0.41	159-176

* See Section 240-3.

Table 346-10. Radius of Conduit Bends

Size of Conduit (in.)	Conductors Without Lead Sheath (in.)
½	4
¾	5
1	6
1¼	8
1½	10
2	12
2½	15
3	18
3½	21
4	24
5	30
6	36

Note: For SI units, 1 in. = 25.4 mm (radius).

Exception: For field bends for conductors without lead sheath and made with a single operation (one shot) bending machine designed for the purpose, the minimum radius shall not be less than that indicated in Table 346-10, Exception.

Table 346-10. Exception, Radius of Conduit Bends

Size of Conduit (in.)	*Radius to Center of Conduit (in.)*
½	*4*
¾	*4½*
1	*5¾*
1¼	*7¼*
1½	*8¼*
2	*9½*
2½	*10½*
3	*13*
3½	*15*
4	*16*
5	*24*
6	*30*

Note: For SI units, 1 in. = 25.4 mm (radius).

Table 346-12(b)(2). Supports for Rigid Metal Conduit

Conduit Size (in.)	Maximum Distance Between Rigid Metal Conduit Supports (ft)
½-¾	10
1	12
1¼-1½	14
2-2½	16
3 and larger	20

Note: For SI units, 1 ft = 0.3048 m (supports).

Table 370-16(a). Metal Boxes

Box Dimension in Inches, Trade Size, or Type	Minimum Capacity (in.^{3})	Maximum Number of Conductors*						
		No. 18	No. 16	No. 14	No. 12	No. 10	No. 8	No. 6
4 x 1¼ round or octagonal	12.5	8	7	6	5	5	4	2
4 x 1½ round or octagonal	15.5	10	8	7	6	6	5	3
4 x 2⅛ round or octagonal	21.5	14	12	10	9	8	7	4
4 x 1¼ square	18.0	12	10	9	8	7	6	3
4 x 1½ square	21.0	14	12	10	9	8	7	4
4 x 2⅛ square	30.3	20	17	15	13	12	10	6
4¹¹⁄₁₆ x 1¼ square	25.5	17	14	12	11	10	8	5
4¹¹⁄₁₆ x 1½ square	29.5	19	16	14	13	11	9	5
4¹¹⁄₁₆ x 2⅛ square	42.0	28	24	21	18	16	14	8
3 x 2 x 1½ device	7.5	5	4	3	3	3	2	1
3 x 2 x 2 device	10.0	6	5	5	4	4	3	2
3 x 2 x 2¼ device	10.5	7	6	5	4	4	3	2
3 x 2 x 2½ device	12.5	8	7	6	5	5	4	2
3 x 2 x 2¾ device	14.0	9	8	7	6	5	4	2
3 x 2 x 3½ device	18.0	12	10	9	8	7	6	3
4 x 2⅛ x 1½ device	10.3	6	5	5	4	4	3	2
4 x 2⅛ x 1⅞ device	13.0	8	7	6	5	5	4	2
4 x 2⅛ x 2⅛ device	14.5	9	8	7	6	5	4	2
3¾ x 2 x 2½ masonry box/gang	14.0	9	8	7	6	5	4	2
3¾ x 2 x 3½ masonry box/gang	21.0	14	12	10	9	8	7	4
FS - Minimum internal depth 1¾ single cover/gang	13.5	9	7	6	6	5	4	2
FD - Minimum internal depth 2⅜ single cover/gang	18.0	12	10	9	8	7	6	3
FS - Minimum internal depth 1¾ multiple cover/gang	18.0	12	10	9	8	7	6	3
FD - Minimum internal depth 2⅜ multiple cover/gang	24.0	16	13	12	10	9	8	4

Note: For SI units, 1 in.^{3} = 16.4 cm^3.

* Where no volume allowances are required by Sections 370-16(b)(2) through 370-16(b)(5).

Table 370-16(b). Volume Allowance Required per Conductor

Size of Conductor (AWG)	Free Space Within Box for Each Conductor (in.^{3})
18	1.50
16	1.75
14	2.00
12	2.25
10	2.50
8	3.00
6	5.00

Note: For SI units, 1 in.^{3} = 16.4 cm^3.

APPENDIX B

PLANS

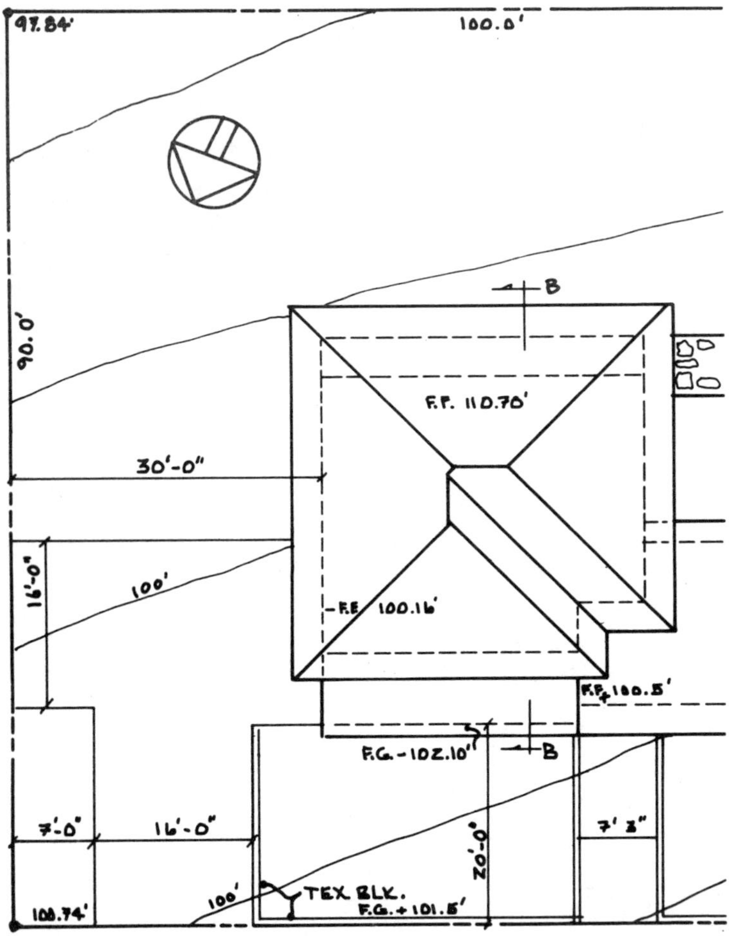
97.84'
100.0'
B
90.0'
F.F. 110.70'
30'-0"
16'-0"
100'
F.F. 100.16'
F.F. 100.5'
F.G. - 102.10'
B
7'-0"
16'-0"
7' 3"
20'-0"
100'
TEX. BLK.
F.G. + 101.5'
100.74'

PLOT PLAN

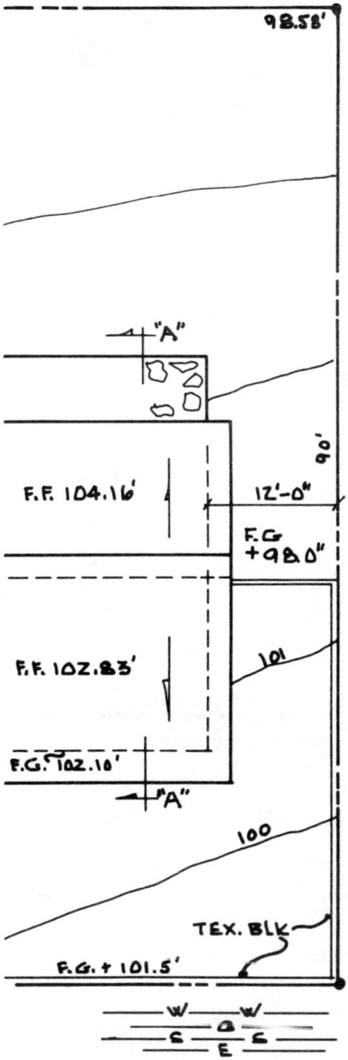

LEGAL DESCRIPTION:
LOT 29 RANCHO SUBDIVISION.
ACCORDING TO MAP #5760
SAN COUNTY.

NOTES:
CONFIRM DEPTH OF SEWER IN STREET BEFORE START OF WORK.
ALL UTILITIES UNDERGROUND IN STREET.
ALL FRAMING DIMENSIONS ARE STUD LINE DIMENSIONS.

WHEN ROUGH GRADING STOCK PILE TOP SOIL AT 'SE' CORNER OF LOT FOR FINE FINISH GRADING.

ALL UTILITIES IN STREET

SCALE

SHEET 1

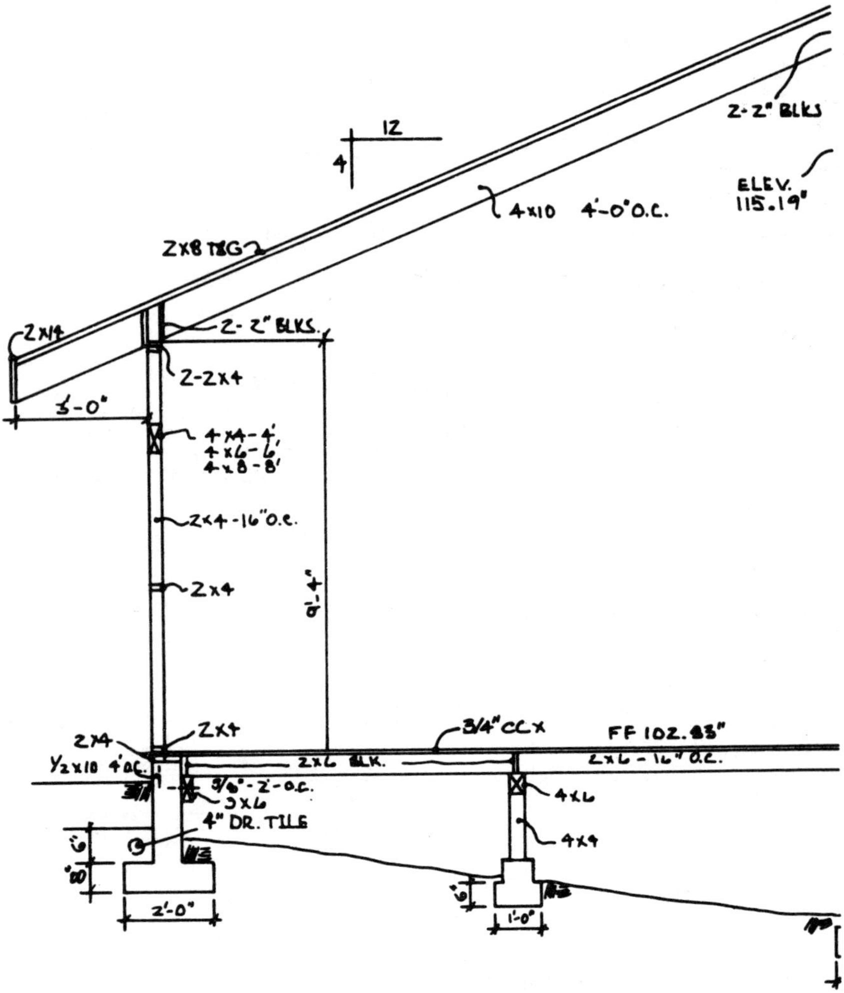

TYPICAL SECTION A-A

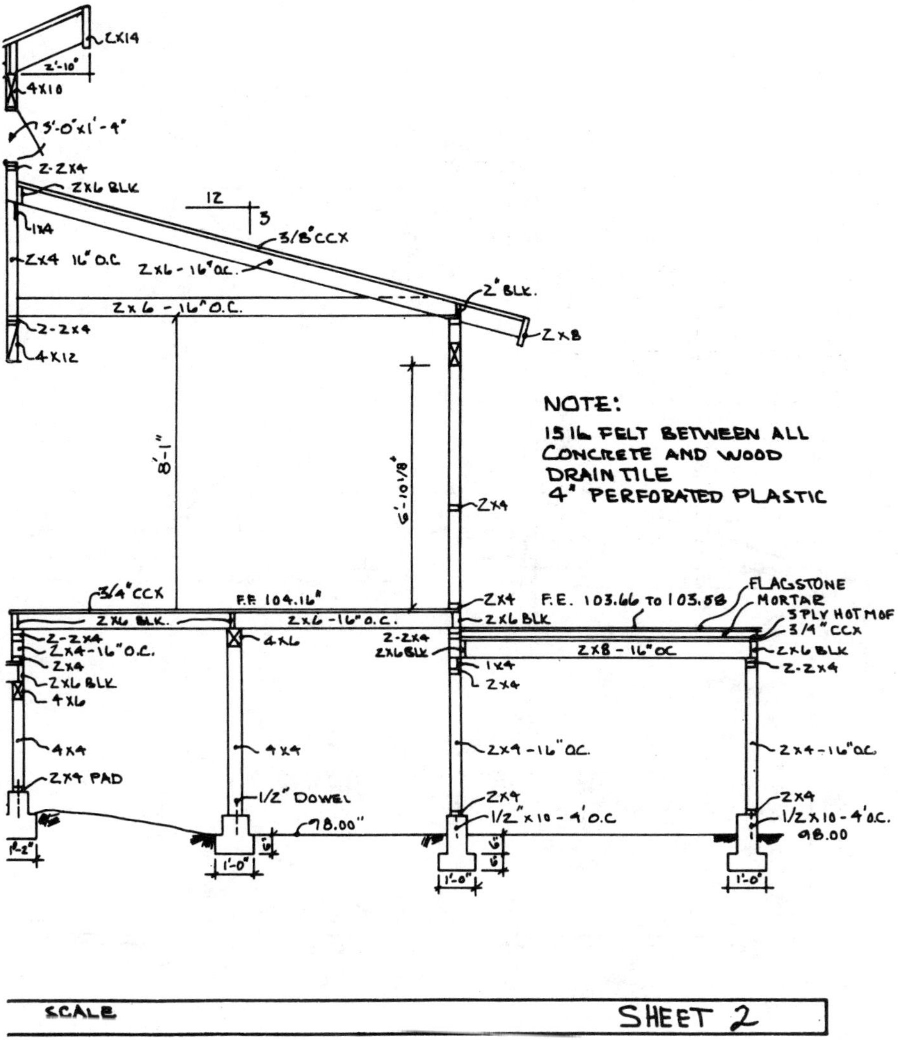
2X14
2'-10"
4X10
3'-0"x1'-4"
2-2X4
2X6 BLK
12
3
1X4
3/8" CCX
2X4 16" O.C.
2X6 - 16" O.C.
2" BLK.
2X6 - 16" O.C.
2-2X4
2X8
4X12
NOTE:
15 lb FELT BETWEEN ALL
CONCRETE AND WOOD
DRAIN TILE
4" PERFORATED PLASTIC
8'-1"
6'-10 1/8"
2X4
3/4" CCX
F.F. 104.16"
2X4
F.E. 103.66 TO 103.58
FLAGSTONE
MORTAR
3 PLY HOT MOP
3/4" CCX
2X6 BLK.
2X6 - 16" O.C.
2X6 BLK
2-2X4
4X6
2-2X4
2X6 BLK
2X8 - 16" OC
2X6 BLK
2X4-16" O.C.
2X4
1X4
2-2X4
2X6 BLK
2X4
4X6
4X4
4X4
2X4 - 16" OC.
2X4 - 16" OC.
2X4 PAD
1/2" DOWEL
2X4
1/2" X 10 - 4' O.C
2X4
1/2 X 10 - 4' O.C.
98.00"
98.00
1'-2"
1'-0"
6"
6"
1'-0"
1'-0"
SCALE
SHEET 2

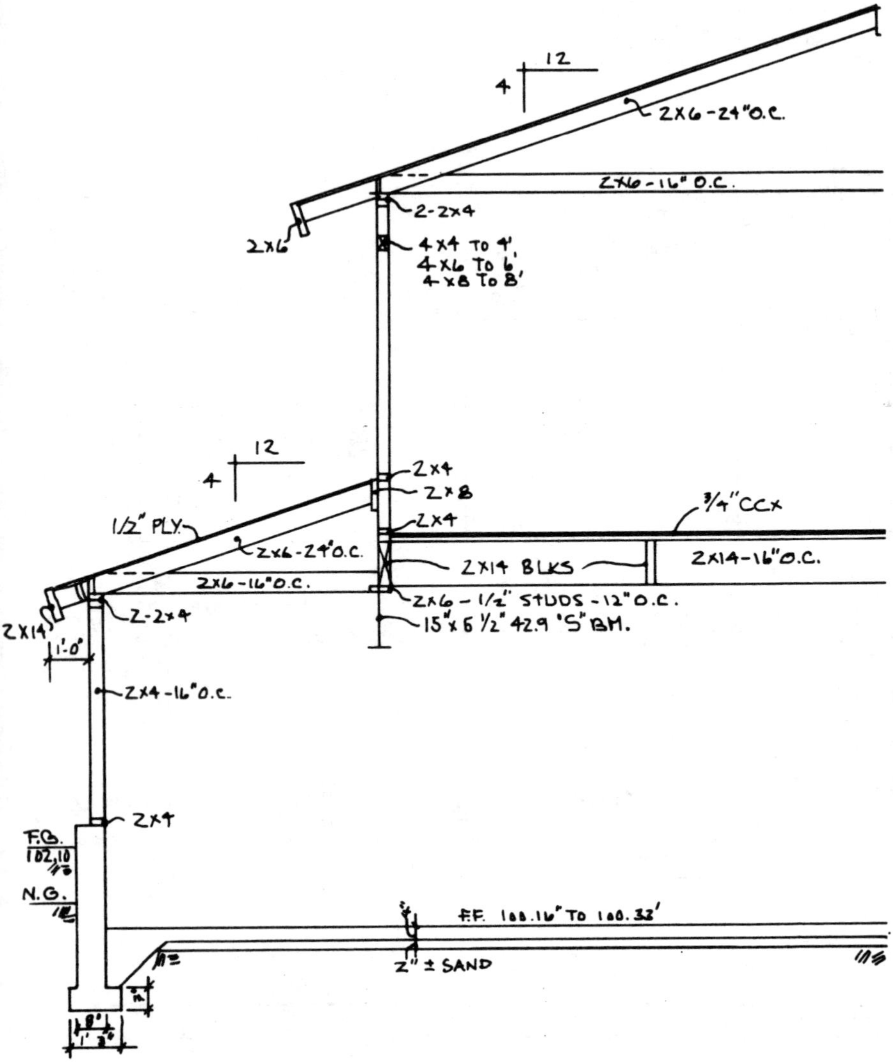

TYPICAL SECTION B-B

SCALE

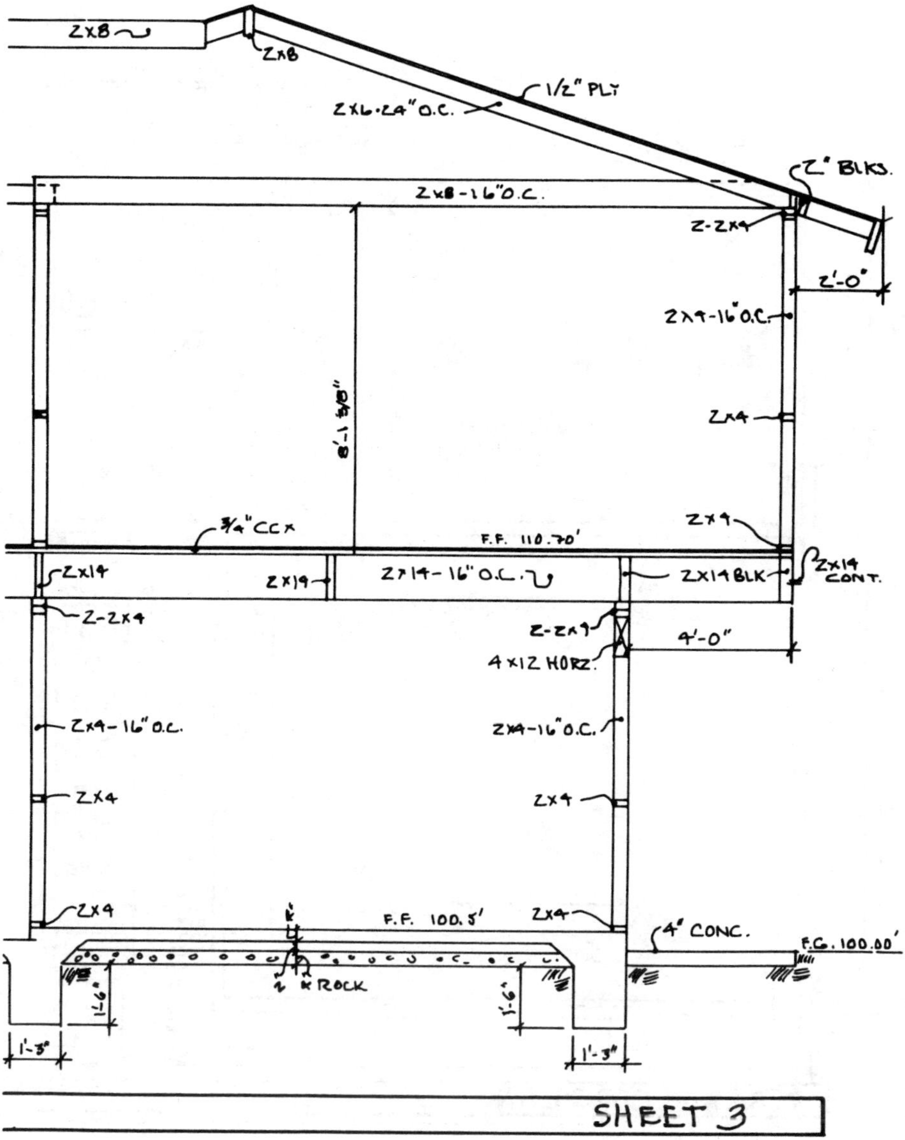
2X8
2X8
1/2" PLY
2X6-24" O.C.
2" BLKS.
2X8-16" O.C.
2-2X4
2'-0"
2X4-16" O.C.
2X4
8'-1 5/8"
3/4" CCX
2X4
F.F. 110.70'
2X14
2X14
2X14-16" O.C.
2X14 BLK
2X14 CONT.
2-2X4
2-2X4
4'-0"
4X12 HORZ.
2X4-16" O.C.
2X4-16" O.C.
2X4
2X4
2X4
F.F. 100.5'
2X4
4" CONC.
F.G. 100.00'
ROCK
1'-6"
1'-6"
1'-3"
1'-3"
SHEET 3

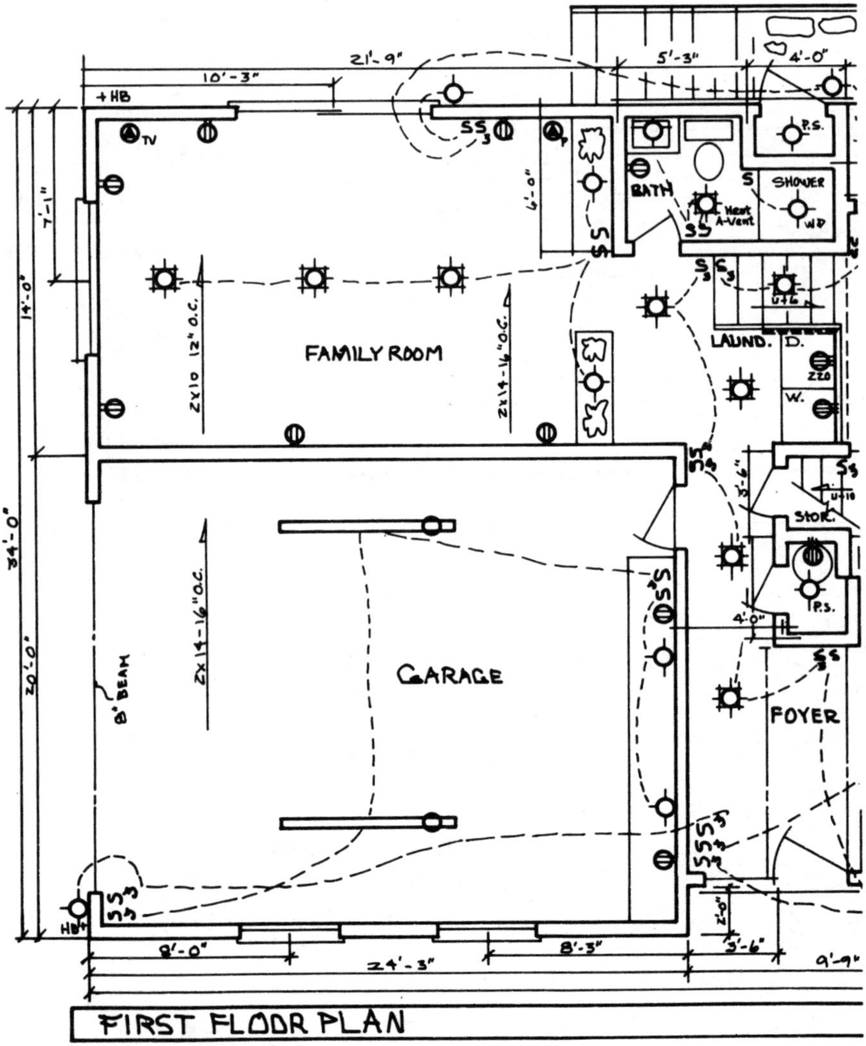
FAMILY ROOM
GARAGE
BATH
SHOWER
LAUND.
STOR.
FOYER
FIRST FLOOR PLAN

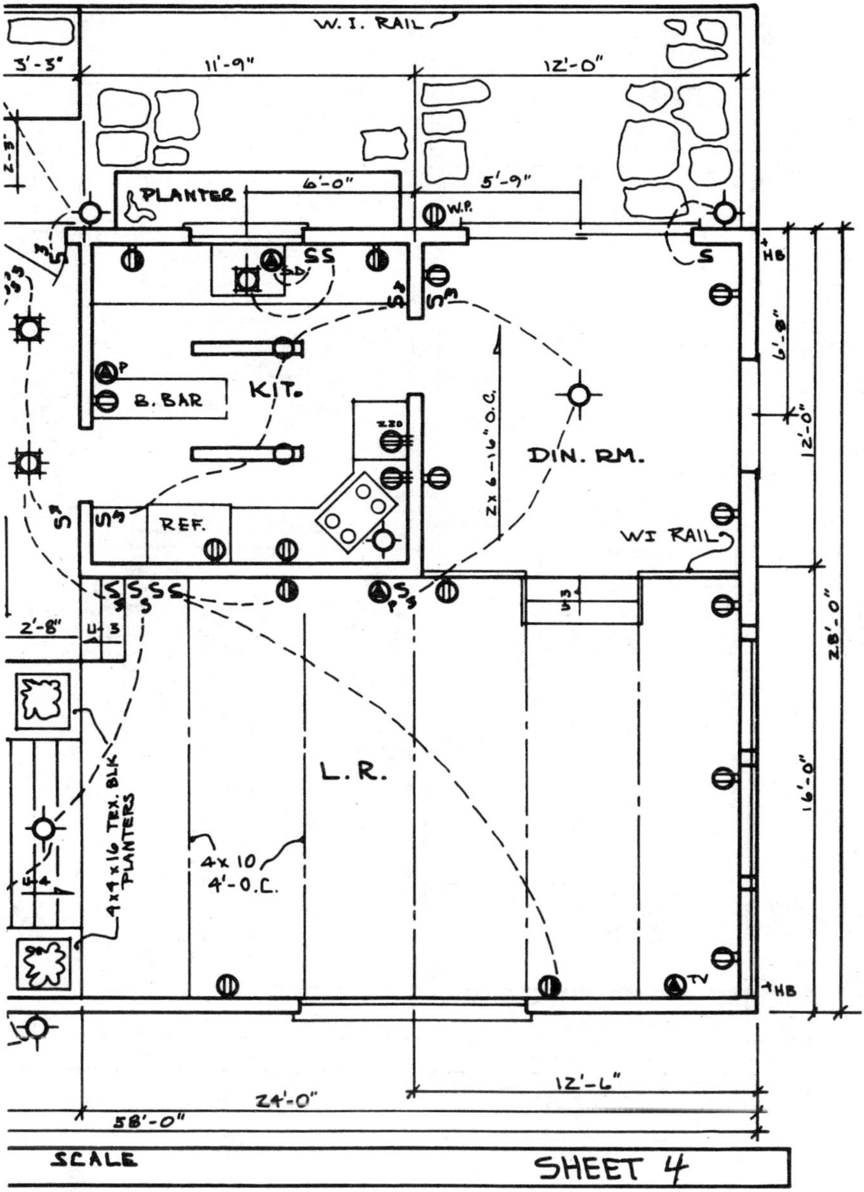
W. I. RAIL
3'-3"
11'-9"
12'-0"
PLANTER
6'-0"
5'-9"
W.P.
SS
HB
KIT.
B. BAR
DIN. RM.
2x6-16" O.C.
REF.
WI RAIL
L.R.
2'-8"
4x4x16 TEX. BLK PLANTERS
4x10
4'-O.C.
TV
HB
6'-8"
12'-0"
28'-0"
16'-0"
12'-6"
24'-0"
58'-0"
SCALE
SHEET 4

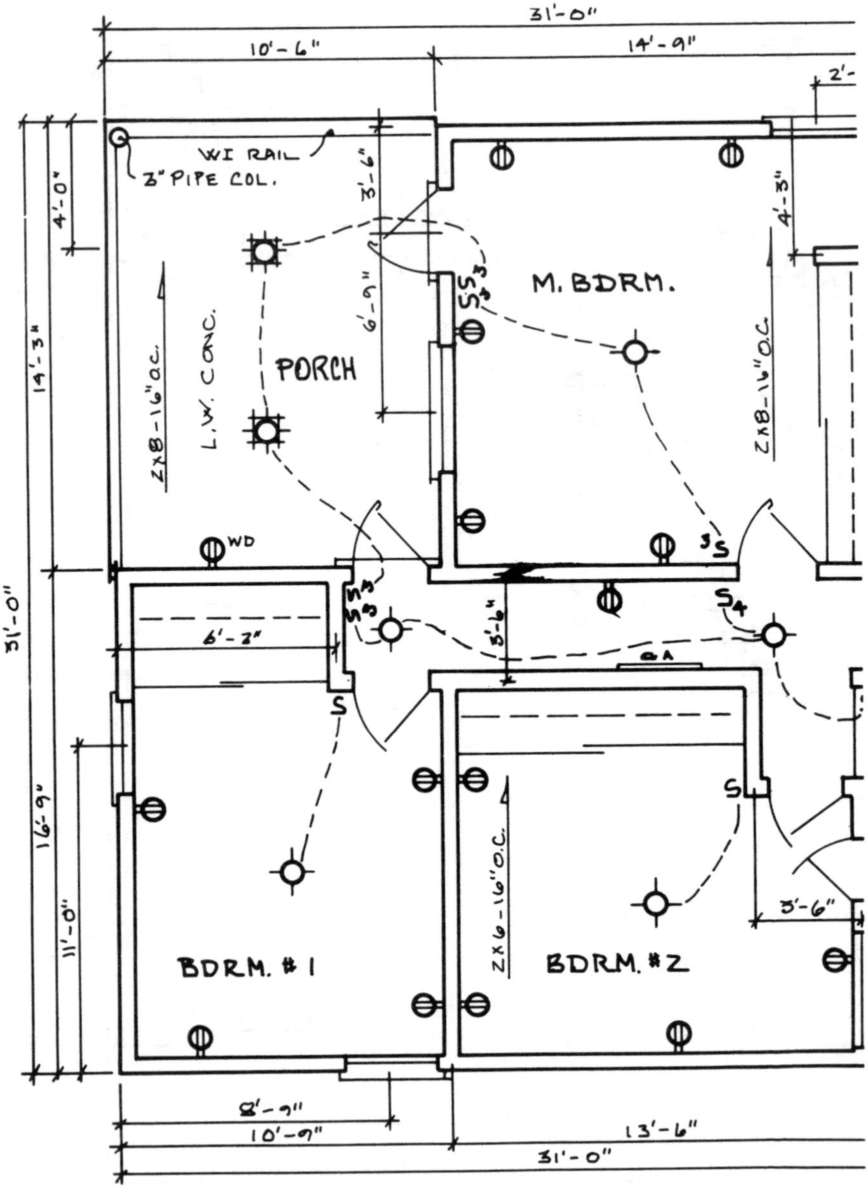

SECOND FLOOR PLAN

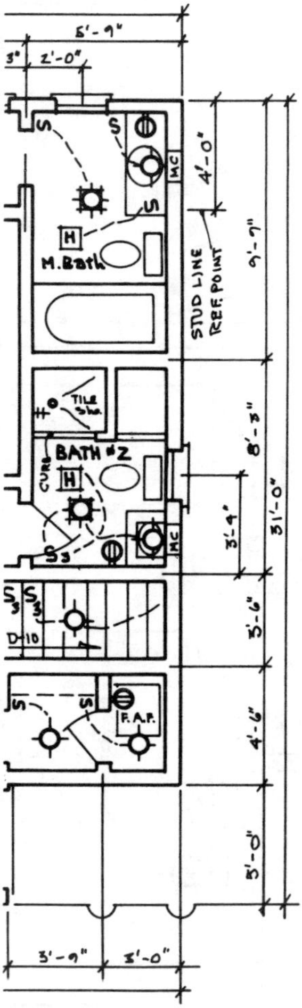

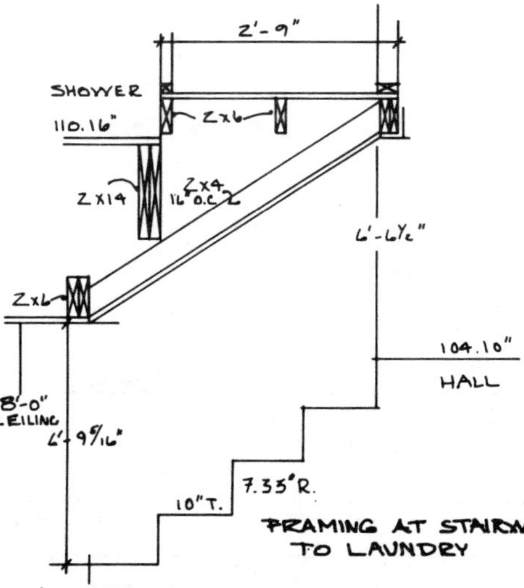

FRAMING AT STAIRWAY
TO LAUNDRY

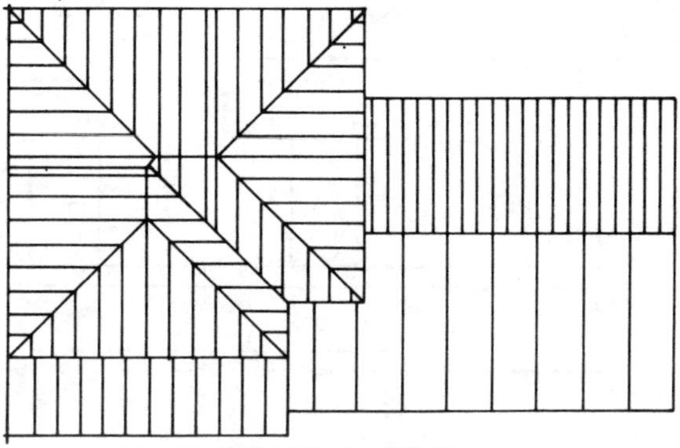

ROOF FRAMING PLAN
SCALE 3/32" = 1'-0"
NO OVERHANG SHOWN

SCALE SHEET 5

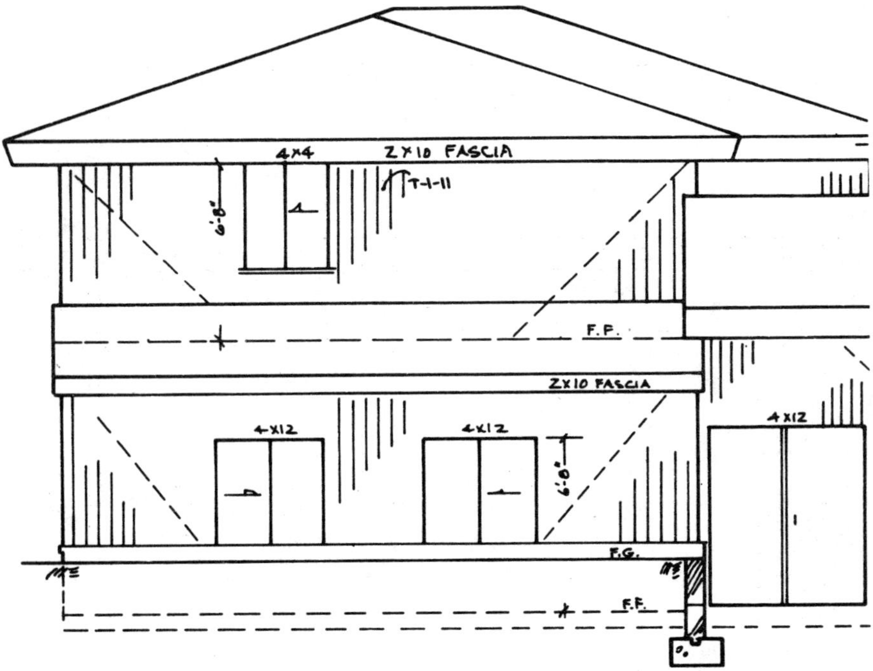

NORTH ELEVATION

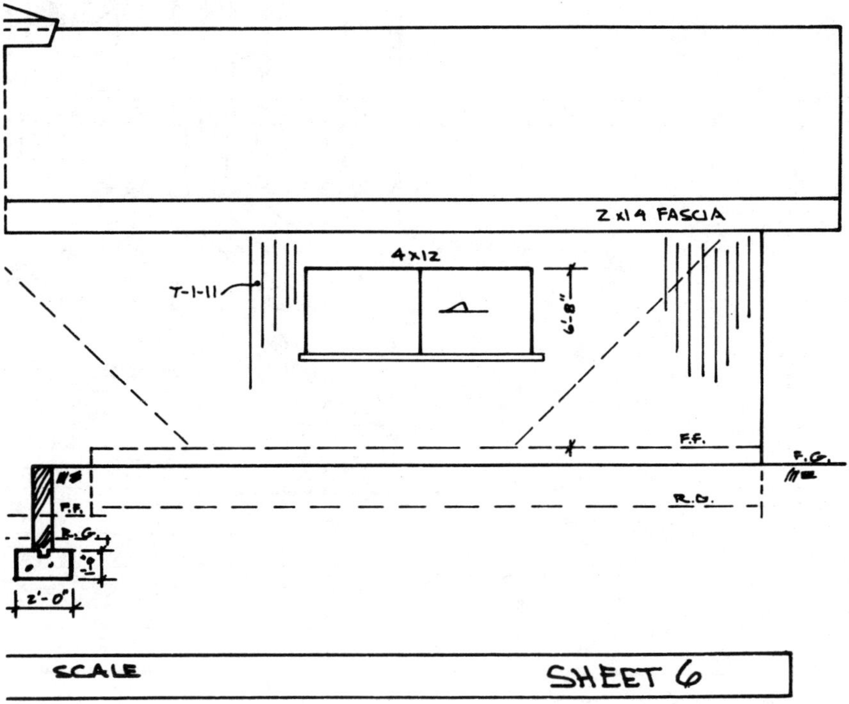
2x14 FASCIA
4x12
T-1-11
6'-8"
F.F.
F.G.
R.G.
2'-0"
SCALE
SHEET 6

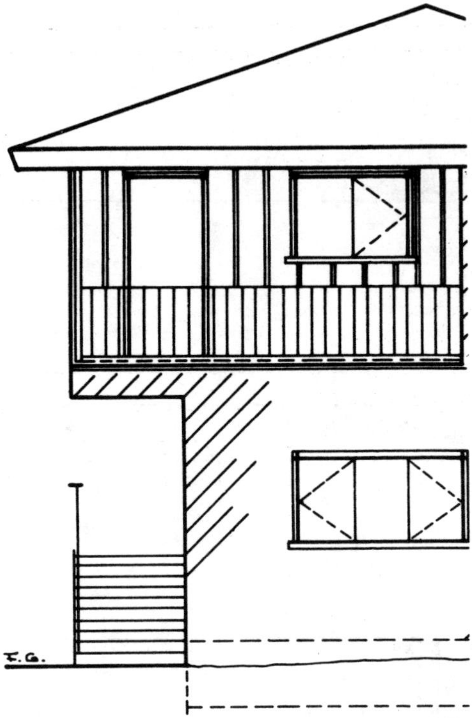

EAST ELEVATION SCALE

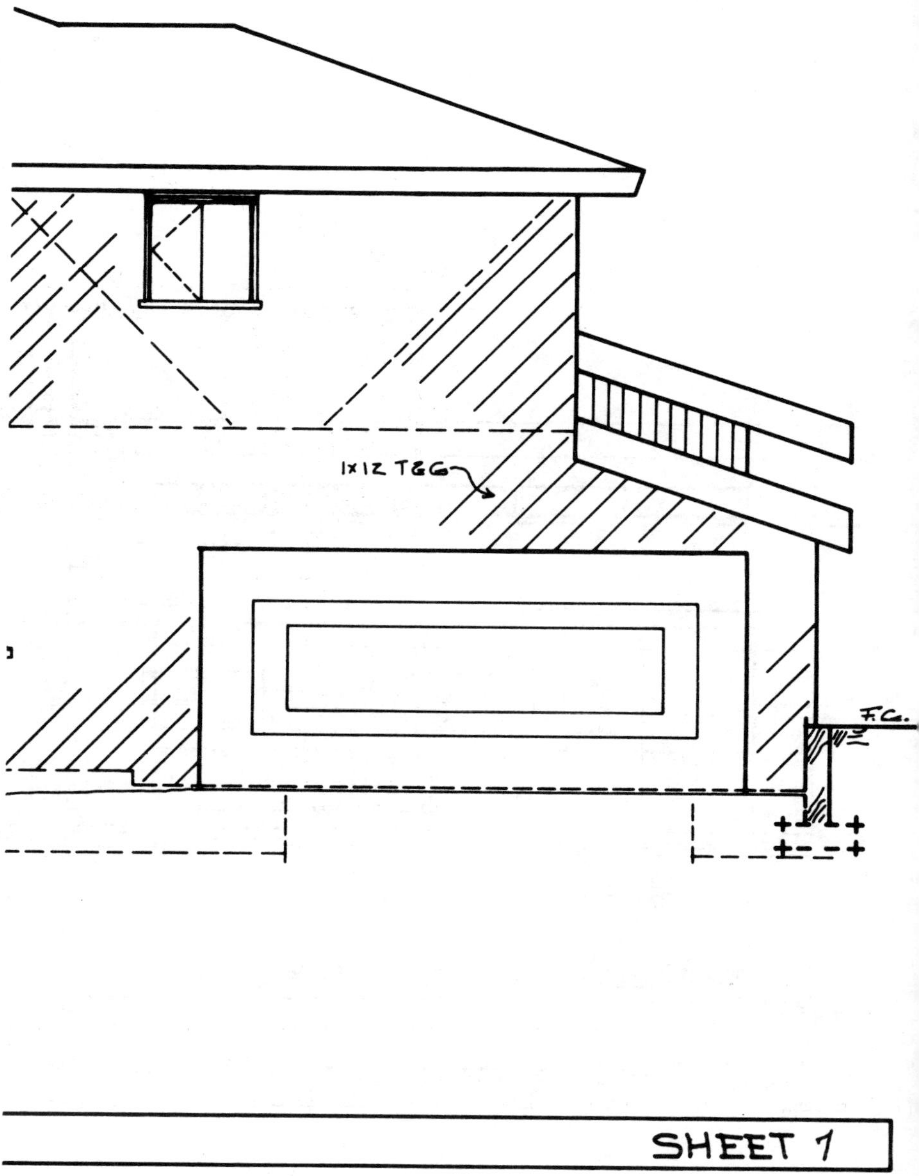
1x12 T&G
F.G.
SHEET 1

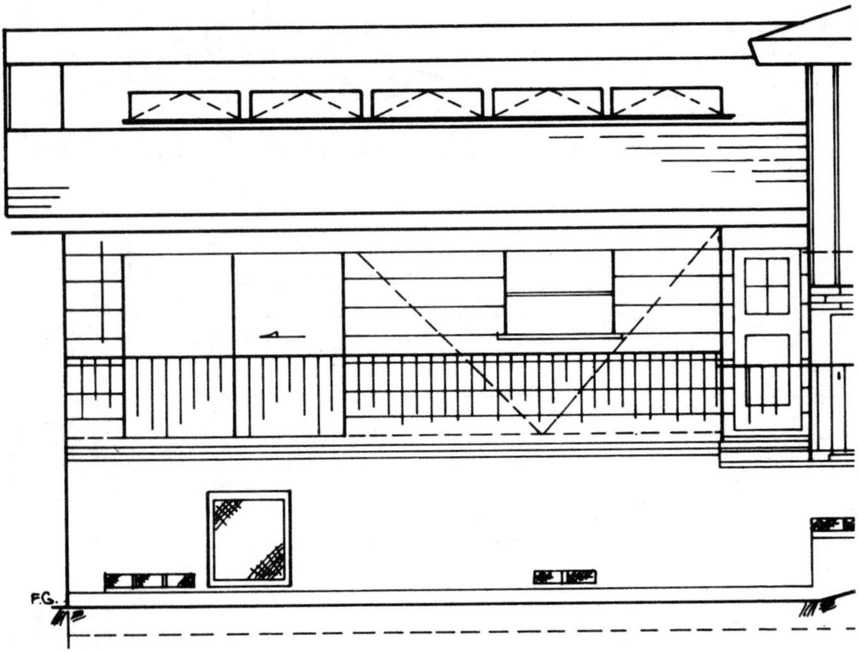

SOUTH ELEVATION

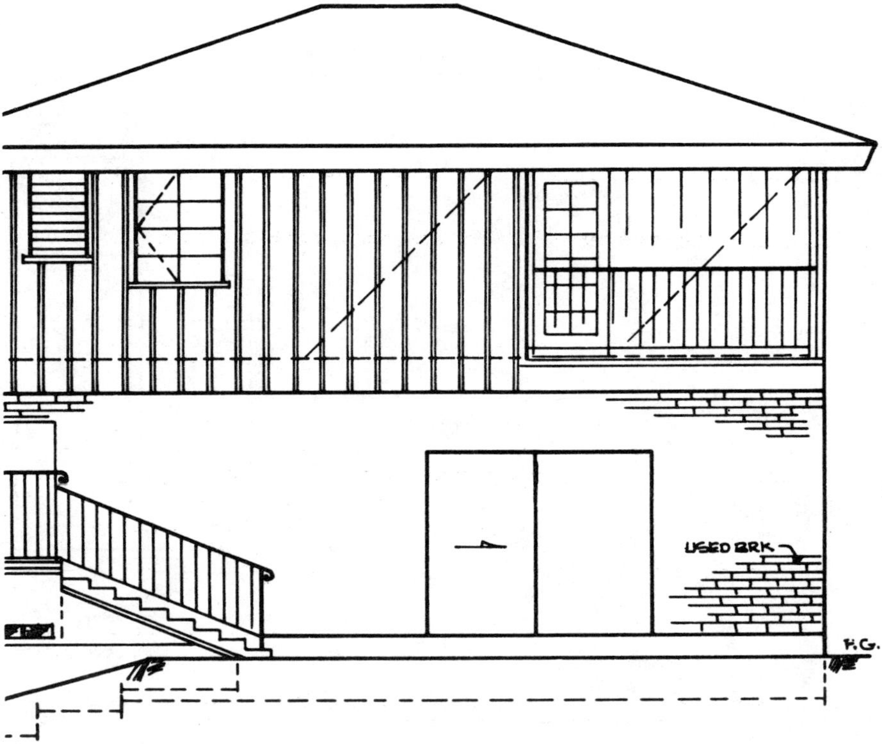

SCALE SHEET 8

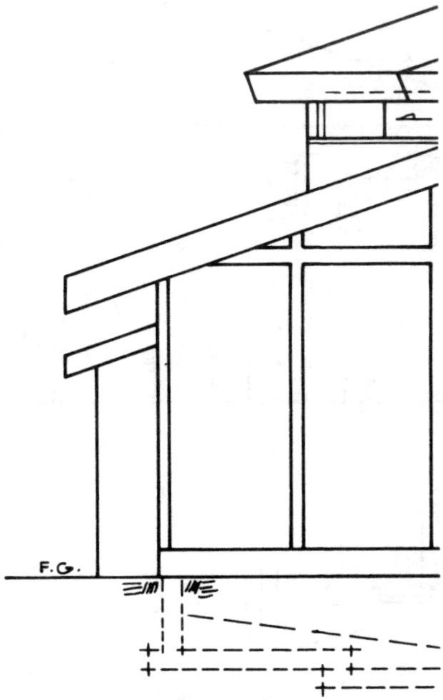

WEST ELEVATION

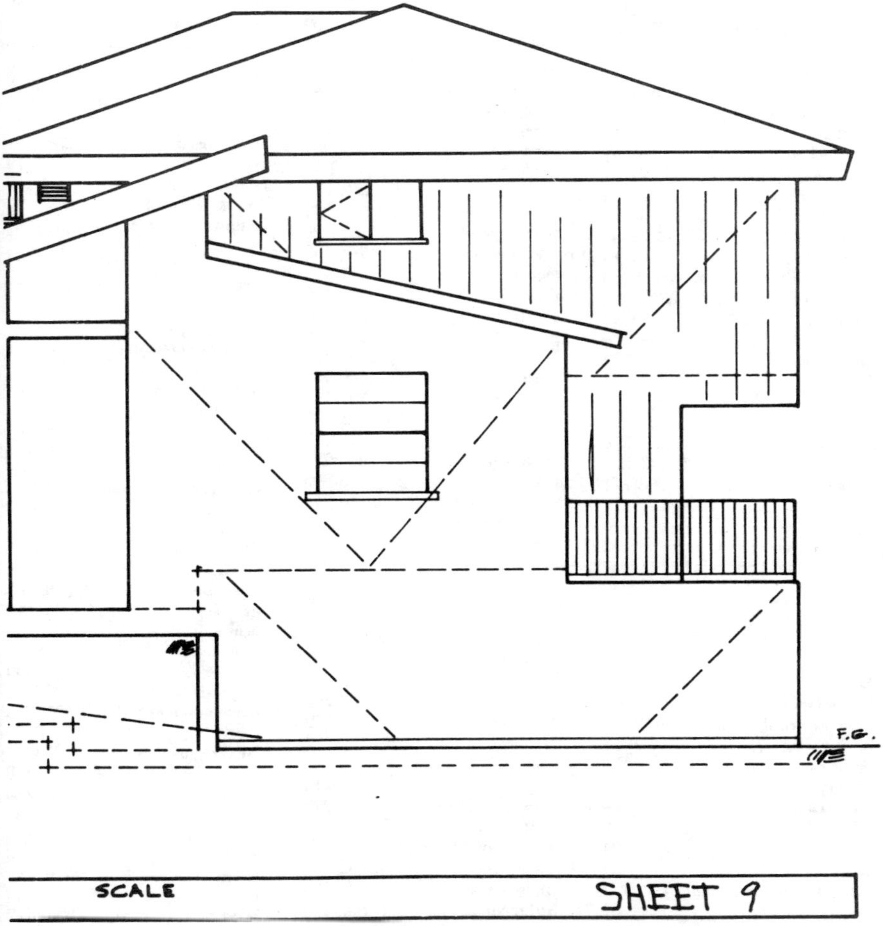
F.G.
SCALE
SHEET 9

Index

T

U

V

W

XYZ

Practical References for Builders

Commercial Electrical Wiring

Make the transition from residential to commercial electrical work. Here are wiring methods, spec reading tips, load calculations and everything you need for making the transition to commercial work: commercial construction documents, load calculations, electric services, transformers, overcurrent protection, wiring methods, raceway, boxes and fittings, wiring devices, conductors, electric motors, relays and motor controllers, special occupancies, and safety requirements. This book is written to help any electrician break into the lucrative field of commercial electrical work.
320 pages, $8\frac{1}{2}$ x 11, $27.50

National Electrical Estimator

This year's prices for installation of all common electrical work: conduit, wire, boxes, fixtures, switches, outlets, loadcenters, panelboards, raceway, duct, signal systems, and more. Provides material costs, manhours per unit, and total installed cost. Explains what you should know to estimate each part of an electrical system. Includes a CD-ROM with an electronic version of the book with *National Estimator*, a stand-alone *Windows*™ estimating program, plus an interactive multimedia video that shows how to use the disk to compile construction cost estimates.
520 pages, $8\frac{1}{2}$ x 11, $47.75. Revised annually

Illustrated Guide to the 1999 *National Electrical Code*

This fully-illustrated guide offers a quick and easy visual reference for installing electrical systems. Whether you're installing a new system or repairing an old one, you'll appreciate the simple explanations written by a code expert, and the detailed, intricately-drawn and labeled diagrams. A real time-saver when it comes to deciphering the current *NEC*.
384 pages, $8\frac{1}{2}$ x 11, $38.75

Electrical Blueprint Reading Revised

Shows how to read and interpret electrical drawings, wiring diagrams, and specifications for constructing electrical systems. Shows how a typical lighting and power layout would appear on a plan, and explains what to do to execute the plan. Describes how to use a panelboard or heating schedule, and includes typical electrical specifications. **208 pages, $8\frac{1}{2}$ x 11, $18.00**

Electrician's Exam Preparation Guide

Need help in passing the apprentice, journeyman, or master electrician's exam? This is a book of questions and answers based on actual electrician's exams over the last few years. Almost a thousand multiple-choice questions — exactly the type you'll find on the exam — cover every area of electrical installation: electrical drawings, services and systems, transformers, capacitors, distribution equipment, branch circuits, feeders, calculations, measuring and testing, and more. It gives you the correct answer, an explanation, and where to find it in the latest *NEC*. Also tells how to apply for the test, how best to study, and what to expect on examination day.
352 pages, $8\frac{1}{2}$ x 11, $32.00

Construction Forms & Contracts

125 forms you can copy and us — or load into your computer (from the FREE disk enclosed). Then you can customize the forms to fit your company, fill them out, and print. Loads into Word for *Windows*™, *Lotus 1-2-3*, *WordPerfect*, *Works*, or *Excel* programs. You'll find forms covering accounting, estimating, fieldwork, contracts, and general office. Each form comes with complete instructions on when to use it and how to fill it out. These forms were designed, tested and used by contractors, and will help keep your business organized, profitable and out of legal, accounting and collection troubles. Includes a CD-ROM for *Windows*™ and Mac. **400 pages, $8\frac{1}{2}$ x 11, $39.75**

Markup & Profit: A Contractor's Guide

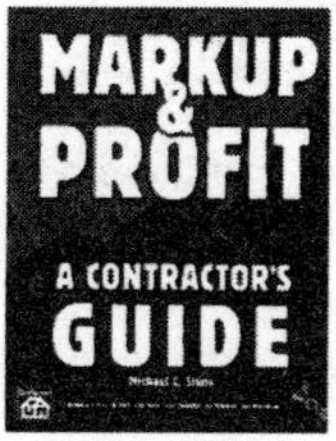

In order to succeed in a construction business, you have to be able to price your jobs to cover all labor, material and overhead expenses, and make a decent profit. The problem is knowing what markup to use. You don't want to lose jobs because you charge too much, and you don't want to work for free because you've charged too little. If you know how to calculate markup, you can apply it to your job costs to find the right sales price for your work. This book gives you tried and tested formulas, with step-by-step instructions and easy-to-follow examples, so you can easily figure the markup that's right for your business. Includes a CD-ROM with forms and checklists for your use. **320 pages, 8½ x 11, $32.50**

Estimating Electrical Construction

Like taking a class in how to estimate materials and labor for residential and commercial electrical construction. Written by an A.S.P.E. National Estimator of the Year, it teaches you how to use labor units, the plan take-off, and the bid summary to make an accurate estimate, how to deal with suppliers, use pricing sheets, and modify labor units. Provides extensive labor unit tables and blank forms for your next electrical job. **272 pages, 8½ x 11, $19.00**

CD Estimator

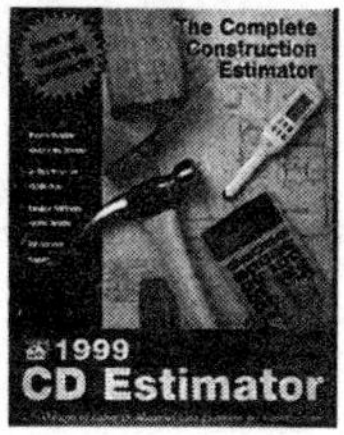

If your computer has *Windows*™ and a CD-ROM drive, *CD Estimator* puts at your fingertips 85,000 construction costs for new construction, remodeling, renovation & insurance repair, electrical, plumbing, HVAC and painting. You'll also have the *National Estimator* program — a stand-alone estimating program for *Windows*™ that *Remodeling* magazine called a "computer wiz." Quarterly cost updates are available at no charge on the Internet. To help you create professional-looking estimates, the disk includes over 40 construction estimating and bidding forms in a format that's perfect for nearly any word processing or spreadsheet program for *Windows*™. And to top it off, a 70-minute interactive video teaches you how to use this CD-ROM to estimate construction costs. **CD Estimator is $68.50**

1996 *National Electrical Code* Interpretive Diagrams

Based on the 1996 *NEC*, this new edition visually and graphically explains and interprets the key provisions of the *National Electrical Code* as seen through the eyes of an electrical inspector. All drawings have been carefully prepared and reviewed by inspectors and the Codes and Standards Committee. Here you'll find approved installations of just about every type of electrical part. **270 pages, 5 x 8, $34.95**

Residential Electrical Estimating

A fast, accurate pricing system proven on over 1000 residential jobs. Using the manhours provided, combined with material prices from your wholesaler, you quickly work up estimates based on degree of difficulty. These manhours come from a working electrical contractor's records — not some pricing agency. You'll find prices for every type of electrical job you're likely to estimate — from service entrances to ceiling fans. **320 pages, 8½ x 11, $29.00**

Contractor's Guide to QuickBooks Pro

This user-friendly manual walks you through QuickBooks Pro's detailed setup procedure and explains step-by-step how to create a first-rate accounting system. You'll learn in days, rather than weeks, how to use QuickBooks Pro to get your contracting business organized, with simple, fast accounting procedures. On the CD included with the book you'll find a full version of QuickBooks Pro, good for 25 uses, with a QuickBooks Pro file preconfigured for a construction company (you drag it over onto your computer and plug in your own company's data). You'll also get a complete estimating program, including a database, and a job costing program that lets you export your estimates to QuickBooks Pro. It even includes many useful construction forms to use in your business. **320 pages, 8½ x 11, $39.75**

NO POSTAGE
NECESSARY
IF MAILED
IN THE
UNITED STATES

BUSINESS REPLY MAIL

FIRST CLASS MAIL PERMIT NO. 271 CARLSBAD, CA

POSTAGE WILL BE PAID BY ADDRESSEE

Craftsman Book Company
6058 Corte del Cedro
P.O. Box 6500
Carlsbad, CA 92018-9892

NO POSTAGE
NECESSARY
IF MAILED
IN THE
UNITED STATES

BUSINESS REPLY MAIL

FIRST CLASS MAIL PERMIT NO. 271 CARLSBAD, CA

POSTAGE WILL BE PAID BY ADDRESSEE

Craftsman Book Company
6058 Corte del Cedro
P.O. Box 6500
Carlsbad, CA 92018-9892